URBAN CARNIVORES

URBAN CARNIVORES

ECOLOGY, CONFLICT, AND CONSERVATION

Edited by

STANLEY D. GEHRT

SETH P. D. RILEY

BRIAN L. CYPHER

THE JOHNS HOPKINS UNIVERSITY PRESS | BALTIMORE

© 2010 The Johns Hopkins University Press
All rights reserved. Published 2010
Printed in the United States of America on acid-free paper
9 8 7 6 5 4 3 2

The Johns Hopkins University Press
2715 North Charles Street
Baltimore, Maryland 21218-4363
www.press.jhu.edu

Library of Congress Cataloging-in-Publication Data

Urban carnivores : ecology, conflict, and conservation / edited by Stanley D. Gehrt,
Seth P. D. Riley, Brian L. Cypher.
    p. cm.
  Includes bibliographical references and index.
  ISBN-13: 978-0-8018-9389-6 (hardcover : alk. paper)
  ISBN-10: 0-8018-9389-5 (hardcover : alk. paper)
  1. Carnivora—Ecology. 2. Urban animals—Ecology. I. Gehrt, Stanley D. II. Riley,
Seth P. D. III. Cypher, Brian L.
  QL737.C2U73 2009
  599.7'1756—dc22   2009018179

A catalog record for this book is available from the British Library.

Photography Credits (chapter facing pages)
Chapter 1: National Park Service
Chapter 2: Jeremy Bruskotter
Chapters 3 and 9: Don Hunford
Chapter 4: John Harrison
Chapter 5: Christine Van Horn Job
Chapter 6: Paul Cecil
Chapter 7: Max McGraw Wildlife Foundation
Chapter 8: Jerry W. Dragoo
Chapter 10: Martin Schwab
Chapter 11: National Park Service
Chapter 12: Sue Mandeville, Cat Goods, Inc.
Chapter 13: Beate Ludwig
Chapter 14: Stephen W. R. Harrison
Chapter 15: Humane Society of the United States
Chapter 16: Francesca Ferrara
Chapter 17: USGS Western Ecological Research Center

Special discounts are available for bulk purchases of this book. For more information, please contact
Special Sales at 410-516-6936 or specialsales@press.jhu.edu.

The Johns Hopkins University Press uses environmentally friendly book materials, including
recycled text paper that is composed of at least 30 percent post-consumer waste, whenever
possible. All of our book papers are acid-free, and our jackets and covers are printed on
paper with recycled content.

To my parents, Earl and Jody, and my wife, Jane, for their love and support
SDG

To Elise Kelley, for her love, support, and sacrifices, during the writing of this book and always; and to Eric York, a great colleague and a great friend who is greatly missed
SPDR

To Ellen Cypher for her ever-present support, encouragement, companionship, and love
BLC

CONTENTS

PREFACE

Why urban carnivores? The field of urban ecology is bursting at the seams, and it does not seem necessary at this point to justify the necessity of focusing attention toward urban areas. But what is unique about carnivores in urban areas as opposed to wildlife in general?

Forman (2008) pointed out that conservationists typically focus their attention on rural or remote areas, where people are scarce, whereas urban planners are focused on cities, where people are concentrated. As urbanization and its consequences for conservation accelerate, an integration of the disciplines is needed. In many ways, studying the order Carnivora encourages linkage between rural/remote and urban areas. Some species exhibit movements that traverse the urban-rural gradient, such that policies and management of those species in one area may well influence the other. For some Carnivora, urban centers may serve as population sinks on the landscape, whereas for others, cities represent refugia and possibly even sources of migrants for precarious populations located outside urban areas.

Members of Carnivora present both similar, and unique, conservation and management problems compared with other wildlife species. For some species, issues such as fragmentation, conservation of native habitat, and roads are paramount, just as they are for many other wildlife species. But for others, especially those that occupy the top of the trophic pyramid, there are issues unique among wildlife species, the most important probably being the relationship between those top carnivores and humans. In the United States, cities are relatively new and, largely through persecution, have been devoid of wild carnivores for most of their existence. However, in recent decades, the number of larger carnivore species appearing in cities has apparently increased. This increase could be the result of efforts to conserve habitat quality and protect predators in remote/rural areas. Likewise, urban planners' emphasis on green spaces may provide opportunities for carnivores to move from rural/remote sites into the urban landscape.

These points are certainly solid arguments for the study of urban carnivores. But we would argue that the most important reason for a focus on Carnivora is visceral. It is undeniable that members of the Carnivora elicit strong feelings in people—fascination, admiration, fear, hate—unlike any other wildlife group. Hans Kruuk (2002) has suggested that our feelings toward carnivores and resulting management policies may be a manifestation of ancestral predator-prey relationships, developed during a time when we were prey of the Carnivora. In other words, our feelings toward Carnivora were created eons ago and lie deep within each of us.

This book began as a symposium at the 2006 Annual Conference of The Wildlife Society in Anchorage, Alaska. There was a certain irony in presenting a symposium on urban carnivores in the least-populated state in the United States. During that conference, we solidified the concept of an edited volume, and we began recruiting contributors. Our goal for this project was to invite researchers that have conducted long-term, extensive research on a particular carnivore species within an urban environment. We have found that rarely does all information from long-term studies get presented in journal articles, so we thought that researchers with firsthand accounts would be able to provide insight that isn't currently available in existing literature. To that end, we encouraged contributors with considerable experience to go beyond the data, when appropriate, and share anecdotes or professional opinions that may at times be provocative or stimulating. We also welcomed the presentation of data that has not been previously published, both to give a fuller picture of the species or population studied and because many studies represent either long-term efforts, where new information is constantly generated, or they are recently completed projects.

ACKNOWLEDGMENTS

All chapters were reviewed independently by one to three peers. We thank the following for serving as reviewers:

Philip Baker, University of Reading; Erin Boydston, U.S. Geological Survey; Jeremy Bruskotter, Ohio State University; Kevin Crooks, Colorado State University; Stephen DeStefano, USGS Cooperative Research Units; David Drake, University of Wisconsin; Mark Elbroch, University of California–Davis; Paul Flournoy, Ohio State University; Matthew Gompper, University of Missouri; Todd Gosselink, Iowa Department of Natural Resources; Stephanie Hauver, Ohio State University; Graziella Iossa, University of Bristol; Roland Kays, New York State Museum; Paul Krausman, University of Montana; Serge Larivière, Cree Hunters and Trappers Income Security Board; David Macdonald, Oxford University; Suzie Prange, Ohio Department of Natural Resources; Howard Quigley, Craighead Beringia South; Katherine Ralls, Smithsonian Institution; Timothy Roper, University of Sussex; Benjamin Sacks, California State University–Sacramento; Ray Sauvajot, National Park Service; Alastair Ward, Central Science Laboratory, York, England; Paige Warren, University of Massachusetts; Gregory Warrick, Center for Natural Lands Management; Alison Willingham, Ohio State University.

We thank V. Burke for giving us the opportunity to work with the Johns Hopkins University Press, and E. Rodgers and V. Burke for stewarding us through this process. J. L. Gehrt provided invaluable assistance with the bibliography and spot-checking the manuscript. We thank Lena Lee, data manager and GIS specialist at Santa Monica Mountains National Recreation Area, for her expertise and extensive efforts in producing and refining maps and figures for six different chapters. SDG wishes to thank the Max McGraw Wildlife Foundation, including Charlie Potter (President and Chief Operating Officer), for support while he morphed from a small-town Kansas boy to an urban wildlife ecologist. His research also has benefited immeasurably through the collaborations with Chris Anchor and the Forest Preserve District of Cook County, and Dr. Dan Parmer (deceased) and Dr. Donna Alexander, Cook County Animal and Rabies Control. SPDR would like to thank the staff of Santa Monica Mountains National Recreation Area, the world's best national park, for their support during the writing of this book and of all the wildlife projects at the park.

The CSUS Endangered Species Recovery Program provided logistical support for BLC during preparation of this book.

Funds from the Western Section of The Wildlife Society, the San Joaquin Valley Chapter of The Wildlife Society, the Max McGraw Wildlife Foundation, and Ohio State University Extension helped to defray publication costs.

Last, we certainly do not pretend that this book covers the whole field of urban carnivores. There are many species, populations, cities, and subjects that remain to be addressed. However, we hope that this volume will give the reader a good sense of the research that has been conducted on urban carnivores to date and of the management and conservation issues that arise. And we hope it will engender further interest in this exciting and growing field.

CONTRIBUTORS

PHILIP J. BAKER
Lecturer in Vertebrate Conservation
School of Biological Sciences
University of Reading
Whiteknights
Reading
Berkshire RG6 6AS
United Kingdom

PAUL BEIER
Professor of Conservation Biology
and Wildlife Ecology
School of Forestry
University of Northern Arizona
Flagstaff, Arizona 86011
USA

ERIN E. BOYDSTON
Research Ecologist
U.S. Geological Survey
Western Ecological Research Center
Irvine, California 92602
USA

KEVIN R. CROOKS
Associate Professor
Department of Fish, Wildlife, and
Conservation Biology
Colorado State University
Fort Collins, Colorado 80523
USA

PAUL D. CURTIS
Associate Professor
Cornell University
Ithaca, New York 14853
USA

BRIAN L. CYPHER
California State University–Stanislaus
Endangered Species Recovery Program
Turlock, California 93382
USA

DANIEL J. DECKER
Professor and Director
Human Dimensions Research Unit
Department of Natural Resources
Cornell University
Ithaca, New York 14853
USA

STEPHEN DeSTEFANO
Unit Leader and Research Professor
U.S. Geological Survey
Massachusetts Cooperative Fish
and Wildlife Research Unit
University of Massachusetts
Amherst, Massachusetts 01003-9285
USA

JERRY W. DRAGOO
Mephitologist and Research Associate
Division of Mammals
Museum of Southwestern Biology
University of New Mexico
Albuquerque, New Mexico 87131-0001
USA

TODD K. FULLER
Professor
Natural Resources Conservation
University of Massachusetts
Amherst, Massachusetts 01003-9285
USA

STANLEY D. GEHRT
School of Environment and Natural Resources
The Ohio State University
Columbus, OH 43210-1095
Max McGraw Wildlife Foundation
Dundee, Illinois 60118
USA

TODD E. GOSSELINK
Forest Wildlife Research Biologist
Iowa Department of Natural Resources
Chariton Research Station
Chariton, Iowa 50049
USA

JOHN HADIDIAN
Director, Urban Wildlife Programs
The Humane Society of the United States
Washington, D.C. 20037
USA

STEPHEN HARRIS
School of Biological Sciences
University of Bristol
Woodland Road
Bristol, BS8 1UG
United Kingdom

JAN HERR
Service de la Conservation de la Nature
Administration de la Nature et des Forêts
L-2453 Luxembourg
Luxembourg

GRAZIELLA IOSSA
School of Biological Sciences
University of Bristol
Woodland Road
Bristol, BS8 1UG
United Kingdom

LISA M. LYREN
Ecologist
U.S. Geological Survey
Western Ecological Research Center
6010 Hidden Valley Road
Carlsbad, California 92011
USA

SUZANNE PRANGE
Research Biologist
Ohio Department of Natural Resources
Waterloo Wildlife Research Station
Athens, Ohio 45701
USA

SETH P. D. RILEY
Wildlife Ecologist, National Park Service
and Adjunct Assistant Professor,
University of California, Los Angeles
Santa Monica Mountains National Recreation Area
Thousand Oaks, CA 91360

RICHARD ROSATTE
Senior Research Scientist
Ontario Ministry of Natural Resources
Wildlife Research and Development Section
Trent University
Peterborough, Ontario K9J 7B8
Canada

RAYMOND M. SAUVAJOT
National Park Service
Santa Monica Mountains National Recreation Area
Thousand Oaks, California 91360
USA

WILLIAM F. SIEMER
Research Specialist
Human Dimensions Research Unit
Department of Natural Resources
Cornell University
Ithaca, New York 14853
USA

KIRK SOBEY
Ontario Ministry of Natural Resources
Rabies Research Unit
Trent University
Peterborough, Ontario K9J 7B8
Canada

CARL D. SOULSBURY
School of Biological Sciences
University of Bristol
Woodland Road
Bristol, BS8 1UG
United Kingdom

TIMOTHY R. VAN DEELEN
Assistant Professor
University of Wisconsin, Madison
Madison, Wisconsin 53706
USA

PAIGE S. WARREN
Assistant Professor
Natural Resources Conservation
University of Massachusetts
Amherst, Massachusetts 01003-9285
USA

PAULA A. WHITE
Museum of Vertebrate Zoology
University of California, Berkeley 94720
USA

HEATHER WIECZOREK HUDENKO
Graduate Research Assistant
Human Dimensions Research Unit
Department of Natural Resources
Cornell University
Ithaca, New York 14853
USA

URBAN CARNIVORES

1

The Urban Ecosystem

STANLEY D. GEHRT

T HE TIMING of this book coincides with a landmark period in the
history of humans on this earth. On one level, the current human
population, estimated at more than 6 billion (United Nations [UN]
2006), continues to set new, daily landmarks; with each day, the human
population reaches unprecedented levels. However, the landmark specifically relevant to this book and, perhaps not as well known to the general
public, is that, for the first time in human history, the world's population
is now dominated by people living in cities rather than rural populations
(figure 1.1). Currently, an estimated 3 billion people reside in urban areas,
and this will increase to 5 billion by 2030 (UN 2006). Some of the human
population growth will occur in the 2 dozen megacities across the earth,
but most of the population growth will occur in the many smaller cities
of 250,000–500,000 people (Forman 2008). This historic transformation
apparently occurred around 2008, and projections are that the worldwide
population will remain dominated by urbanites as the population stabilizes
for the long term (United National Population Fund [UNFPA] 2007). *Homo
sapiens can now be called an urban species, and we have entered the urban century*
(Forman 2008).

The increasing urbanization of the world has garnered the attention
of the scientific community, but this interest in urban areas took time.
Although studies of urban flora occurred in Europe as early as the 1600s
(Sukopp 2002), the concept of cities as ecosystems in their own right, and
the development of urban ecology among European ecologists, did not
occur until the 1970s (Sukopp 2002). In North America, ecologists intent

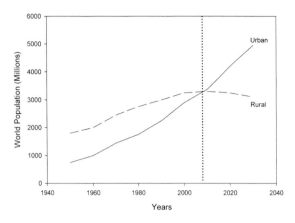

Figure 1.1. World human population trends (United Nations Population Fund, 2007).

on exploring the mysteries of the natural world understandably avoided cities and their surrounding areas, given that they are considered antitheses of natural systems. Consequently, our understanding of the ecology of urban landscapes lags behind most systems, giving us this rich irony: urban systems, the areas in which most ecologists live and work, remain frontiers for ecology and conservation. This is especially true for mammalian carnivores, often considered symbols of wilderness.

Although the field of urban ecology does not have a long history, it has been making up for lost time. There are now journals exclusively dedicated to an urban focus, such as *Urban Ecosystems, Landscape and Urban Planning,* and *Urban Forestry.* Traditional journals have dedicated special sections to urban issues. Likewise, urban issues have been the subjects of special symposia at professional conferences, and some conferences are completely dedicated to urban research. High-profile Long-Term Ecological Research projects funded by the National Science Foundation have been dedicated to the study of urban ecology in the Baltimore and Phoenix metropolitan areas. Currently, the discipline of urban ecology has progressed beyond a focus within urban areas, such as studies of species within cities, to focus on urban areas ecologists recognize as ecosystems in their own right (Grimm et al. 2000). That is, urban ecologists are interested in ecosystem processes within and among urban areas (Shochat et al. 2006), including how urban ecosystems interact with rural or natural landscapes (Pickett et al. 2001; Kaye et al. 2006). The results emanating from this burst of research strongly suggest we are only beginning to

uncover the mysteries of cities and how they affect our world.

WHAT IS URBAN?

This book addresses the relationship between urban areas, urbanization, and urban residents and species within the mammalian order Carnivora. While it is a rather simple process to identify carnivore species (despite a tremendous range of characteristics among species; see chapter 2), coming to a consensus about identifying an urban system, and whether an animal is living in an urban area, is not as easy. McIntyre et al. (2000) provided a review of different interpretations of "urban," particularly among different disciplines. Some definitions included human population size or density for a particular area, such as >2,500 people or >620 individuals/km^2 (U.S. Census Bureau), or >20,000 people in one area (e.g., UN). Alternatively, other ecologists have used energy consumed by area, such as >100,000 kcal/m^2/yr. Most interesting, social scientists may define an urban area by density-dependent relationships among people, or a psychologist may use emotional responses by subjects to define urban (Ulrich et al. 1991). In other cases, general definitions are used, such as "a dense agglomeration of people and firms" (Glaeser 1998).

Because different carnivore species may interpret urban areas differently, not to mention urban biologists, we have avoided restricting authors in this book to a single definition. However, the various references to "urban" throughout this book are consistent with a general description of an urban area as a large grouping of people and associated structures comprising at least one town or city.

CHARACTERISTICS OF URBAN AREAS

What are the characteristics of an urban area? Large cities have an inner core (see glossary; table 1.1) of nearly complete development and impervious surfaces, with a lack of evident greenness. The inner core is surrounded by a region of variable levels of development and is most commonly dominated by suburbs. This outer region is often highly variable, ranging from intense development to preserved green spaces. In larger metropolitan regions, the outer ring may contain pockets of smaller "inner cores" of city centers, providing a complex landscape-level urban-rural gradient.

Table 1.1. Glossary of terms associated with urban ecosystems

Anthropogenic	Entity or process associated with humans.
City	A center of high human population density within an urban ecosystem (Kaye et al. 2006).
Ecological footprint	The total land area required to meet the demands of its population in terms of consumption and waste assimilation (Rees and Wackernagel 1996).
Exurban	Similar to periurban. Human developments on the periphery of urban areas. Also, residences in the transition zone between rural and suburban fringes.
Heat island	Temperature difference between the urban core and surrounding rural areas.
Natural	A habitat or relationship that resembles a system without modification by human actions.
Periurban	Similar to exurban. Human developments on the periphery of urban areas. Also, residences in the transition zone between rural and suburban fringes.
Suburban	Human-dominated areas, primarily residential or commercial, ringing and interacting with an urban core.
Synanthropic	Species that thrive in urban areas.
Urban	An area of human residence, activity, and associated land area developed for those purposes. Usually defined by a threshold human density.
Urban core	Center of high population density and activity within the urban ecosystem, with noticeably high levels of impervious surfaces.

Figure 1.2. The Chicago metropolitan area, Illinois, United States. Max McGraw Wildlife Foundation.

Most urban areas of cities worldwide with 250,000 to 10 million people have a radius of about 70–80 km (Forman 2008). Cities tend to be oriented near bodies of water (i.e., rivers, lakes, oceans). Urban areas are incredibly dynamic, especially in the outer ring where land conversion is never-ending and can be dramatic. The conversion of rural land or natural habitat for urban use can outpace population growth. For example, much of my research has focused on the Chicago metropolitan area (figure 1.2), which is one of the world's megacities with a human population exceeding 9 million. In that area, the human population increased 4% over a 20-year period (1970–1990), whereas land consumption for urban purposes increased 46% during the same period (Northeastern Illinois Planning Commission 1992).

The effects of urbanization on the landscape are profound. Natural landscapes are fragmented, isolated, and degraded (Marzluff 2001); soil composition and hydrology is altered (Booth and Jackson 1997); energy flow and nutrient cycling is modified if it exists at all (McDonnell et al. 1997; Grimm et al. 2000); and floral and faunal communities are simplified and homogenized (Blair 2004).

CHARACTERISTICS OF THE URBAN ECOSYSTEM

The urban ecosystem primarily comprises the same abiotic and biotic components and their interactions that characterize most ecosystems. Figure 1.3 is a conceptual model outlining these components in the form of spheres (following Marzluff et al. 2008), including the following abiotic spheres: atmosphere (air), hydrosphere (water), and pedosphere (soil). The biotic sphere includes flora and fauna and their interactions, such as trophic interactions (between primary producers and consumers) and drivers influencing interactions, such as predation, parasitism, and competition. The critical aspect of the urban ecosystem is the anthroposphere, which represents human influences through socio-politico-economic actions. All spheres interact with each other to varying degrees, but the center of the ecosystem is the anthroposphere and the relatively strong interactions between that sphere and all others.

Although general patterns have been identified for urban ecosystems, it is important to note that no two cities are identical, with the same population density, same pattern of development, same topography, or

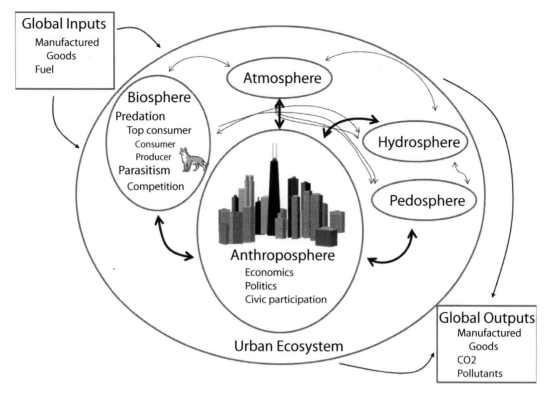

Figure 1.3. Conceptual model of the urban ecosystem organized into abiotic and biotic spheres (adapted from Marzluff et al. 2008). Arrow widths reflect the relative importance of the anthroposphere sphere.

even the same socioeconomic characteristics. Indeed, I have come to think of cities as having their own personalities, much like people that look similar but behave quite differently. These different "personalities" may translate to variations in ecological processes, especially with differences in area or human population size. For example, plant species richness may vary based on city size or history (Pickett et al. 2001), and avian predators may respond to urban areas differently, depending on the relative size of the city. Sorace and Gustin (2009) observed that specialist and generalist avian predator distributions differed among 27 cities in Italy and that some specialized predators occurred in urban centers of small towns but not large cities. Keeping in mind this inherent variability among urban systems, I will proceed to give a brief overview of the characteristics of urban ecosystems.

To illustrate some basic principles of the urban ecosystem within the context of this volume, let's consider the urban ecosystem from the perspective of a carnivore. Let us construct a hypothetical animal, perhaps from a species similar to a coyote: medium sized

(12–15 kg), with a penchant for small, vertebrate prey but that may take advantage of human foods (e.g., refuse, pet food, ornamental fruit) when available. As our carnivore travels along the rural-to-urban gradient, what are they likely to experience?

Climate

Our hypothetical carnivore may be affected to some degree by an altered climate, even if it does not directly notice the increase in temperature near the urban core. Urban ecosystems typically have an elevated ambient temperature as a product of the "heat island effect." The combination of factors, including impervious surfaces that absorb sunlight and release heat, the production of particulates (that help contain the heat), and energy from combustible fuels, results in an island of raised temperatures in and around the urban core. Temperature differences between the heat island and surrounding rural areas are small for cities <250,000 (Berry 2008) but can be more substantial (6°–12°C) in larger cities (Kaye et al. 2006). In large cities, the heat island can have profound effects on vegetation phenology and primary productivity. Field data have revealed that flowering dates are four to seven days earlier in urban areas than in adjacent rural landscapes (Roetzer et al. 2000), and the urban growing season is on average 15 days longer for cities >10 km² in size in the eastern United States (Zhang et al. 2004). Remote sensing indicates a city's climate "footprint" may be 2.4 times greater than that of the city limits, and the larger the city, the greater the extended greenup effect in rural areas (Zhang et al. 2004). Depending on local environmental characteristics and city size, heat island effects can extend 10 km into the surrounding rural landscape. In addition to higher temperatures, other general patterns for climate include a greater preponderance of contaminants, less sunshine, more cloud cover, more precipitation (but less snow near the core), higher frequency of thunderstorms, lower relative humidity, and lower mean wind speed in urban compared with rural areas (Berry 2008). Our carnivore may not be directly affected by these climate changes but may be indirectly affected if plant productivity affects mast production (if omnivorous) or prey abundance.

Water

Depending on the rural matrix, our carnivore may notice the altered hydrology that occurs in urban ecosystems. If our urban area is surrounded by an arid environment, the carnivore may notice an increase in available surface water due to artificial watering of the landscape, with subsequent greening of vegetation compared with the rural landscape. In addition, there is nearly always a significant water body (river, lake, etc.) associated with a city and sometimes with individual neighborhoods in the form of water retention ponds. Hydrologic processes are typically altered substantially to accommodate human needs. Probably not noticed by our carnivore is that urban streams are typically channelized and modified to reduce flooding and provide areas for development. However, semiaquatic carnivores (e.g., mink, river otter) may be aware of accelerated flow rates and simplified structures of urban streams (Paul and Meyer 2001; Riley et al. 2005). Because of impervious surfaces, water travels farther in urban areas than in rural areas. In Baltimore, Maryland, increasing impervious surfaces to only 10%–20% of overall surface area will double surface runoff compared with a forest system (Kaye et al. 2006). An extreme example of altered hydrology would be in the Chicago metropolitan area (figure 1.2), where engineers actually reversed the flow of the Chicago River

in 1871 for waste management (Greenberg 2002). The river now flows southwest into the Illinois River rather than into Lake Michigan where the mouth of the river occurs.

Light

One physical property of larger urban areas that might be noticed by our carnivore is the urban glow at night. As a native Kansan (an area of low human population), one of the first things I noticed after beginning work in the Chicago metropolitan area was that even on the most clear night stars were no longer visible because the urban landscape produced a tremendous amount of light. Fifteen years later, I am still amazed that I can see the urban glow from 60 miles away, depending on night conditions. As our animal moves closer to the urban core, cycles of light associated with moon phases may become nonexistent.

Urban light, sometimes termed "light pollution," consists of two types: (1) astronomical light pollution and (2) ecological light pollution (Longcore and Rich 2004). Astronomical light pollution refers to the interference with our ability to see the night sky, such as celestial bodies and changes of light from moon phases. Ecological light pollution includes direct glare or fluctuations of artificial light that may come directly or indirectly from building lights, street lights, vehicle lights, or sky glow (astronomic light pollution). Artificial light has been shown to affect the navigation of some wildlife, by confusing them (migration by birds at night), repelling them (e.g., some spiders, rodents, mountain lions), or attracting them (e.g., some arthropods, amphibians; reviewed in Longcore and Rich 2004). The response by some species to artificial light can be dramatic. We have all seen the dramatic response some insects have in their attraction to mercury vapor lights.

Artificial lights have been shown to alter predator-prey relationships for some reptiles and their prey (Schwartz and Henderson 1991), as well as those moths attracted to lights and the bats that prey on them (Svensson and Rydell 1998). Crows roosts in urban areas tend to be oriented near lights, presumably to avoid owl predation (Gorenzel and Salmon 1995).

Some small mammals, the likely prey of our carnivore, tend to reduce their activity and movements on bright nights (Gilbert and Boutin 1991; Lima 1998), which would either suggest the bright urban sky would be detrimental to our carnivore or to the prey (since the prey could not wait for a "dark" night). However, it is sometimes misleading to extrapolate animal be-havior from natural systems to predictions for urban areas, as both predator and prey may acclimate their activities and senses to a bright night. I am unaware of any research that has demonstrated an effect of "urban light pollution" on the predator-prey dynamic involv-ing mammalian carnivores and their prey.

Noise

Humans are noisy, and there is no doubt that carni-vores would notice the increase in noise pollution as they move toward the city core. Urban noise may come from point sources or may have a linear pattern (i.e., roads), can be reflected vertically by buildings, or have daily fluctuations in intensity (Katti and Warren 2004). The noise from human activities can be substantial (Rheindt 2003). One of my long-term study areas, the Ned Brown Forest Preserve in Cook County, Illinois, is situated near O'Hare International Airport and is bordered on two sides by six- to eight-lane interstate highways with traffic volumes averaging >100,000 vehicles/24 h and is bisected down the middle by a four-lane state highway (see Prange et al. 2003, 2004). In some parts of the park, my technicians and I have to shout, despite standing next to one another, to be heard above the airplanes overhead and vehicle traf-fic below. Although we have not formally measured it, we have noticed that background noise volumes in the park decline during evening, which is a common phenomenon (Katti and Warren 2004), and have varied across days as the flight patterns from O'Hare change from day to day. I have often wondered how animals that rely on sounds adjust to living in this area. Recent research from other sites suggests that at least some songbirds adjust the frequencies of their calls in re-sponse to urban background noise (Slabbekoorn and Peet 2003). The possibility exists that, in addition to habitat fragmentation and road barriers, background noise may represent another challenge for wildlife in urban settings (Katti and Warren 2004).

Habitat Fragmentation

A noticeable characteristic of the urban ecosystem is the fragmentation of natural habitats. Our carnivore will quickly notice this pattern as they seek refuge from people and room to hunt in habitats with which they are most comfortable. Likely, they are sensitive to some degree to the distance they must travel through the ur-ban matrix between patches (Crooks 2002), although they may begin using patches of modified habitats

when natural habitats become difficult to find. Indeed, it is possible that this feature, at this point, dictates more than any other whether our carnivore decides to continue into the urban ecosystem. Our carnivore may notice that small patches have fewer potential prey spe-cies than larger patches (e.g., birds; Chace and Walsh 2006) and that urban habitat patches may have fewer native species than similar patches in rural landscapes (Knapp et al. 2008). Despite a general relationship be-tween urban landscapes and biotic diversity, urbaniza-tion tends to create homogenization of species across the larger landscape (McKinney 2006).

Complicating things further for our carnivore is that roads often separate patches of habitat. All ter-restrial wildlife must face the challenge of negotiating roads if they are to persist in the urban ecosystem. Our carnivore may have noticed one or more conspecifics dead along these roads, as carnivores are vulnerable to vehicle collisions (e.g., Smith-Patten and Patten 2008). Carnivores may avoid the dangers of roads by treat-ing roads as barriers and avoid them (Gehrt 2004b), or they may learn traffic patterns (however, this has not been verified for any species to my knowledge). Over the years of radio-tracking hundreds of raccoons and coyotes in the Chicago area, my technicians and I have opportunistically observed individuals crossing roads, and some behaviors suggest some adjustments for traf-fic patterns (such as delaying the crossing until there is a gap in traffic, rushing across roads rather than a normal trot, stopping in medians to wait for a gap in traffic, etc.). We have occasionally observed raccoons congregate on the side of a road and then quickly cross the road as a group once there was a break in traffic, and coyotes crossing half of a multilane interstate, sit and wait in the median, and then quickly crossing the other half.

Roads may be a paradox for some carnivore species because they inhibit movements across their surface, but roads also may serve as movement corridors into the urban matrix. For species that cannot adjust to roads, the urban ecosystem may represent an eco-logical "sink" to overall populations (e.g., black bears; Beckmann and Lackey 2008). Our carnivore may learn the dangers of vehicles and survive to continue the journey, at least temporarily.

Biota

Our carnivore will likely notice a change in the types of plants and animals it encounters in the city, and its ability to adjust to these changes will be important to its success in the urban ecosystem. Contrary to our intuition that urban landscapes are detrimental to wildlife species, overall biotic diversity is often relatively high in urban landscapes (Ricketts and Imhoff 2003; Pickett et al. 2008), which has been observed for various taxa and spatial scales in the United States and Europe (Dobsonet al. 2001; McKinney 2002; Wania et al. 2006). Such diversity occurs because of a propensity for human settlements to exist in areas of inherent native diversity, the influx of alien species, or the tremendous heterogeneity of habitat types characteristic of urban development.

Flora

Biotic diversity may be highest at the urban fringe because of a mix between urban-sensitive species and synanthropic species (Alberti and Marzluff 2004; Pickett et al. 2008). Floral diversity is relatively high in urban landscapes, typically due to an influx of exotic species (Pyšek et al. 2004) and to remnant native communities. For example, exotic species comprised an average of 40% of the plants in 54 Central European cities (Pyšek 1998). Floral diversity reflects the individual characteristics of cities, such as size and history of development. Number of plant species tends to correlate positively with number of human inhabitants (Klotz 1990 in Pickett et al. 2001). Likewise, larger old cities tend to harbor more plant species than similar-sized younger cities (Sukopp 1998 in Marzluff et al. 2008). However, long-term monitoring of vegetation over 30 years in Plzen, Czech Republic, revealed an overall decline in floral diversity and an increase in homogenization across the city (Pyšek et al. 2004).

Canopy cover may increase or decrease with urbanization, the direction of change depending on the rural landscape. For example, urban canopy may only be approximately 31% of rural areas in regions of nearly continuous forest (e.g., mesic forest regions), whereas the urban forest canopy may be significantly greater than adjacent rural landscapes in prairie-savanna or desert regions (Nowak et al. 1996; Pickett et al. 2001). In addition, the composition of the forest habitat in urban areas is usually relatively diverse owing to exotic species, including subcanopy species.

Recent research has revealed that the enhanced biological richness in urban landscapes may extend beyond the physical boundaries of the city, much like the patterns observed for urban climates and resulting primary productivity. Outside Berlin, Germany, floral richness, as measured by seed deposition, was greater along motorways exiting the city than roads entering the city (von der Lippe and Kowarik 2008). Plant dispersal of exotic species is facilitated by human activities such as seed deposition by vehicles.

Fauna

Animal diversity is largely homogenized within and among urban areas (Chace and Walsh 2006; McKinney 2006). However, the relationship between urbanization and species diversity may vary based on the surrounding habitat matrix, taxonomic group under consideration, size of the metropolitan area, and pattern of urbanization (Alberti and Marzluff 2004; Jokimäki et al. 2005). For example, bat diversity actually increases with urbanization in the Chicago metropolitan area because urban areas contain more water and woodland habitat than the surrounding rural landscape (which is dominated by row crop agriculture; Gehrt and Chelsvig 2003, 2004). However, in most cases, it appears that native diversity or richness declines as urbanization fragments and increasingly isolates natural habitats (Stratford and Robinson 2005; Knapp et al. 2008).

Our carnivore may notice that some species from rural areas have disappeared, whereas others have become quite common. They will also notice species they may not have seen before. In addition to exotic plants, urban ecosystems have exotic animals. Exotic species are typically commensal species and are those closely associated with human developments, such as the house mouse or Norway rat, which may serve as prey. While exotic species increase in abundance in urban areas, their distributions may be highly variable within the urban matrix. For example, the three most common exotic bird species in the Baltimore, Maryland, system were not spatially correlated, and each was associated with different urban habitats (Pickett et al. 2008).

Rodents and some avian species may serve as potential sources of food, or our carnivore may otherwise ignore them. Our carnivore will likely notice an increase in domestic cats and dogs that may serve as competitors, prey, predators (at least for small carnivores), or hosts of pathogens that can be transmitted across species (Beck 1973; Bradley and Altizer 2006). Indeed, these domestic carnivores may be the most common carnivores found in some parts of the urban ecosystem (chapter 12). Thus, these exotic species may present an increasing challenge or opportunity for our carnivore as they attempt to avoid, to ignore, or to depredate these strange exotics.

Unseen Dangers

Vehicles are likely the most obvious challenge to survival in urban landscapes for most mammalian carnivores; however, other challenges to survival are more subtle, such as disease. Urbanization results in an increase in the types and concentrations of pollutants in the air, ground, and water. The direct or indirect effects of pollutants, including heavy metals, are probably relatively minor compared with other challenges of living with people; however, this needs more study. Frequent exposure to contaminants may lead to a compromised immune system, but this has received little attention (Bradley and Altizer 2006). Carnivores may be exposed to poisons distributed in the urban environment intentionally, such as rat poison (chapter 7), or unintentionally, such as antifreeze (S. D. Gehrt, unpublished data).

Exposure to macro- or microparasites may increase in urban systems (Bradley and Altizer 2006), and this may be particularly important for mammalian carnivores given the plethora of parasites they can harbor. Human activities often cause changes in the distribution and abundance of carnivore species, which may alter the dynamics of pathogens in their populations, or pathogens may spill over across species. Prevalence of sarcoptic mange in red foxes (Baker et al. 2000; Gosselink et al. 2007), canine parvovirus in gray foxes (Riley et al. 2004), or canine distemper in raccoons and striped skunks (Gehrt 2005) may be higher in urban landscapes. Prevalence of heartworm was nearly an order of magnitude greater for urban coyotes than for rural coyotes in Illinois (chapter 7).

Sociopolitical Boundaries

Within urban ecology, biological ecology and social ecology blend together so that they cannot completely be partitioned (Grimm et al. 2000; Pickett et al. 2001). Social differentiation, a fundamental aspect of social ecology, has important implications for the allocation of resources and development across the landscape. Social differentiation recognizes that resources are not allocated evenly and occurs over spatial scales (Machlis et al. 1997). Interactions between carnivores and human residents are certainly influenced by social differentiation. For example, human responses to some carnivore species will likely vary, depending on the resources or political influence available to those individuals.

Our carnivore undoubtedly fails to notice shifts in sociopolitical dynamics as it moves from the rural landscape into the urban ecosystem. However, I have saved this feature for last because of its potential importance, as it may have as much, or more, influence on the future success of our animal, despite its intangible quality, than any physical changes to the environment from urbanization. Nearly all species of Carnivora are classified as furbearers by states and provinces across North America. Thus, in rural areas, there is a legal harvest during the furbearer season and for some species a hunting season. For example, nearly seven thousand coyotes were legally harvested in Illinois during the 2007–2008 fur harvest season (Illinois Department of Natural Resources, unpublished data), and this activity represents the primary cause of death for rural coyotes in the state (Van Deelen and Gosselink 2006). However, within the Chicago metropolitan area, which spans all or portions of six counties, there are few, if any, places where hunting or trapping is legal.

The sociopolitical factors become more complex for our carnivore in the metropolitan landscape, where intangible changes take place across municipal boundaries, especially with regard to their responses to the presence of carnivores. At least 128 municipalities are packed into Cook County, Illinois, which is the center of the Chicago metropolitan area, and they vary with respect to nuisance wildlife programs and their tolerance toward carnivores. On another level, sociopolitical factors occur at neighborhood levels as well, with neighborhoods varying considerably in their response to carnivores and other wildlife. At both spatial scales, policies and programs are dynamic, so the sociopolitical landscape is a constantly shifting mosaic of multiple spatial scales that can have a dramatic effect on life or death for some Carnivora. Some of the radio-collared coyotes from our research in the Chicago area (chapter 7) have had home ranges that spanned portions of five different cities, each with different politics and resources for carnivore management. One could argue that the rural-urban shifts in sociopolitical dynamics are more extreme with greater implications for Carnivora than for any wildlife group (Kellert et al. 1996; Kruuk 2002).

CONCLUSION

This is a brief introduction to urban ecosystems. I have not addressed soil dynamics, carbon cycles, contaminants, nutrient flows, and refuse (these are addressed in detail in other sources), but my goal was to provide some background and to get the reader thinking about urban principles from a carnivore's perspective. Each urban ecosystem is unique, with different sizes and patterns of development, but they all share this characteristic: humans drive the system (figure 1.3). Indeed, the challenge for urban ecology is to determine to what extent "human drivers" alter traditional ecological processes (Pickett et al. 2008). More specifically, can ecological principles developed from natural systems be applied to urban ecosystems, or are human drivers so dominant and unique that new principles must be identified for urban systems (Kaye et al. 2006). Now that urban ecologists are focusing on as well as within urban ecosystems, identifying human drivers and their relative roles should produce new surprises.

It is probably safe to assume that cities were seen as refuges for people from mammalian carnivores during most of human history. However, as urbanization continues to consume land in this increasingly urban world, mammalian carnivores will seek to exploit our neighborhoods to varying degrees or face disappearing from the landscape altogether. As we move forward into this century of urbanization, it will be interesting to see whether *Homo sapiens* are able to find levels of coexistence in their own backyard with animals that, for all of our history, have served as our competitors and, for larger species, our predators.

2

Carnivore Behavior and Ecology, and Relationship to Urbanization

TODD K. FULLER

STEPHEN DESTEFANO

PAIGE S. WARREN

WHAT IS A CARNIVORE?

Ever since there have been animals, there have been carnivores; that is, organisms that kill and eat animal species and thus live a carnivorous lifestyle. There are carnivorous plants, intertidal anemones, and insects, and in the vertebrate world, any number of fish, snakes, and birds are carnivorous. Even among mammals a variety of species are part-time (e.g., bonobos; Hohmann and Fruth 2008), mostly (e.g., bats; Bonato et al. 2004), or even obligate (e.g., tiger quolls; Belcher et al. 2007; or killer whales; Baird et al. 2006) carnivores. The carnivores we are concerned with here are some of the species that compose the mammalian order Carnivora.

Here we introduce the variety of carnivoran species that occur in the world and try to make clear how diverse a group they are. We consider their taxonomic relationships and give examples of the tremendous variation in carnivore geographic distribution, morphology, ecology, and behavior, including their interactions with prey and with other predators. Finally, we introduce the consequences of this variation in biology with respect to interactions with humans and particularly with urban environments. Specific comparisons between urban and rural populations will be made in the species chapters (4–12) and between species, genera, and families in the taxonomic comparisons chapter (13). Here we provide the basic knowledge of the family and how its members may respond to urbanization.

TAXONOMIC CONSIDERATIONS

The origin of the order Carnivora dates to between 60 and 55 million years ago (mya), and these earliest carnivorans were miacids and viverravids (Flynn 1996). By about 25–35 mya, the extant families of Carnivora had differentiated and are identifiable morphologically through fossil remains and chemically through genetic analyses. An additional family that had differentiated from one of these groups about 20 mya, the Amphicyonidae, or "bear dogs," died out about 6–9 mya (Hunt 1996). Within the extant taxa, some species are basically monophyletic (e.g., red or lesser panda, although they are most closely related to the raccoon family) and have no current ancestors with any relationship much closer to them than to any other carnivoran. However, others (e.g., cats in the "Ocelot" lineage) have many closely related or sister species with which they share a common ancestor perhaps less than 1 mya (Johnson et al. 2007).

Of the approximately 270 living species in 11 families classified as Carnivora (Wozencraft 1989; but see Dragoo and Honeycutt 1997; Macdonald 2006), 34 or so species are marine carnivorans in 2 families, the Otariidae (sea lions and walruses) and the Phocidae (seals). The other 9 families are considered terrestrial and related as 2 suborders: the Feliformia, including cats (Felidae), civets and genets (Vivveridae), mongooses (Herpestidae), and hyenas (Hyeanidea); and the Caniformia, including dogs (Canidae), bears (Ursidae), raccoons and coatis (Procyonidae), and weasels, otters, and badgers (Mustelidae). Also, recent genetic studies have consistently demonstrated that skunks and stink badgers descend from a common ancestor and together form a separate familial lineage (Mephitidae) that diverged before the split between Mustelidae and Procyonidae (Dragoo and Honeycutt 1997).

Among these terrestrial groups, the Hyaenidae has the smallest number of species ($n = 4$) and has the least size diversity (about fivefold difference in mass between the aardwolf and the spotted hyena), while the Mustelidae number about 55 (Wozencraft 1989) and vary 1,000-fold in size (least weasel vs. sea otter). There are no ursids or procyonids in Africa, though they occur on all other continents except Antarctica (where there are no terrestrial carnivorans) and Australia (where only introduced canids and felids occur; Hunt 1996). Viverrids, herpestids, and hyenids are absent from North and South America but occur in Africa and Asia (and introduced mongooses occur on some Caribbean and Hawaiian islands). Procyonids and skunks occur only in North and South America, and stink badgers occur only in Java. Canids and felids have worldwide distributions on all continents except Antarctica.

Overall carnivore diversity has been assessed quantitatively in Africa (Mills et al. 2004), the Neotropics (Loyola et al. 2008), and worldwide (Sechrest et al. 2002). In Africa, where 70 species occur, carnivoran diversity is not highest in the biologically diverse rain forests of Central Africa as one might expect; rather, the "hot spots" (top 10% richest areas) for carnivoran species are in East Africa and the northern parts of southern Africa (Mills et al. 2004). In the Neotropics, carnivore phylogenetic diversity is greatest in rain forest ecoregions, whereas for large-bodied carnivores, the tropical Andes are important (Loyola et al. 2008). Worldwide, about 89% of carnivore species occur in at least 1 of 25 previously identified hot spots (Sechrest et al. 2002).

DISTRIBUTIONAL VARIATION OF SPECIES

Aside from domestic dogs and cats, red foxes have the widest geographic variation of any carnivore (70 million km^2; Macdonald and Reynolds 2004), but several other species range across multiple continents. For example, brown bears live throughout the Holarctic, striped hyenas occur across Africa and Eurasia, and pumas range from the southern tip of South America into the Rocky Mountains of western Canada (Macdonald 1992; Jackson and Nowell 1996).

In contrast, a number of species have restricted ranges. Seven of the eight carnivore species found on Madagascar (e.g., the fossa, a viverrid) are endemic to the island; the other, the small Indian civet, was introduced there (Hawkins and Racey 2008). Ethiopian wolves are an endangered endemic species restricted to isolated Afro-alpine regions (mountain "islands") of Ethiopia (Sillero-Zubiri and Marino 2004). Island foxes occur only on six small islands off the west coast of California in the United States (Roemer et al. 2004), and Iriomote cats, the most endangered felid species (a total population of about 100 individuals), are inhabitants of a single island in the western Pacific Ocean about 200 km east of Taiwan (Jackson and Nowell 1996).

Many carnivoran species have experienced major reductions in their geographic range due to persecution and habitat reduction by humans (e.g., African wild dogs; Woodroffe et al. 2004), but a number of species' ranges have been expanded as a result of introductions (Boitani 2001) or human-caused landscape changes. For example, red foxes are common in Australia but were introduced there (Macdonald and Reynolds 2004). The small Indian mongoose was introduced to several islands in the Caribbean and Hawaii where it also has thrived (Hoagland and Kilpatrick 1999). Introduced raccoon dogs now have a wider geographic range in northern and eastern Europe than their natural range in the Asian Far East (Kauhala and Saeki 2004). Finally, coyotes have greatly expanded their range in North America, not through introduction but rather on their own and likely as the result of land conversion and the removal of wolves since 1900 (Moore and Parker 1992).

On every continent of the world except Antarctica, many species of carnivorans are found; plant and animal prey of all sorts are distributed similarly. This also means that many species are well adapted to extremes in temperature, precipitation, and topography. For example, polar bears (Ferguson 2000) spend much of their lives on the Arctic ice pack. The habitat of Ruppell's fox typically includes sand and stone desert across the Sahara of northern Africa and the Middle East (Cuzin and Lenain 2004). Binturongs, a viverrid also known as the Asian bearcat, thrive in the rain forests of Southeast Asia (Grassman et al. 2005). Alternatively, snow leopards live throughout the high, cold mountain ranges of Central Asia (McCarthy and Chapron 2003), and meerkats (or suricates) inhabit all parts of the relatively flat and hot Kalahari Desert in Botswana and South Africa (Clutton-Brock et al. 2001).

The degree of disparity in habitat use within species also varies. Aside from those species whose absolute geographic range, and thus variation in habitat use, is extremely small, there are other species, some of which have rather large geographic ranges, whose habitat preferences are still very narrow. Black-footed ferrets, for example, live only in the grasslands of the Great Plains of North America, where they are obligate predators on prairie dogs, a ground-dwelling rodent (Miller et al. 1996). Similarly, giant pandas live only in bamboo forests of Asia (Schaller et al. 1985). Conversely, wolves range from arctic tundra in Siberia to deserts in the Middle East to rain forests in western Canada (Fuller 2004), and leopards occur in African deserts, Indian jungles, and northern forests in the Russian Far East (Jackson and Nowell 1996).

VARIATION IN ECOLOGY AND BEHAVIOR

Although most differences in species' distributions have a historic, evolutionary genesis, the consequential strategies for continued survival have resulted in tremendous variation in carnivore ecology and behavior. What follows is a cursory review of some of these contrasts.

Morphology and Physiology

Body sizes (weights/masses) of the Carnivora range from the 100-g least weasel to the 800-kg polar bear (Gittleman 1989b). Body weight is not correlated with habitat, activity cycle, or latitudinal gradients, but there are significant differences in body weight among insectivorous, herbivorous, and carnivorous species, and, among predatory carnivores, prey size and diversity increase with body weight (Gittleman 1985).

One can easily imagine that, given the badger's digging, the otter's fishing, or the cheetah's chasing, the skeleton and thus body proportions of carnivores also vary tremendously (Ewer 1973). Ambulatory bears and digging badgers have plantigrade (heal-on-ground) stances and relatively shorter legs than, for example, cursorial canids, which are digitigrade (toe-runners; Taylor 1989). It also follows that the variation in dental adaptations of carnivores is great as well; felids are often pure meat eaters with large cutting carnassial teeth, whereas frugivorous/plant-eating bears or shell-crunching otters have broad, grinding molars, and bone-crushing spotted hyenas have stout conical premolars and molars (Van Valkenburgh 1989).

In general, the digestive system of carnivores is simple, especially for those species (e.g., cats) consuming mainly easily digested meat; their gut length is only four times their body length (Ewer 1973). Some species have relatively long guts (e.g., sea otters), and others (e.g., spotted hyenas) produce copious hydrochloric acid to help dissolve bones (Macdonald 1992). The caecum, or gut pocket, of carnivores is usually absent or relatively small, though in canids it is prominent (Ewer 1973).

Some carnivores, like the canids, are typically seasonal breeders that regularly come into estrus (spontaneous ovulators), mate, and produce pups fifty to seventy days later (Ewer 1973). Others, like American mink and striped skunks, are induced ovulators (eggs released after behavioral, hormonal, or physical stimulation; Larivière and Ferguson 2003), with similar-length gestation periods. However, some ursids and mustelids and both "pandas," though spontaneous ovulators, have delayed implantation, or embryonic diapause and, as a result, have gestation periods lasting 95–365 days (Mead 1989).

Highly efficient sense organs seem essential for predators, both for capturing prey and for interacting successfully with one another. The eyes of nocturnal species usually have a relatively large anterior chamber with a large, curved lens and a highly convex cornea, as well as the presence of one or more layers of reflecting cells, the tapedum lucidum, outside the receptor layer of the retina (Ewer 1973). Carnivores also often have binocular vision, as well as a nictitating membrane, further enhancing visual acumen. The potential for carnivores to hear well is reflected in the facts that red foxes need to listen for rodents, aardwolves for termites, and hyenas for prey calling in distress (Mills et al. 2001). From a social perspective, it is not uncommon to hear wolves howl, lions roar, dholes whistle, hyenas whoop, or African wild dogs twitter to maintain communication among their kind (Peters and Wozencraft 1989). Similarly, olfaction plays a key role in prey capture (Cushing 1985) and carnivore social organization (Gorman and Trowbridge 1989).

Population Ecology

The rate at which an animal population could potentially increase is a result of several factors, all of which vary widely among carnivore species. The first age of reproduction of many species is 1-year-old, but for other large species, such as black bears, it may be as old as 6 years (Kasworm and Thier 1994). Similarly, number of offspring produced per reproductive event (litter size) varies from only 1 (e.g., giant panda) to 19 in arctic foxes (Ovsyanikov 1993). Most small- or medium-sized carnivores, especially in temperate areas, produce a litter every year after reaching maturity, but bears only reproduce every 2, 3, or 4 years, depending on species and location (Dahle and Swenson 2003). In monogamous species (e.g., red foxes), every adult of breeding age could potentially breed; for some social species, however, reproductive suppression of subordinates can result in a substantial portion of adults not producing offspring in a given year (e.g., >80% for dwarf mongooses; Creel 1996). Given this variation in parameters, it is easy to understand that populations of monogamous species with an early age of first reproduction, a large litter size, and young that disperse and breed quickly have a high potential of increase (e.g., 1.5- to 2.0-fold / year for wolves; Mitchell et al. 2008).

Conversely, some species with low reproductive rates (old age at first reproduction, small litters, long intervals between litters) have a low potential rate of increase (e.g., 5%–10% / year for brown bears; Garshelis et al. 2005).

Within species, the density of carnivores varies from one to over three orders of magnitude, and though obviously affected by intensity of mortality, a species' density is ultimately determined by the availability of food (Fuller and Sievert 2001); in fact, a comprehensive analysis of the entire order of carnivores indicated that, in general, 10,000 kg of prey supports about 90 kg of a given species of carnivore, irrespective of body mass (Carbone and Gittleman 2002). Food affects a species' density because relatively more abundant food resources result in better physical condition and thus affect reproduction by lowering first age of reproduction, increasing litter size, and/or lowering mortality rate of juveniles (Fuller and Sievert 2001). Poor food resources result in increased mortality of adults, larger home ranges, and increased rates of movement (Fuller and Sievert 2001).

Food Habits

Not all Carnivora are carnivorous. In fact, most are omnivorous. In addition to animal prey, some vegetation (often high-calorie fruits or nuts) is also consumed. Felids are usually considered the most obligate of carnivores (Ewer 1973), and canids, bears, viverrids, and perhaps procyonids are the most omnivorous and have the most generalized diet. Many carnivores scavenge (DeVault et al. 2003), yet cats rarely do so (but see Bauer et al. 2005); canids do so fairly frequently (e.g., Schrecengost et al. 2008), and hyenas do so quite often (e.g., brown hyenas; Mills and Mills 1978). Some species are mostly myrmecophagous in that they specialize in feeding on ants and termites (e.g., aardwolf; Richardson 1987; Kruuk and Sands 1972; and sloth bear; Joshi et al. 1997). Similarly, giant pandas are bamboo specialists (Schaller et al. 1985), and black-footed ferrets eat only prairie dogs (Miller et al. 1996). Within a species, diet can vary greatly, depending on the resources available (e.g., Iriarte et al. 1990), and for many carnivore species, what they eat depends on "what they can get" (Ewer 1973). Even within an area, diet may change as prey availability changes (e.g., O'Donoghue et al. 1998; Corbett and Newsome 1997).

Overall, it is doubtful that there are animal prey species that some type of carnivore does not eat. Size certainly doesn't matter; wolves kill 1,000-kg bison (Smith et al. 2000), tigers kill 1,500-kg gaur (Karanth and Sunquist 1995), and lions even hunt young elephants (Loveridge et al. 2006), but bat-eared foxes eat termites (Clark 2005). Massive brown bears feast on army cutworm moths (*Euxoa auxiliaries*; White et al. 1998); and 300-g stoats prey on >2,000-g lagomorphs (McDonald et al. 2000). Skunks eat bird eggs and nestlings, insects of various sorts, and apparently even clams (Cuyler 1924), and various viverrids and other small carnivores eat fish, reptiles, amphibians, and crustaceans (Ewer 1973; Wang and Fuller 2003).

Behavior

The daily activity patterns of carnivores range from almost exclusively nocturnal (e.g., large-spotted genets; Fuller et al. 1990), to crepuscular (active mostly at dawn and dusk; e.g., African wild dogs; Fuller and Kat 1990), to almost exclusively diurnal (e.g., cheetahs; Caro 1994). Species often are active throughout the day and night but at varying levels (e.g., spotted hyenas; Kolowski et al. 2007). Activity patterns within a species may also vary with levels and kinds of human activities (e.g., coyotes; Kitchen et al. 2000; red foxes; Baker et al. 2007) or with the composition of competing carnivore species (e.g., jackals; Fuller et al. 1989). With regard to seasonal activity patterns, some carnivore species have highly regularized breeding and denning dates (e.g., wolves; Fuller 1989a), but others have no fixed breeding season (e.g., lions; Haas et al. 2005). Home ranges vary in size between winter and summer for some species (e.g., wolves; Fuller 1989b), largely in relation to changes in prey distribution and thus changes in habitat use (e.g., Koehler 1991).

Most small, nocturnal carnivores and all felids (except for lions and domestic cats) are solitary except for females with offspring (raising them alone) and during mating season (Bekoff et al. 1984; Eisenberg 1989). This does not mean that they do not communicate through vocal olfactory or even visual signals but rather that they don't regularly travel together or interact directly. Many species in other taxa, however, have evolved various levels of sociality from monogamous parings (e.g., maned wolves; Rodden et al. 2004) to groups, clans, packs, or prides of >25 members (e.g., white-nosed coati; Gompper 1997). Additional group members may provide extra provisioning for offspring, increase hunting efficiency, or provide increased protection (Bekoff et al. 1984; Gittleman 1989b). Intraspecific variation in social organization, primarily due to differences in food resources and habitat (Bekoff et al. 1984), is at least

as large as that among species (e.g., golden jackals; Macdonald 1979a).

Carnivores hunt for and kill prey in different ways, constrained by inheritance expressed by morphological specializations and dependent on prey type and behavior (Sunquist and Sunquist 1989). Searching for prey likely always involves traveling through patches of habitat where success has previously occurred, regardless of capture technique. For many carnivores hunting among thick vegetation or other cover, prey are typically captured from ambush or a stalk and short rush or chase (Sunquist and Sunquist 1989). Such hunting may be carried out singly (e.g., a weasel or snow leopard) or in large groups (e.g., lions). Conversely, groups of wolves, African wild dogs, and spotted hyenas employ long-distance chases of prey, some covering as much as 20 km (Mech and Korb 1978). Still, foxes pounce on mice they hear through the snow, bears turn over logs and lick up ant pupae, and many individual carnivores hunt in several ways during the same foraging bout, depending on the food types that are available.

An individual's home range size (the area regularly used by an individual) is determined by several factors. First, range size increases with metabolic needs, and thus larger individuals, all else being equal, have larger ranges (Gittleman and Harvey 1982); it also follows that ranges of groups are larger than for individuals. Furthermore, trophic level affects range size; in particular, meat eaters of a given body mass have larger ranges, and intraspecific variation in feeding patterns causes variation in range size (Gittleman and Harvey 1982). Within a species, range size increases in proportion to prey density (e.g., bobcats; Knick 1990), and distribution of patches of food within a range also affects its size (e.g., Cape clawless otters; Somers and Nel 2004).

Ranges of some carnivores overlap considerably among individuals, depending largely on resource density and distribution (e.g., leopards; Bailey 1993; Marker and Dickman 2005) and genetic relatedness (e.g., black bears; Moyer et al. 2006). Many carnivores have individual- or group-specific defended ranges (i.e., territories) that often do not overlap among the same sex but do overlap between sexes (e.g., fishers; Arthur et al. 1989). Where prey distribution is stable, these territories are often stable, but under other circumstances, they drift (e.g., red foxes; Doncaster and Macdonald 1991) or either move with migrating prey (e.g., wolves; Walton et al. 2001) or are fixed but temporarily left by individuals to find prey (e.g., spotted hyenas; Hofer and East 1993).

Community Ecology

Aside from interacting with others of their own species and with prey species, carnivores must regularly deal with other species of carnivores (e.g., chapter 14). In circumstances in which food habits do not overlap, resources are distributed widely, and size differences are small, competition is relatively low, and these interactions are relatively benign. In other circumstances, carnivore species may compete intensively, and the results usually are manifested in differences among species with respect to size, diet overlap, habitat-specific abundances, activity patterns, and behavioral dominance (e.g., Fuller et al. 1989; Johnson et al. 1996; Fedriani et al. 2000). Under some circumstances, interspecific killing (Palomares and Caro 1999) takes place, and this can lead to species population regulation and restriction in distribution (e.g., gray foxes; Farias 2000; Farias et al. 2005).

Infectious diseases of carnivores are emerging as an important factor in population viability (Murray et al. 1999). In addition to outbreaks that may decimate specific populations (e.g., lions; Roelke-Parker et al. 1996) and change community structure, noninfectious diseases may act as factors modifying resistance and susceptibility to other diseases (Funk et al. 2001). Some disease risk assessments have been carried out for large, endangered carnivore species (e.g., Ethiopian wolves; Sillero-Zubiri and Macdonald 1997), but small carnivores have been largely neglected, as have disease interactions among carnivore species (Funk et al. 2001).

Carnivores interact with humans in a variety of ways (Clutton-Brock 1996). Historically, many carnivore species were considered competitors for prey, a natural resource in the form of fur and food, or pests because of depredation on livestock (or humans) or as carriers of disease. As a consequence, many species were persecuted relentlessly, and their populations and distribution reduced dramatically. They have served as totems, icons, entertainment, and political "footballs," and some have been domesticated and now number in the millions.

More recently, the role that carnivores play in ecosystem process and stability has been investigated and debated (Ray et al. 2005), giving them yet another, perhaps essential, perceived value. Thus, the need to conserve carnivores, even in the face of perpetual conflict with many, has led to wide-ranging discussion of options and strategies for management (Sillero-Zubiri and Laurenson 2001). Aside from the practical reasons for learning more about carnivores to manage them efficiently and reasonably, there is still the point that they are amazing and wondrous organisms that humans unabashedly admire once they have observed them in natural circumstances (Macdonald 2001).

EFFECTS OF URBANIZATION

So, given the amazing variation in carnivore biology and ecology, how might the relatively recent effects of human urbanization affect carnivores? Clearly, the specifics of such interactions will be covered in detail in the chapters that follow. But in terms of general expectations, we might consider the following.

There are exceptions to every rule in ecology, and attempts to generalize within a large and varied taxonomic group such as carnivores often fall short of reality. However, having acknowledged this, it seems clear that carnivores that do well in urban and suburban areas often share several common characteristics. First, it seems like most of the carnivores that inhabit urban and suburban ecosystems tend to be small to medium sized (see chapter 13). The largest carnivores in North America, such as wolves, grizzly bears, and mountain lions, are not usually found within highly developed human landscapes, though they are sometimes nearby (Dickson and Beier 2007; chapter 11). Many urban and suburban carnivores have relatively high reproductive potential; that is, they are capable of having and raising large numbers of young and could be classified as r-selected ("fast"-reproducing) species. Coyotes, raccoons, and red foxes fit this description. Females can breed at an early age, can have relatively large litters, and can breed every year, thus potentially producing a large number of offspring in their lifetimes. The black bear, however, is a key exception to these generalizations. It is a large-bodied species with low reproductive potential increasingly inhabiting suburban environments (Lyons 2005).

Perhaps most important, the carnivores that are most well adapted to urban and suburban environs seem to be the ones that are diet generalists (Nilon and Pais 1997; Pickett et al. 2001); that is, they are not picky about what they eat and consume plant or vegetable matter, human refuse, and living or dead animals, depending on availability. A generalist diet, in fact, can lead to many negative interactions between people and carnivores. Intentionally feeding carnivores table scraps, unintentionally feeding carnivores by not properly containing garbage or leaving out pet food, or mistakenly feeding carnivores by leaving birdfeeders out year-round often leads to human intolerance of carnivores (DeStefano and DeGraaf 2003; Ditchkoff et al. 2006).

What else then, beyond life history, defines the success of urban carnivore species? Perhaps behavioral characteristics such as a tolerance for or lack of avoidance of humans account for the success of many urban carnivore species. Tolerance for humans may be an intrinsic trait ("Species X is more afraid of people than species Y"), but it is more likely a trait with the capacity for plasticity and adaptation both intra- and interspecifically. Synurbanization (Adams et al. 2005; Ditchkoff et al. 2006) is thought to involve an increase in animals' tolerance and even affinity for humans. Many examples of variation in tolerance can be found in this book. For example, some populations of coyotes are more willing to occupy urban lands than are others (Gompper 2002; chapter 7).

Shifts in the behavior of populations or species might occur as a consequence of phenotypic plasticity (e.g., behavioral flexibility) or through genetic evolution, given sufficient opportunity for selection to act. Contemporary evolutionary change (often thought of as occurring within human lifetimes) is generally facilitated by an increase in the strength of selection (e.g., from a disturbance or other environmental change; Stockwell et al. 2003) as well as by genetic isolation of the affected population (Slabbekoorn and Peet 2003). For many carnivores experiencing habitat fragmentation from suburbanization, conditions are likely ripe for facilitating evolutionary responses to the novel urban environments they are encountering (Hendry et al. 2008).

Perhaps even more likely is the possibility that humans both directly and indirectly facilitate adaptation (either phenotypic or genetic) of urban carnivores. People provide food resources to carnivores in all sorts of ways, whether directly (as noted earlier) or indirectly (e.g., providing food and cover to prey species like rabbits and deer). These elevated food resources generally lead to increased densities for species able to occupy urbanized lands (McKinney 2002; Shochat et al. 2006). In addition, urban environments frequently ameliorate climate effects (Kanda et al. 2005; Parris and Hazell 2005) and may even provide refugia for some species from predators (Faeth et al. 2005). Some species are already showing behavioral shifts in response to human provisioning (e.g., black bears with reduced hibernation intervals; Beckmann and Berger 2003a). Finally, changes in human behavior, such as reduction in hunting pressures, have been suggested as leading to relaxing previously strong selection pressures on species to avoid humans (Gompper 2002; Geist 2008). All of these factors, from intrinsic behavioral flexibility to altered selection pressures from humans, may lead to surprising shifts in tolerance of humans. This makes us wonder whether there are any species, given adequate resources, truly lacking the capacity to adapt to urban or suburban environments.

CONCLUSION

The aim of this chapter is to illustrate the biological and ecological diversity of carnivore species. Combined with the often amazing capacity for adaptation in carnivores, this diversity makes it difficult to identify more than generalizations regarding their responses to urbanization. We suggest that one project for readers of this book is to extract and to synthesize facts from the growing body of research presented here and then define new, more generalizable predictions about the vulnerability or success of carnivores in response to global urbanization. As understanding grows of the mechanisms limiting the distribution of carnivores in urban settings, we may also find ways to facilitate the success of the more vulnerable species in the face of growing human influences.

3

Urban Carnivore Conservation and Management

The Human Dimension

HEATHER WIECZOREK HUDENKO

WILLIAM F. SIEMER

DANIEL J. DECKER

THE COMPLEX relationship between humans and carnivores has ranged from conflict to mutual benefit and has been evident in culture for millennia. Carnivores have been both maligned and revered in Western culture and institutions. These extremes in human perspectives about carnivores have been evident in bounties and eradication programs as well as legislative protection of endangered species. Despite a long history of attention from wildlife professionals, carnivore management and conservation remain a magnet for controversy. Wildlife managers involved in such controversy ask, "How do we make sense of the diverse human sentiment toward carnivores? Can we understand what drives it? And, can we influence human-carnivore relationships positively?" Good questions, but the answers do not come easily, especially when the context for conservation and management varies widely and is constantly changing.

Modern wildlife conservation and management, especially of carnivores, typically has focused on public lands, protected areas, and rural, agricultural landscapes. As urban land development has increased over the past half-century, however, human-carnivore interactions in urban, in suburban, and in exurban areas have taken on increasing significance. The characteristics of human-carnivore interactions in these more altered and densely populated landscapes range widely. In particular, human-carnivore interactions that result in negative impacts to people and companion animals are of growing concern to citizens and wildlife professionals. In some locations, the need for effective management is urgent.

In this chapter, we posit that the core of society's interest in specific carnivores, as well as their support for carnivore conservation and management activities, is related to the way humans are, or expect to be, affected by a particular species. Interactions with carnivores can affect humans in myriad ways. Put simply, the focus of conservation and management activities is on preserving or enhancing positive impacts and mitigating or eliminating negative impacts in an effort to strike an ecologically and socially sustainable balance. We believe that striving for such a balance is essential to the future of human-carnivore coexistence. This is true across landscapes, including urban environments. (In this chapter, we will use "urban" to refer to urban, suburban, and exurban environments.)

Fundamentally, how individuals perceive impacts is a matter of human cognition and behavior. Understanding human perceptions, attitudes, beliefs, and behaviors related to interactions with carnivores can be achieved through human dimensions research (Decker et al. 2001, 2004). Integrating and applying knowledge about the biological and human dimensions of carnivore conservation and management is the work of wildlife professionals and policy makers. Their ability to be successful with this is affected significantly by the quality of insight they have to inform their decisions and actions.

This chapter provides an overview of theoretical and empirical literature relevant to the human dimensions of carnivore conservation and management. The literature specific to human dimensions of urban carnivores is thin. Indeed, the social science literature on humans' beliefs, attitudes, and behaviors with respect to carnivores and their management is not extensive even when rural and wildland landscape contexts are considered. The limitations of current literature notwithstanding, we have organized the main threads of available empirical research and theory that may improve insight about human-carnivore interactions in urban environments.

IMPACTS

Interactions among carnivores, people, and habitat characteristics yield a multitude of effects. Wildlife-related effects recognized as important by citizens, communities, managers, or other stakeholder groups are called "impacts" (Riley et al. 2002). To be an impact, an effect must be both perceived (recognized) and evaluated as important by a stakeholder. Effects that go unnoticed or that are deemed unimportant are not considered impacts (i.e., are not effects that warrant management attention).

Impacts resulting from interactions between humans, carnivores, and habitats may be positive or negative, depending on stakeholders' perceptions. Given the array of effects associated with a human-carnivore interaction, an individual stakeholder may evaluate a single event or interaction as having both positive and negative impacts. Similarly, a particular effect may be interpreted as a positive impact by one stakeholder and as negative by another. Carnivore impacts on ungulate species offer a good example. The possibility of carnivores reducing ungulate populations or changing movement patterns is often viewed as a positive impact by ecologists because they believe these alterations have broad-scale ecosystem benefits (Smith, Peterson, and Houston 2003). Conversely, some stakeholders may view these same changes as a negative impact because it results in fewer opportunities to observe or to hunt affected species (Pate et al. 1996; Enck and Brown 2002). Similar issues arise in urban areas. For instance, bowhunters in urban regions of New York state express concern about coyotes reducing white-tailed deer numbers and therefore hunter satisfaction, but gardeners and homeowners view this same effect as a tremendous benefit that protects their cultivated and ornamental plants (Wieczorek Hudenko et al. 2008). The practical implication of such disparate perspectives is that wildlife managers must weigh the concerns of multiple stakeholders.

Wildlife professionals engage in management activities to create a variety of outcomes, such as conservation of species or habitats, maximization of benefits derived from wildlife (e.g., wildlife viewing, hunting), or alleviation of problems related to wildlife (Siemer and Decker 2006). An outcome regarded as important by stakeholders (i.e., an impact) often is the driving force behind wildlife management decisions and practices (Riley et al. 2002). Wildlife professionals ideally may strive to avoid human-carnivore conflict and promote coexistence in urban areas by first identifying and prioritizing impacts and then selecting a suite of management actions that maximize positive impacts and minimize negative impacts. In practice, however, weighting various impacts associated with human-carnivore interactions can be complex, controversial, and political. Complexity notwithstanding, a first step toward impact management simply is to identify and to clarify the full range of impacts that exist in a given situation.

The range of possible effects associated with carnivore presence and human-carnivore interactions can be grouped into a few broad categories. Human dimensions research in support of impact management for black bears suggests that a set of five categories is useful for classifying impacts associated with any urban carnivore: (1) ecological, (2) economic, (3) health and safety, (4) psychological, and (5) social (Siemer and Decker 2006; table 3.1). In any specific urban carnivore context, it is likely that multiple effects are important to various stakeholders and thus become impacts of management significance. Examples of such impacts follow.

Ecological Impacts

Ecological impacts may include a carnivore's valued effects on species and ecosystems, and the value humans place on sustaining viable carnivore populations. Carnivores are often considered keystone predators, and their presence may promote biodiversity (Wilmers et al. 2003). When a carnivore habitat becomes urbanized, it may endanger certain populations (Kautz et al. 2006), and it may allow others to flourish and perhaps even provide ecosystem services beneficial to humans (Gompper 2002; Gehrt 2006). Human dimensions studies have shown the importance many stakeholders place on these various ecosystem effects. For example, studies often document the value substantial numbers of people put on biodiversity (Kellert 1996), the existence of wildlife generally (Loomis and White 1996), and presence of carnivores specifically (Bright and Manfredo 1996; Fulton et al. 1996).

The importance North Americans place on these positive ecological impacts is reflected in laws, policies, and wildlife regulations that protect and conserve carnivores. At the federal level, carnivore conservation has been addressed primarily through statutes aimed at wise management of the lands on which carnivores exist (e.g., the National Forest Management Act of 1976; the National Environmental Policy Act of 1970). For example, section 2(a) of the Endangered Species Act (ESA) of 1973 declares that economic growth and development can contribute to the extinction of plant and animal species that have ecological and other values to the nation and its people. The ESA is arguably "the most significant law in any nation designed to maintain biodiversity" (Schwartz et al. 2003, 55) and has played a central role in the conservation of several carnivores, including the gray wolf, the grizzly bear, the Florida panther, and the black-footed ferret. At a state level, the importance that stakeholders place on biodiversity and species conservation is reflected in policies and regulations that limit taking of carnivores. For instance, states often seek to maintain population viability of native carnivore species by affording them protection as game species (i.e., by limiting take through regulated hunting and trapping seasons) or by designating a species as nongame or state threatened / endangered (i.e., protected from all taking).

Carnivores may also create negative ecological impacts. Expanding or introduced carnivore populations may prey on other species, fundamentally altering food webs and nutrient distribution or even endangering prey populations. Introduced Arctic foxes in the Aleutian Islands, Alaska, for example, decimated seabird populations to such a degree that wildlife managers implemented fox eradication programs (Williams et al. 2003).

Economic Impacts

Economic impacts cover positive and negative monetary effects associated with interactions among humans and carnivores. Positive impacts may include revenue generated from wildlife viewing or other tourism activities related to carnivores (Rasker and Hackman 1996; Montag et al. 2005). Property damage or livestock depredation from carnivores can create negative economic impacts (Knowlton et al. 1999). Negative economic impacts associated with livestock depredation were one of the driving forces in carnivore management in the United States during much of the twentieth century (Conover 2002), and those impacts remain important in rural agricultural areas today. Negative economic impacts associated with carnivore conservation also can become a major consideration in urban areas where land use is restricted to maintain or to restore habitat for a threatened species (e.g., San Joaquin kit fox, Florida panther).

Health and Safety Impacts

Health and safety impacts associated with carnivore species often include illness, injury, or death of humans or their companion animals. Human dimensions research in this area has focused on documenting the number and distribution of health and safety events (Herrero and Fleck 1990; Conover 2002; Herrero 2005), or characterizing the circumstances under which unsafe interactions occur (Herrero 1985; Mansfield and Charlton 1998). Although injury to people from human-carnivore interactions is rare in both rural and urban areas of North America, even at a low frequency such

Table 3.1. Effects of human-carnivore interactions, and examples of related research

Effects categories	Examples within category	Related research
Ecological Effects on other wildlife, wildlife habitats, or ecological systems that result from interactions between carnivores, other wildlife, humans, and the land.	• Anthropogenic threats to long-term carnivore population viability. • Biodiversity / ecosystem benefits provided by carnivores. • Ecosystem services provided to humans in the form of predation on nuisance wildlife.	• Contingent valuation study showing coyotes have existence value to residents of New England (Stevens et al. 1994). • Statewide survey showing a majority of New York state residents believed restoring wolves to the Adirondack region would reduce rodent populations and balance the deer population (Enck and Brown 2002). • How human alterations to landscapes affect distribution of coyotes, red fox, and raccoon along an urban-rural gradient in northern Illinois (Randa and Yunger 2006). • Effects of human recreation on spatial and temporal activity patterns of bobcat in an urban nature reserve in California (George and Crooks 2006).
Economic Monetary effects produced as a consequence of carnivore presence or human interaction with carnivores.	• Costs of damage to commercial property. • Costs for prevention and treatment of disease transmitted from a carnivore vector species to pets, livestock, or other wildlife. • Cost of damage to residential property. • Economic activity associated with carnivore-related recreation (hunting, viewing, photography).	• Economic impact of coyote depredation on sheep (Knowlton et al. 1999) • The wolf-viewing experience in the Lamar Valley of Yellowstone National Park (Montag et al. 2005). • British residents' willingness to pay to control transmission of bovine tuberculosis from badgers to cattle (White and Whiting 2000).
Health and safety Effects carnivores exert on human health or safety.	• Human and companion animal injuries and deaths related to carnivore interactions with humans. • Diseases transmitted from a carnivore vector species to humans, pets, livestock, or other wildlife.	• Modeling the relationship between various human activities and two types of mountain lion depredation (i.e., depredation on domestic sheep and pets; Torres et al. 1996). • Cougar attacks on humans in California (Mansfield and Charlton 1998). • Raccoons as a vector for rabies transmission (MacInnes and LeBer 2000).

Psychological
Positive or negative effects on the psychological well-being of individuals or stakeholder groups.
	• Personal satisfaction associated with recreation activities (hunting, viewing, photography).
	• Personal/psychological effect of
	  • commercial property damage.
	  • residential property damage.
	  • injury or loss of pets, hobby animals.
	  • injury or loss of livestock.
	  • perceived threat to health, safety.
		• Psychological benefits associated with common wildlife (including foxes and badgers) in Trodheim, Norway (Bjerke and Ostdahl 2004).
		• Recreation motivations and satisfactions associated with furbearer trapping in the northeast (Daigle et al. 1998).
		• Perceived threat to humans posed by coyotes in Chicago metropolitan area (Miller et al. 2001a).

Social Social effects associated with interactions among humans, where carnivores are the reason for the interaction.	• Conflict between stakeholder groups over carnivore species • Concern about loss of a way of life among some stakeholders who suffer negative economic consequences due to existence of a predator population.	• Understanding social controversy over river otter management in Missouri (Hamilton et al. 2000; Goedeke 2005) • Deconstructing conflict between stakeholders in controversy about wolf management in Yellowstone National Park (Wilson 1997).

events can become a focal point for management consideration in urban areas. Land development near urban areas and expanding carnivore populations may increasingly bring people and carnivores into contact. This could increase the likelihood of negative impacts from human-carnivore interactions.

In contrast to the possible physical risks, social aspects of health and safety threats also have the potential to develop into a management priority in urban contexts. Media coverage of negative encounters with carnivores in urban settings may amplify risk perception among stakeholders. Some mass media research suggests that reports about negative human-wildlife encounters make up a large percentage of wildlife-related news coverage, and reports about wildlife conflicts may be more common in urban than in rural newspapers (Corbett 1992, 1995). Moreover, studies suggest that cycles in wildlife-related news stories peak right after a dramatic event such as a wildlife-related human fatality (Wolch et al. 1997; Gore et al. 2005).

Despite the focus on risk often associated with carnivores, positive health and safety impacts are also possible. It has been documented that stakeholders believe a reduction in urban ungulate populations may reduce the incidence of ungulate-vehicle collisions, thereby reducing the health and safety risks to humans (Whittaker et al. 2001). Urban carnivores could provide one possible mechanism for limiting ungulate populations. Some urban communities take actions to reduce populations of Canada geese or white-tailed deer because they believe that doing so may reduce the risk of animal-human disease transmission in their communities (Smith et al. 2001; Decker et al. 2004). Again, certain urban carnivores, like coyotes, are known to prey on fawns and goose eggs (Gehrt 2006), potentially affecting long-term population trends and therefore generating positive health and safety impacts.

Psychological Impacts

Positive feelings of enjoyment or appreciation and negative emotions such as anxiety or fear may all be associated with the presence of a carnivore species. These effects fall under the category of psychological impacts and can be either beneficial or detrimental. Psychological impacts may influence stakeholder tolerance of, and attitudes toward, particular species. For example, negative emotions may be related to decreased support for carnivore conservation, such as wolf reintroduction (Bright and Manfredo 1996).

Inquiries of human attitudes toward carnivore species are relatively numerous in the literature, but often descriptive. In general, these studies characterize public sentiment toward a species or how attitude strength or direction differs by age, education, gender, place of residence, or other traits (Kellert 1985; Stevens et al. 1994; Pate et al. 1996; Bjerke et al. 1998; Harrison 1998; Miller et al. 2001; Williams et al. 2002; Zinn and Pierce 2002; Bjerke and Østdahl 2004; Bjurlin and Cypher 2005; Casey et al. 2005). Results from these studies have revealed variation in attitudes toward predators across a range of sociodemographic and interest groups, and some inquiries find conflicting results (Williams et al. 2002). A few studies have provided insights about the relation between psychological impacts and carnivore acceptance (Zimmermann et al. 2001; Kleiven et al. 2004), indicating that higher levels of fear about a particular carnivore are correlated with lower overall acceptance of the species.

Social Impacts

Social impacts result from positive or negative interactions between people in response to a wildlife-related issue. They can be a catalyst or barrier to public action, and therefore social impacts can have far-reaching consequences for carnivore conservation and management. In northern New Jersey, for example, bitter social conflict between stakeholders over black bear management has produced a level of divisiveness and distrust that hobbles progress toward management of a burgeoning population of black bears (a brief description of that conflict appears in Siemer et al. 2007). Conversely, human-carnivore interactions can also stimulate positive social impacts that build community capacity to address wildlife management issues. For instance, concern in Louisiana about the consequences for property owners of listing the Louisiana black bear as threatened, led to the formation of a coalition of more than 60 organizations. The coalition is credited with creating a landowner information and training network that built support for the bear and encouraged the creation of suitable habitat. This effort was so successful that the Louisiana black bear population began to recover (Black Bear Conservation Committee 2006).

Specific human dimensions research about social impacts has focused on characterizing and understanding the perspectives of different stakeholder groups. For example, Wilson (1997) characterized the conflicting perspectives of ranchers and animal protectionists in a wolf management controversy. Hamilton et al. (2000) and Goedeke (2005) described the competing claims made by interest groups in a controversy about river otter management. Understanding social perspectives may help wildlife professionals anticipate and manage wildlife issues. Anticipating social conflict may be especially important for an urban setting when carnivore populations expand or restoration activities are considered.

Secondary Impacts

The various impacts described relate to the direct, or primary, impacts of carnivore presence and human-carnivore interactions. Management of these primary impacts can also create secondary effects that are often important to stakeholders, thereby creating secondary impacts. Such secondary impacts are often generated by a management intervention designed to conserve or to control a species. Reactions to expected collateral effects of carnivore reintroduction (Curt 1999; Enck and Brown 2002) or community response to lethal carnivore control efforts (Teel et al. 2002; Martinez-Espiñeria 2006) may become significant secondary impacts. Particularly in an urban setting, the secondary effects created when managers take action may be more consequential than the primary impacts they intend to manage. Highly contentious public issues associated with opposition to traps and trapping in urban spaces may lead to strife in a community and limit agencies' ability to implement management activities (Andelt et al. 1999).

THEORIES AND CONCEPTS TO UNDERSTAND AND MANAGE IMPACTS

Many theoretical frameworks are helpful to understand human-wildlife relationships and the resulting impacts. Several lend themselves to application in a human-carnivore context. A number of human dimensions concepts relate basic social science theory to wildlife issues and can illuminate how interactions with carnivores may influence people's beliefs, attitudes, and behaviors. For example, the conceptual framework of impact dependency uses an established psychological theory that explores human behavioral motivations, Maslow's hierarchy of needs (Maslow 1943), to provide insight into how people translate effects into impacts and prioritize impacts. Another social science construct, risk perception, explains details

about people's appraisal of risk situations (Slovic 1987), which is information pertinent to understanding human response to and evaluation of health and safety impacts.

Wildlife acceptance capacity (Decker and Purdy 1988), examined within a rubric of norms theory, is particularly useful in characterizing tolerance of impacts across stakeholder groups, defining public acceptability of carnivore management actions in urban areas, and identifying level of public agreement about potential management policies (Manfredo et al. 1995). All of these ideas are useful in the carnivore management context where the presence of carnivore species can create a variety of impacts; however, human cognitions about, and responses to, these impacts are not well understood. We briefly describe these concepts and describe how they relate to the human-carnivore interface.

Impact Dependency

Behavioral motivations shape how an individual responds to impacts and his or her acceptance of management actions to affect impacts. A theory developed by Maslow (1943) to explain motivations for human behavior, called a "hierarchy of needs," helps frame human responses to carnivores. The hierarchy describes the prioritization of needs that are common to most humans. Maslow asserted that human actions were generally motivated by a desire to meet these needs based on their importance for human existence, from the most basic requirements for survival to higher-order aspirations. This theory suggests that basic physiological needs, such as food and shelter, must be met before (or are of greater priority than) higher-order needs, such as self-esteem or feelings of belonging. As the various needs are met, a human's behavioral motivations move up the hierarchy, directed toward fulfilling higher-order needs.

The concept of impacts can be related to Maslow's hierarchy of needs through a framework termed *impact dependency* (Decker, Jacobson, and Brown 2006), where a "need" is affected by a human-wildlife interaction. Impacts resulting from interactions between humans, carnivores, and habitat may affect different needs in Maslow's hierarchy (figure 3.1). For example, promoting conservation goals might be considered a self-actualizing activity because it presumably creates a sense of purpose and fulfillment. A wildlife nuisance issue, such as a bear or raccoon raiding curbside trash-cans, would affect esteem because it can affect an indi-

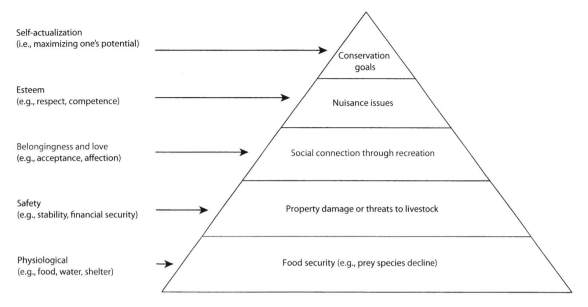

Figure 3.1. Maslow's hierarchy of needs and examples of wildlife impacts that may affect each need level (adapted from Maslow 1943).

vidual's ability to feel competent to control the problem. Engaging in wildlife-related recreation activities can foster social connections and feelings of affiliation, which would correlate with the "belongingness and love" echelon. Economic impacts of carnivores, such as property damage or threats to livestock, may influence economic security, a safety-level need in Maslow's hierarchy. Finally, a carnivore that threatens food security through prey population reductions in rural subsistence communities, or human safety via attacks on people, would create an impact that correlates with Maslow's most basic category—physiological needs.

The impact dependency framework may help explain seemingly contradictory results from studies of human responses to carnivores and aid our understanding of human-carnivore interactions in urban environments. If impacts from carnivores are perceived by stakeholders as affecting or threatening needs lower in the hierarchy—that is, affecting more basic needs—then acceptance of the impact, and thus tolerance for the animal, can be expected to be low. This idea may explain why people's primary wildlife management concern is often impacts on human health and safety (Reiter et al.1999).

Impact dependency suggests that tolerance of wildlife is less about acceptance of the species or its population density and more about acceptance of the *impacts* that result from living with the species. Empirical research on urban wildlife species (e.g., deer, geese, moose, and beaver) demonstrates that wildlife acceptance capacity is correlated with perceptions of wildlife-related impacts (Loker et al. 1999; Whittaker et al. 2001; Jonker et al. 2009). As discussed previously, media coverage and public attention may make impacts from carnivores a particular focus in urban areas.

Researchers have been exploring connections between carnivores and impacts through a variety of mechanisms such as acceptance, tolerance, and desired management. It appears that many of these studies have found empirical support for application of the impact dependency hypothesis. Research on gray wolf reintroduction shows that anticipated impacts from wolves are more predictive of support for reintroduction than are positive attitudes about wolves (Lohr et al. 1996). Similarly, severity of carnivore behavior has been linked more closely to acceptance of carnivores than to attitudes or sociodemographic characteristics (Kleiven et al. 2004). A more recent inquiry specifically related acceptance of carnivores (expressed through management preferences) with the human need that was affected by a carnivore's behavior (i.e., impacts incurred; Decker et al. 2006). Acceptance of impacts decreased as the significance of an impact increased (i.e., as impacts affected more basic needs) and the researchers suggested that acceptance might be "impact dependent."

North American studies about management alternatives for carnivores specifically in urban areas also support the impact dependency hypothesis. Surveys among urban residents about desired management options for various carnivore species identify the connection between tolerance for a species and impacts. Acceptance of lethal management options often depends on the type of encounter: whether the carnivore has been sighted in a residential area, caused damage to property, killed a pet, or threatened human health or safety (Whittaker et el. 2001). Such impact dependency has been found for mountain lions, coyotes, raccoons, and a variety of other carnivore species (Manfredo et al. 1998; Zinn et al. 1998; Miller et al. 2001; Casey et al. 2005). Although many factors influence the way humans relate to carnivores, studies of impact-dependent human responses are consistent in their conclusions.

Risk Perception

Interactions between humans and carnivores at times result in negative impacts, which can become a focus of public concern. Perception of risk is therefore an important element in understanding human responses to carnivores. Carnivores are often perceived as a "hazard," or threat, to humans and their domesticated animals. When evaluating a hazard, most people form risk perceptions based on intuitive risk judgments, unlike the technical assessments made by experts (Slovic 1987). Risk perception is a broad cognitive process influenced by a number of interrelated characteristics, such as whether the risk is familiar, certain, controllable, equitable, known, voluntary, observable, immediate, and manageable (Slovic 1987). Understanding two classes of risk perceptions, cognitive and affective, may provide insight into stakeholder views of wildlife species (Gore et al. 2005). Cognitive risk perception is the perceived probability of a threat (Renn 1992), for example, potential for injury from a carnivore. Affective risk perception includes feelings related to the threat (Sjöberg 1998; e.g., concern or worry about the threat carnivores pose). People evaluate perceived risk with context specificity, and such perceptions can shape their decisions and behaviors. Risk perceptions can influence people's attitudes and beliefs, as well as their acceptance of various wildlife species (Decker et al. 2002). Kellert (1985) demonstrated that risk perception variables influence attitudes toward carnivores.

Carnivore presence in urban areas can pose risks, such as disease transmission or injury to people and pets. A number of specific risk characteristics may influence people's risk perception associated with particular carnivore species. For instance, studies show that familiarity (Gore et al. 2006), certainty (Zimmermann et al. 2001), and controllability (Kleiven et al. 2004) influence risk perceptions about grizzly bear, gray wolves, European lynx, and wolverine. These studies suggest a link between risk perceptions and attitudes about and tolerance for carnivores.

In general, studies examining risk perception related to carnivore species indicate that if a wildlife-related risk is unfamiliar, uncertain, novel, or uncontrollable, it will likely lead to a higher perceived risk. All of these factors can be associated with a carnivore perceived by residents as "new" to an area. Carnivores might be considered new as human development expands outward into natural areas and people encounter certain species for the first time, or as certain carnivore populations continue to expand and use urban environments, thereby interacting with people more frequently. With both phenomena occurring regularly, it is not surprising that urban carnivore management is drawing much attention.

THE IMPORTANCE OF RISK COMMUNICATION

Understanding stakeholders' risk perceptions and communicating with stakeholders to increase their understanding of actual threats are important tools for impact-based management. An excellent example of this is a proposal to restore bobcats to Cumberland Island National Seashore that generated significant opposition from hunters and other stakeholders (Warren et al. 1990). These groups were concerned about the negative effect bobcat might have on white-tailed deer and turkey hunting, and media coverage amplified perception of this risk. Opposition waned only after managers used a communication campaign to convey anticipated impacts and alleviate concerns about potential risks to human health and safety and recreation interests. A decade after bobcat reintroduction, a survey revealed that hunters using Cumberland Island National Seashore held neutral (not negative) attitudes toward bobcats on the island (Brooks et al. 1999). Managers reflecting on this experience encourage others to research stakeholder risk and impact perceptions and to use this information to guide communication efforts (Warren et al. 1990).

WILDLIFE ACCEPTANCE CAPACITY

Wildlife acceptance capacity was originally developed to refer to the maximum wildlife population level, or its associated impacts, that a stakeholder group is willing to tolerate (Decker and Purdy 1988). Stakeholders' perceptions of the costs and benefits associated with a wildlife species, such as its impacts on humans and the environment, shape overall wildlife acceptance capacity (Carpenter et al. 2000). Variables thought to influence wildlife acceptance capacity include risk perceptions, attitudes toward and beliefs about a species, sociodemographic traits (e.g., gender, age, education), experience with a species, particular impacts (positive and negative) created by the species, perceptions of wildlife population trends, and attitudes about management (Decker and Purdy 1988). It follows that people with different stakes have different acceptance capacities for various wildlife species (Carpenter et al. 2000). This variation may occur because wildlife interactions, such as those experienced with carnivores, affect different levels of needs, per Maslow's hierarchy, and therefore may vary in importance.

Norm theory (Sherif 1936; Sherif and Sherif 1969; Fishbein and Ajzen 1975) provides the theoretical underpinning for wildlife impact acceptance (Lischka et al. 2008). Norms are beliefs about: (a) what constitutes "normal," or typical, behavior in a given circumstance or (b) what individual or collective behavior people should exhibit in a given circumstance (Cialdini et al. 1990). For example, if an urban resident believes that typical behavior among most people is to refrain from harassing coyotes when they encounter them around their home, and this resident therefore avoids chasing coyotes from their backyard, then the person is reflecting a norm. Studies often find that many people believe wildlife agencies should be allowed to take lethal action to protect human safety, which also is a normative belief. Much of the normative research in human dimensions of wildlife management is based on the work of Jackson (1965), who proposed a model to describe norms by means of an "impact-acceptability curve" (see Vaske and Whittaker 2004 for a review of norm theory applications in a natural resource context).

Similar to the concept of "range of tolerable conditions" found in the norms literature (Vaske and Whittaker 2004), the acceptance capacity concept holds that an individual's acceptance threshold is situation specific and depends on the severity of impacts experienced or anticipated from a human-wildlife interaction. For instance, songbirds are likely to provide numerous benefits to stakeholders with few negative impacts, generating a relatively high acceptance capacity. Other species capable of causing damage to property or injury to people, such as black bear or coyotes, may be perceived as having greater costs associated with their presence. However, large carnivores may be perceived as providing great benefit because of their aesthetic appeal or their role in balanced ecosystems. Consequently, predicting a stakeholder's acceptance capacity for large carnivores is less intuitive. One might safely assume that the various stakeholders within a community would have diverse acceptance capacities for these species because they are capable of significant positive and negative impacts. Indeed, studies have found conflicting results about acceptance and tolerance for large carnivores. Examples of these inquiries are discussed next.

Some variables influencing acceptance capacity, such as risk perception, may change over time as people and carnivores coexist in the same area. The longer carnivores live near developed landscapes, the more experience residents are likely to have with the species. Experience could occur through direct encounters or through vicarious experience passed through social networks and the media. Regardless of the mechanism, this greater richness of experience is likely to increase familiarity with, knowledge of, and certainty about the carnivore, all of which are presumed to lower risk perception (Slovic 1987). In fact, some studies have found that acceptance in certain communities actually increases over time with greater exposure to carnivores (Decker and O'Pezio 1989; Harrison 1998; Kleiven et al. 2004; Bjurlin and Cypher 2005).

A review of 13 attitude surveys, including both urban and rural residents, demonstrated a similar possibility of potential increased carnivore acceptance over time (Zimmermann et al. 2001). On the basis of their review, Zimmermann and colleagues proposed that a sequence of attitude change occurs over time. Across the attitude studies, negative attitudes toward a carnivore species increased as the animal's range expanded toward the people surveyed. Negative attitudes peaked when the species established residence near people, and then over time, as experience with the animal accumulated, negative attitudes declined. Eventually, after prolonged exposure to sharing a geographic area with a carnivore species, negative attitudes were even lower than when the species was newly colonizing areas far from the people surveyed. This reduction of negative attitudes over time is likely to exert a positive influence on acceptance capacity for the referent species. It may be that similar circumstances would occur specifically for a carnivore in an urban area, but particular variables (e.g., resident background [urban or rural] or land-use patterns [distribution of resources]) may make the urban context unique.

In contrast to the idea that acceptance capacity for carnivores may increase with experience, other studies show that acceptance capacities may remain constant or even decrease with experience (Ericsson and Heberlein 2003). Recall that risk perception can be a key determinate of acceptance capacity. Certain risk perception variables, like control, may not improve with greater carnivore experience. A perception of lack of control, perhaps reinforced by actual experience, can decrease acceptance of certain carnivores (Treves and Karanth 2003). Even if an individual has low cognitive risk perceptions and understands that the threat of injury from a carnivore is a low probability event, affective risk perceptions based on the perceived threat may be significant enough to modify acceptance capacity (Riley and Decker 2000a). It may be possible that acceptance declines over time because of conflict with a carnivore (Riley and Decker 2000b). In the host of factors influencing acceptance capacity, it may be that stakeholder perceptions of impacts can outweigh other variables and contribute more substantially to acceptance capacity.

The role of acceptance capacity and norms is particularly relevant to management of carnivores in urban areas, where intervention options are closely scrutinized and often contentious. The unique aspects of the urban context will influence acceptable management actions. In general, norms for acceptability of management actions vary by species, depending on the animal's popular image, perceived abundance, and impact potential. For example, coyotes are often portrayed as scavengers or pests, capable of creating significant negative impacts, such as threats to pets or people (McIvor and Conover 1994). As a result, killing a coyote is more acceptable than killing individuals of many other species (Wittmann et al. 1998), but lethal control often is not feasible in areas of dense human population and may be unacceptable to urban residents who are not experiencing negative impacts (Zinn et al. 1998). Severity of impacts is likely to drive acceptance of management actions. Understanding norms for ac-

ceptance of actions is crucial for managers hoping to resolve the challenges of carnivore management in the urban context.

INFORMATION NEEDS TO SUPPORT CARNIVORE CONSERVATION AND MANAGEMENT IN URBAN AREAS

Wildlife managers and policy makers continuously seek improved insight about issues for which they are required to make decisions. Such information is needed for both the ecological and human dimensions of any wildlife issue. One way to organize the litany of research needs is to focus inquiry around the elements of management. In table 3.2, we present general and carnivore-specific human dimensions information needs for recognized aspects of management.

The scope of information needs presented in table 3.2 is broad and multifaceted. It includes some relatively straightforward needs, which can be addressed by practicing wildlife professionals through routine techniques for data collection. For example, practitioners working in urban contexts may benefit from designing and implementing processes to describe their wildlife context and identify effects that their various stakeholders regard as impacts worthy of management attention. Such situation analyses and impact identification processes could be useful to professionals making decisions about urban carnivore conservation and management programs.

Table 3.2 also outlines information needs that present greater challenges and will require the attention of human dimensions specialists or interdisciplinary teams of researchers and wildlife professionals. We believe that four topics: (1) understanding impact dependency, (2) advancing practices for weighting different stakes, (3) assessing carnivore-related risk perceptions, and (4) clarifying the role of co-tolerance in human-carnivore interactions in urban environments, should be research priorities and offer opportunities for scholarship to advance carnivore conservation and management in urban areas. We describe each priority briefly.

Impact Dependency

The examples discussed earlier illustrate that researchers have made some effort to improve understanding of the relationships between impact perception and tolerance of carnivores and acceptability of management tools. Application of the impact dependency framework has not been extensive in the context of

Table 3.2. A matrix of human dimensions information needs to support carnivore conservation and management in urban areas

General information need	Related information needs for carnivore conservation and management
Defining goals • Baseline data on public values toward wildlife and how values are changing. • Baseline data on stakeholder-defined impacts.	• Clarify impact perceptions among stakeholders in hot spots for human-carnivore interaction. • Identify value orientation subgroups in urban, suburban, and exurban areas.
Identifying problems and opportunities • General understanding of overall system. • Knowledge of extent and severity of existing problems. • Understanding of bases for conflict between people and wildlife. • Knowledge of opportunities to achieve conservation goals.	• Identify stakeholder characteristics, motivations, experiences with, and attitudes about carnivores. • Identify extent and nature of human-carnivore interactions. • Increase understanding of possible human-carnivore co-tolerance.
Identifying conservation and management objectives • Clarification and critique of objectives (i.e., formative evaluation of program objectives based on understanding of system). • Define acceptable limits (normative standards) for a given situation.	• Identify normative standards for acceptable carnivore behavior. • Identify factors affecting tolerance of carnivores. • Identify carnivore acceptance capacities for key stakeholders. • Identify carnivore-related risk perceptions.
Developing management action alternatives • Baseline data on public acceptance of management alternatives. • Forecast anticipated outcomes from management actions (i.e., secondary effects).	• Assess stakeholders' management preferences. • Identify norms related to lethal and nonlethal management actions.
Evaluating outcomes of conservation and management programs • Quantitative feedback on outcomes of actions.	• Efficacy of programs to promote problem prevention behavior by key groups (e.g., pet owners). • Efficacy of removing problem carnivores as a means to reduce negative impacts. • Efficacy of restoration programs to achieve conservation goals and create sustainable coexistence.

urban carnivores. More research is needed to validate the idea and explore how it may be employed to facilitate human-carnivore conservation and management in urban areas. To refine impact dependency as a conceptual framework, future research must evaluate the hypothesis that acceptance capacity for an urban carnivore will be lower when a stakeholder perceives that the species inhibits their ability to meet a basic need or, conversely, that acceptance capacity for an urban carnivore will be higher when the stakeholder has no basic needs threatened by the carnivore species.

Wildlife professionals may benefit from context-specific research on perceptions of various stakeholders to understand how those stakeholders believe their needs are affected by the presence of an urban carnivore. This information will inform the design of program actions that best address the appropriate issues. To enable practitioners to conduct such data collection, human dimensions specialists will need to develop valid and reliable measures of impact dependency and demonstrate that those measures have practical value to the wildlife agencies and organizations charged with management of urban carnivores.

Weighting Different Stakes

Identifying impacts of interest or concern among stakeholders is necessary but not sufficient to serve the conservation and management decision-making process because primary impacts of human-carnivore interactions vary widely in type and in magnitude, as do the secondary impacts of management interventions. These diverse impacts are not easily aggregated and balanced through formulaic trade-off analysis to guide human-carnivore management strategy. Research and management "experiments" that approach weighting of various impacts (the stakes that create stakeholders) in different ways are needed (Carpenter et al. 2000). Currently, this key human dimensions consideration occurs through professional judgment, stakeholder group processes where consensus is sought, and political lobbying at various levels of government. These processes of persuasion, negotiation, mediation, and consensus building that assist in weighting stakes and balancing impacts in decision making need further inquiry and refinement for carnivore conservation and management.

Carnivore-Related Risk Perception

While risk perception theory may help clarify our understanding of how people regard carnivores, the application of this theory in an urban carnivore context has been limited. Few studies evaluate the full set of characteristics thought to influence risk perception, and little replication has occurred. More research along an urban gradient is needed to examine thoroughly the ways in which carnivore-related risk perceptions influence human-carnivore interactions and thus the impacts associated with carnivore species. Perceptions of risk and risk-related impacts can be expected to influence acceptance of conservation and management actions in urban areas.

Agencies and nongovernmental organizations also would benefit from applied research to guide risk communication and risk management actions. Research is needed to shed light on the role of mass media in risk perception associated with urban carnivore issues. Case studies could lead to improvements in government-nongovernment collaboration to provide risk communication or management.

Causes and Consequences of "Co-tolerance"

Impacts of human-carnivore interactions may result from an interplay between human and carnivore behaviors. Human-carnivore interactions in urban environments are believed to be exacerbated by two primary mechanisms: food conditioning and habituation of animals. Many studies conclude that the development of problematic carnivore behaviors is linked to human behaviors through these two mechanisms (e.g., McNay 2002; Herrero et al. 2005). While not all studies find that food conditioning inevitably leads to human-carnivore conflict, access to anthropogenic food sources facilitates the persistence of carnivore populations in residential areas (Fedriani et al. 2001), which in turn increases the potential for negative interactions (Bounds and Shaw 1994). Research also indicates that unless humans discourage carnivores from using urban habitats, the habituation process will increase the likelihood of an encounter (Kitchen et al. 2000; McNay 2002; Lambert et al. 2006). Although typically examined from a wildlife behavior perspective, food conditioning and habituation are human-induced conditions and should be understood within a theoretical and research framework that includes human dimensions.

Habituation can be observed in both humans and wildlife. Habituation is the waning of a behavioral response following repeated exposure to a nonthreatening stimulus (Bernstein 2006, 195–96). Typically, habituation in wildlife refers to an animal's lack of behavioral fear response to the presence of humans after repeated, nonconsequential encounters (McNay 2002; Herrero et al. 2005). For instance, based on observations of urban coyotes in Banff, Alberta, Gibeau (1998) suggested that habituation of these animals to urban environments was the result of frequent nonthreatening interaction with humans. Habituation in this context generally refers to the tolerance of human presence or activity by wildlife. We postulate that this process is reciprocal; in other words, people also may habituate to the presence or activities of carnivores.

Although studies on food conditioning and habituation of carnivores are useful for understanding carnivore behaviors, they do not provide significant insight into why humans respond to carnivores as they do. Wildlife professionals would benefit from research about the extent to which humans may habituate to or tolerate the presence of carnivores. Just as habituation in carnivores has adaptive advantages, this process in humans with respect to carnivores also may have benefits. Wildlife and humans habituating to each other, referred to recently as "co-habituation" (Zinn et al. 2008), could lead to greater tolerance and successful coexistence of humans and carnivores in urban environments. For example, habituation may reduce people's anxiety and fear associated with living with carnivores or ease concerns about minor nuisance impacts. Of concern, however, is the possibility that human habituation to carnivores conceivably could lead to "overtolerance," creating situations in which carnivores are tolerated in areas where they could pose a threat to the safety of companion animals and people, thereby increasing the risk of a negative encounter. This is a topic in need of inquiry, and it may present significant implications for carnivore conservation and management in urban areas.

CONCLUSION

Understanding human-carnivore relationships is central to urban carnivore conservation and management and ultimately to the possibility of sustainable coexistence. Theories and concepts developed or applied in wildlife management exist that are useful for examining the human component of urban carnivore conservation and management, but they have not been widely applied in this context. Similarly, empirical studies, even of a descriptive nature, are uncommon. Building a knowledge base to support urban carnivore conservation and management efforts will require both specific human dimensions insight and integrated human and ecological / biological research. These needs are escalating as humans and carnivores interact more frequently in urban landscapes. Evidence exists to suggest that, with better human dimensions insight, managers and communities have an opportunity to focus on fostering coexistence between people and carnivores, rather than on mitigating conflict.

4

Raccoons (*Procyon lotor*)

JOHN HADIDIAN

SUZANNE PRANGE

RICHARD ROSATTE

SETH P. D. RILEY

STANLEY D. GEHRT

RACCOONS ARE arguably the most widespread and abundant of all urban carnivores in North America. They were variously reported as living in and around cities from the turn of the twentieth century on (McAtee 1918; Kieran 1959), suggesting they were early to exploit urbanizing habits. Systematic studies to document their presence, however, did not appear until fairly recently. Darrell Schinner and Erick Cauley, under the direction of Jack Gottschang at the University of Cincinnati, conducted the first field research on raccoons in suburban environments in the late 1960s (Schinner 1969; Cauley 1970; Cauley and Schinner 1973; Schinner and Cauley 1974). Other studies have followed; between 20 and 30 publications from 13 urban areas (10 in North America) now describe aspects of the behavior and ecology of urban raccoons (see table 4.1).

Raccoons are medium-sized carnivores (adults typically range from 4 kg to 10 kg) widely distributed throughout North and Central America and found in most metropolitan areas within their range (Gehrt 2003). As introductions, raccoons are also found in cities in parts of Europe and Japan (Hohmann et al. 2002; Ikeda et al. 2004; Bartoszewicz et al. 2008).

Adapted to making use of both vertical as well as horizontal space, raccoons are easily able to scale and use human structures. They are also dietary generalists, tend to be nocturnal and somewhat secretive, and appear flexible in their ability to use a variety of sites for denning and refuge, other traits that commend them to urban living. Their hardiness enables them to survive even severe physical impairment, such as blindness or the loss of a limb (Sunquist et al. 1969; Hadidian and Riley 1996). Beyond this,

Table 4.1. A representative sampling of studies focusing on urban raccoon populations

Study	Metropolitan area	Landscape type
Schinner 1969	Cincinnati, Ohio	Mixed suburban
Cauley 1970		
Cauley and Schinner 1973		
Schinner and Cauley 1974		
Hoffmann and Gottschang 1977		
Hoffmann 1979		
McComb 1981	Eastern Connecticut Specific study locales not reported	Suburban, urban, and industrial
Rosatte 1985	Metropolitan Toronto, Canada	Mixed urban
Rosatte 1986		Urban
Rosatte and MacInnes 1989		Urban and rural
Rosatte et al. 1991		Urban
Rosatte et al. 1992a		Urban
Slate 1985	New Jersey	Mixed suburban
Valentine et al. 1988	Ithaca, New York	Urban
Robel et al. 1990	Manhattan, Kansas	Urban
Hadidian et al. 1991	Washington, D.C.	Urban national park
Hadidian and Riley 1996		
Riley et al. 1998		
Smith and Engeman 2002	Fort Lauderdale, Florida	Urban park
Prange et al. 2003	Chicago metropolitan area	Urban, suburban, and rural parks and preserves
Prange et al. 2004		
Gehrt 2004b		Forest preserve
Prange and Gehrt 2004		
Mosillo et al. 1999		
Randa and Yunger 2006		
O'Donnell and DeNicola 2006	Suburban Connecticut	Suburban towns, Hartford and Tolland counties
Hohmann et al. 2001	Bad Karlshafen, Germany	Urban parks, mixed urban
Hohmann et al. 2002		
Ikeda et al. 2004	Kamakura City, Hokkaido Prefecture, Japan	Urban, suburban, rural areas
Bartoszewicz et al. 2008	Kostrzyn, western Poland	Suburban

Note: This is not intended to be a comprehensive list.

their ability to manipulate objects (Lyall-Watson 1963), intelligence (Whitney 1933), relatively high fecundity (Sanderson 1987), and extended parenting (Stuewer 1943; Johnson 1970; Schneider et al. 1971) may allow raccoons to tolerate challenges associated with urbanization that might imperil more sensitive species.

A relatively rich vocal repertoire in comparison with other procyonids (Sieber 1984) accompanies olfactory capabilities that not only enhance survival, but probably mediate communication within populations in ways we still know very little about (e.g., Ough 1982). Far greater sociality than generally attributed to this "solitary" carnivore is perhaps already suggested by studies that demonstrate neighbor recognition (Barash 1974), ordered social relationships (Sharp and Sharp 1956), long association of mother-cub groups (Sieber 1985; Gehrt and Fritzell 1998a), philopatry (Ratnayeke et al. 2002), and growing evidence of nonfamilial spatial group associations (Gehrt and Fritzell 1998b). Underlying all this is the persona of the raccoon, which combines attractiveness and apparent amiability when young with a wild and untamable nature when grown (Godman 1826; MacClintock 1981; figure 4.1).

POPULATION CHARACTERISTICS

Studies of raccoons in both urban and nonurban habitats indicate they have a positive association with tree cover and water (Stuewer 1943; Johnson 1970; Schinner and Cauley 1974; Hoffmann and Gottschang 1977; Hadidian et al. 1991; Rosatte et al. 1991; Dijak and Thompson 2000). Broadfoot et al. (2001) combined landscape data with demographic and habitat information for raccoons in an urban area of metro Toronto to suggest discrete subpopulations of raccoons that were linked by dispersal to create a metapopulation. Such

Figure 4.1. Garbage and other food attractants are frequently the cause for conflicts with urban carnivores. Photo by Michael Ashdown.

nonrandom distribution would have implications for understanding raccoon interactions with human populations, including the potential for disease transmission (Broadfoot et al. 2001).

Density

Although raccoon populations seem to reach higher average densities in urban areas than elsewhere (Riley et al. 1998; Prange et al. 2003), reported estimates have ranged widely across urban studies (table 4.2). Because density is defined as the number of individuals per unit of space, the determination of the space occupied can have considerable effect on density estimates. Prange et al. (2003) argued for the use of an "effective trapping area" (the total area traversed by all radio-collared raccoons during each season) in estimating raccoon dens-

ities, noting that densities in their study would have been up to 102% higher had the area of the trapping grid been used in estimating density, rather than effective trapping area.

Riley et al. (1998) found raccoon density in an urban park to average 125/km² with an upper estimate of 333/km² along a thin spur of parkland, this being the highest population density reported for the species. Twitchell and Dill (1949) retrieved raccoons by hand from trees in a rural area along a rich bottomland riparian habitat and reported a density estimate of 250 raccoons/km², while Smith and Engeman (2002) removed raccoons from an urban park and reported a density of 238/km². These exceptionally high estimates are all best interpreted cautiously because both methodological (e.g., hand capture or capture and removal as in the second two studies) or landscape/habitat factors (sampling a small peninsula of natural area in a larger urban matrix for the first) may have led to numbers that reflect temporary or seasonal concentrations that are not sustained over long periods. Rosatte et al. (1991) reported microhabitat differences within urban zones, with densities as high as 56 raccoons/km² in forest park areas as opposed to 4/km² in field habitats. Elsewhere, reported densities for urban populations are lower but still usually higher than in either natural or agricultural areas (table 4.2).

Following Rosatte et al. (1991), who speculated that anthropogenic food resources may lead to better physical condition and higher reproductive rates in urban raccoons, Prange et al. (2003) tested higher urban densities against three assumptions: (1) increased survival,

Table 4.2. Urban raccoons: population density, as number of raccoons per km² (range)

Raccoon density	Method	Source	Study area
238	Removal	Smith and Engeman 2002	Urban park
125 (67–333)	Mark-recapture	Riley et al. 1998	Urban national park
111		Schinner and Cauley 1974	Mixed suburban
>100	Mark-recapture	Rosatte 2000	Mixed urban
95–100	Mark-recapture	Hohmann et al. 2002	Urban park
67		Hoffmann and Gottschang 1977	Mixed suburban
55.6	Modified Peterson	Rosatte et al. 1990	Mixed urban
~40–70	Mark-recapture	Prange et al. 2003	Urban, suburban, and rural parks and preserves
36		Slate 1985	Mixed suburban
34–97	Live trap/modeling	Broadfoot et al. 2001	Mixed urban
10–12	Mark-recapture	Rosatte and Lawson 2004	Mixed urban
9–10 (control) 21–22 (treatment)	Removal/live trap (Distemper vaccination)	Schubert et al. 1998	Mixed urban
7–12	Mark-recapture	Rosatte et al. 1992a	Mixed urban
4	Mark-recapture	Rosatte 1985	Mixed urban

(2) increased reproductive rates, and (3) greater site fidelity. They speculated that "superrich" food patches in urban environments may contribute to high-density raccoon populations. The evidence from this study suggested that higher annual recruitment and site fidelity combined with survival to produce higher densities in some parts of the urban landscape. In general, population data for urban raccoons strongly suggest that these habitats support higher population densities, but the determining factors behind these densities remain to be more precisely identified.

Survival and Mortality

A number of studies have focused on survival and mortality in urban raccoons, with general findings that may be true for other urban carnivore populations, too. Survival in Toronto raccoons was about 0.40 to 0.50 annually (Rosatte et al. 1991, 1992a; Rosatte 2000). Riley et al. (1998) reported high survival rates for adults of both sexes (>0.70) for an urban national park in Washington, D.C., with juvenile survival only significantly lower during a rabies epizootic period. They speculated that the absence of hunting and recreational or commercial trapping contributed to higher average survival than in rural areas, while episodic disease outbreaks might be a principal limiting factor in urban environments. One comparison of raccoon survival at urban and rural sites revealed higher survival rates at the urban site for 2 of 3 years, followed by lower survival there during the final year due to disease (Prange et al. 2003). Disease has long been suggested as a limiting factor for raccoon populations (Tevis 1947; Schubert et al. 1998; Broadfoot et al. 2001) and the logic that associates higher population densities with a greater likelihood of disease transmission is suggestive.

However, raccoon survival also varies enough among rural populations that there is reason to question whether living in an urban area confers any special advantage. Clark et al. (1989) reported annual survival among adults of an exploited population ranging from 0.47 to 0.75, while Gehrt and Fritzell (1999) reported on a nonharvested population of raccoons in a rural area with an adult survival rate (>0.84) even higher than those reported for most urban populations. In Manhattan, Kansas, Robel et al. (1990) found that survival was significantly higher in areas of low human disturbance (no trapping and low vehicular traffic) than high disturbance. Rosatte concluded that neither disease nor human intervention had long-term, signifi-

cant consequences for raccoon populations in Ontario because following even high population reduction (83% to 91%) densities returned to precontrol levels within a year (Rosatte 2000; Rosatte et al. 2001, 2007a). The limiting effect of diseases, their periodicity, and the interrelationship between diseases and other mortality factors suggests a complex set of factors may be mediating raccoon survival in urban habitats.

Prange et al. (2003) found more sources of mortality for rural than urban raccoons, with disease and vehicle-related death as the only observed cause of death for urban raccoons over a three-year period in the Chicago metropolitan area. Although road systems and traffic volume were dramatically different between urban and rural sites in that study, mortality from vehicles did not differ between areas, suggesting that adult raccoons adjusted to road traffic in urban areas (Prange et al. 2003). Sources of mortality in Washington, D.C., included rabies, distemper, road kills, euthanasia of "nuisance" animals, gastroenteritis, pneumonia, and attacks by dogs (Riley et al. 1998). Sources of mortality in Toronto included euthanasia, vehicles, attacks by dogs, and diseases such as rabies, canine distemper, and canine parvovirus (Rosatte 2000).

Although a number of studies have reported survival estimates for adult raccoons in urban environments, no study has reported survival estimates for juveniles immediately following their emergence from natal dens. There may be substantial differences in vulnerability to disease, predators, or traffic during this critical period, but, unfortunately, juveniles are typically too small to radio-collar during that time. However, road kill data from the Chicago study over 8 years suggest that juveniles may be particularly susceptible to cars during the summer and early fall of their birth year (figure 4.2).

Reproduction

Raccoons are seasonal breeders, giving birth once a year and typically exhibiting birth peaks in early to late spring, with occasional fall litters suggested for late breeders or females who lost litters early in the season and conceived again (Kaufmann 1982; Sanderson 1987; Gehrt 2003). There is anecdotal information from wildlife rehabilitators that breeding and birth periods in urban areas are extending throughout the year, perhaps because urban microclimates mitigate extreme cold, as well as provide greater availability of food resources, but data to support this have yet to be published.

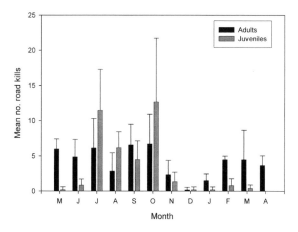

Figure 4.2. Monthly distribution of mean (+standard error) number of raccoons killed on the roads within and bordering an urban park in the Chicago metropolitan area. Road kills (*n* = 557) were tallied weekly during 1995–2001, and classified as juvenile (<1 year) or adult (>1 year). The time sequence begins with May because that is the month that juveniles begin to emerge from natal dens.

In Toronto, breeding occurs during late winter (February–March), with births occurring between April and June (Rosatte et al. 1991). Seventy-three percent of adult female raccoons examined during June and July were lactating or had suckled young as opposed to 94 percent in a rural southern Ontario sample (Rosatte et al. 1991). The estimated median birth date for 103 raccoons in Washington, D.C., was April 29, with the earliest birth occurring on April 2 and the latest on September 12 (J. Hadidian, S. P. D. Riley, and D. Manski, unpublished data).

In a comparison of urban, suburban, and rural raccoon populations, a greater number of juveniles relative to adult females were found at the urban and suburban sites, although the percentage of reproductive females did not differ among these areas (Prange et al. 2003). This suggested that either juvenile survival was higher in these urban and suburban areas or that larger litters were produced. However, in Connecticut, McComb (1981) found that the average number of young per adult or yearling female was almost six times greater in forested and in agricultural habitats than in urban habitats. In Ontario, litters averaged three to four for urban raccoons and four to five for raccoons in rural landscapes (Rosatte et al. 1991; Rosatte 2000). In Scarborough, Ontario, during 1987–1993, although >20% of the raccoon population was removed annually, raccoon density still averaged 12 / km², suggesting that reproductive compensation might have been oc-

curring. Finally, Hoffmann (1979) reported that heavier individuals in a suburban raccoon population reached sexual maturity earlier than lighter individuals.

From the available data, it is hard to tell whether urban raccoons differ from their rural counterparts with respect to basic reproductive parameters. The possibility that reproduction in urban areas might be more productive and only weakly seasonal, with mating and birth events spread out more evenly across the year, is intriguing and potentially significant from a wildlife conflict perspective, and it bears closer scrutiny.

Dispersal and Translocation

Little is known about dispersal in urban raccoons. McComb (1981) argued that the urban raccoon population he studied in Connecticut had to be sustained by dispersal of animals out of forested and agricultural areas, while Rosatte (2000) found that raccoons in Toronto were fairly sedentary, with little dispersal. Far more information is available on the movements of translocated raccoons. Rosatte and MacInnes (1989) reported on the movements of 24 raccoons translocated distances ranging 25–45 km from metropolitan Toronto between August and October 1986. None returned to the point of original capture, with mortality approaching 50% within the first 3 months following release. Mosillo et al. (1999) examined the fate of 76 translocated raccoons trapped as nuisances in suburban Chicago between 1993 and 1994. Three treatment groups were established, including animals trapped in suburban areas and translocated to a rural forest (urban), animals trapped from a wooded area and moved to rural forests (rural), and animals trapped and released within the same rural forest (forest). No differences in mortality were detected among the groups, even though the forest group tended to stay in the release area while animals from the other two engaged in more extensive movements. Clearly, the response of raccoons to translocation can be varied and undoubtedly depends on various influences, ranging from time of year to age, sex, and experience of the individuals moved, among other factors.

LANDSCAPE USE

Habitat / Land Use

Among natural habitats, raccoons generally select mature deciduous woodlands that are near permanent sources of water (Pedlar and Fahrig 1997). In urban areas, it is important to understand the extent to which habitat use falls out between developed versus more natural zones. Several studies in the Toronto metropolitan area have indicated that raccoon use of urban landscapes is preferential toward forested parks and residential areas, followed by cemeteries and field/industrial habitats (Rosatte 1986, 2000; Rosatte et al. 1991, 1992a, 1992b). In Washington, D.C., radio-collared raccoons from two different study areas showed variability in use of the developed and natural landscapes available to them. Some animals stayed entirely within the natural areas of the park; many used both park and residential areas (see example in figure 4.3), and in thin spurs of parkland; some animals stayed consistently in residential areas. However, because trapping occurred mostly within park (natural) areas, raccoons that exclusively used developed areas would be underrepresented. Among a sample of 29 females in the Chicago area, many stayed largely or entirely within forest parks,

which was surprising because the park was closed to visitors (and therefore their refuse) during the winter (Prange et al. 2004). In addition, some adult females ventured out into residential areas during the summer (see next section). Detailed habitat analyses in urban, in suburban, and in rural study sites in the Chicago metropolitan area supported the prediction that raccoon foraging is strongly influenced by the distribution and abundance of human-related food in urban areas (Prange et al. 2004; Bozek et al. 2007), such that many raccoons focused their activity around developed sites (specifically picnic areas) rather than on more natural habitat available to them.

Home Range Size and Shifts

From the first publication reporting radio-tracking data on raccoons (Ellis 1964), a positive association was suggested between high-density populations and

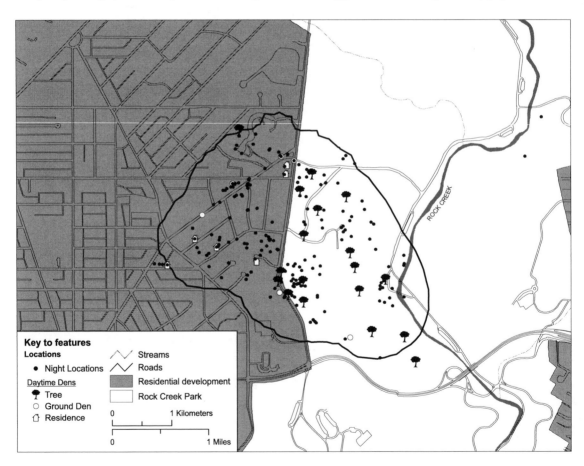

Figure 4.3. Home range (95% adaptive kernel) of one adult female raccoon in Rock Creek Park in Washington, D.C., 1989–1990. The area used comprises approximately 150 ha, about equally divided between an urban neighborhood of single-family residences and the adjacent national park, Rock Creek. Individual points represent nighttime, generally active, locations and daytime resting sites. Resting sites included trees, both in the park and in residential areas; residences, generally chimneys or attics; and ground dens, which were storm sewers under roads or dense patches of undergrowth.

smaller range areas. This, in turn, has been linked to the distribution and quality of important resources, particularly food and dens (e.g., Prange et al. 2003, 2004; Gehrt 2004b). As with density estimates, however, considerable variability in the methodology used to collect and to summarize home range data, as well as likely seasonal and annual variation, suggest caution in interpretation.

Gehrt (2004b) summarized urban home ranges in comparison with those for raccoons from other systems in North America and found a 5–79-ha range of averages for urban raccoons in comparison with 50–300 ha for others. While some of this variability is apparently mediated by landscape differences, home range size can also vary considerably within the same general landscape type. J. Hadidian, S. P. D. Riley, and D. Manski (unpublished data) found that the size of average home ranges (68 ha vs. 24 ha) differed significantly ($t_x = 3.09$, $p < 0.01$) for adult females from two study

areas in Rock Creek Park, Washington, D.C., even though the study locations were no more than 1.6 km apart. Home ranges overlapped considerably wherever multiple raccoons were radio-collared (figure 4.4).

On other continents where raccoons are newly established, similar trends in home range size have been reported. For example, in Kamakura City, Hokkaido Prefecture, Japan, home ranges of raccoons in urban areas averaged 49 ha, compared with 585 ha in adjacent rural, forested habitats (Ikeda et al. 2004). In western Poland, home ranges in a suburban area in the city of Kostrzyn averaged 100 ha, compared with an average of 1,000 ha in an adjacent rural park (Bartoszewicz et al. 2008).

Seasonal variability in home range use and movements are well established for raccoons in most of the habitat types where they have been studied (Kaufmann 1982; Gehrt 2003), but with urban raccoons, trends may not be conclusive. In rural areas, studies

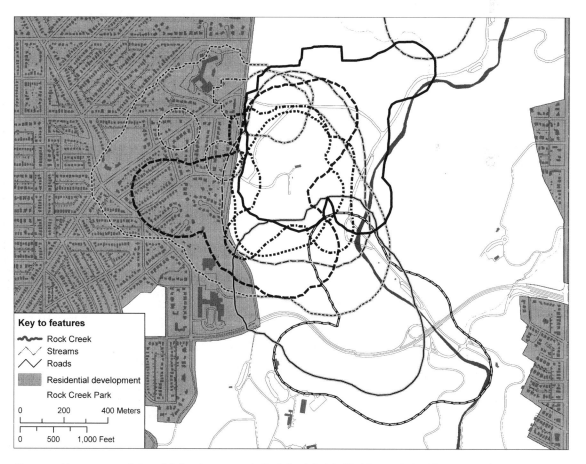

Key to features

- Rock Creek
- Streams
- Roads
- Residential development
- Rock Creek Park

0   200   400 Meters

0   500   1,000 Feet

Figure 4.4. Home ranges of nine adult female raccoons in Rock Creek Park in Washington, D.C., 1989–1990. In this area, some raccoons stayed entirely within the park, while others used both wooded and residential areas. There was also extensive home range overlap within and between sexes.

have shown that raccoons tend to shift activity centers to take advantage of seasonally abundant food resources (Ellis 1964; Turkowski and Mech 1968; Johnson 1970). In urban environments, it is likely they would do the same, except anthropogenic food resources may be more stable and predictable (e.g., feeding stations, Dumpsters). Slate (1985) reported no seasonal change in movements for raccoons studied in suburban New Jersey, while Prange et al. (2004) identified two distinct patterns in their urban study population. Most of the female raccoons they followed ($n = 29$) had small home ranges and stayed within the urban study site year-round. Some, however, shifted their ranges to outlying heavily developed residential/commercial areas with movements of up to 1.6 km in late summer/early fall. Prange et al. (2004) suggested these shifts might be associated with decreasing tolerance of adult females for one another when they began traveling with juvenile offspring, coupled with the highly concentrated nature of anthropogenic resources at picnic areas. Conversely, raccoons at their suburban site exhibited highly stable home ranges across seasons.

Dens

Raccoons make use of a wide variety of sites for denning, ranging from rock ledges, brush piles, and abandoned groundhog burrows, to hollow trees and accessible human structures such as chimneys and storm sewers (figure 4.5; Stuewer 1943; Berner and Gysel 1967; Schinner and Cauley 1974; Rosatte et al. 1987; Hadidian et al. 1991). Where available, hollow trees appear to be preferred, especially when females give birth and raise young (Rabinowitz and Pelton 1986; Endres and Smith 1993). Hadidian et al. (1991), reporting on raccoons from two study areas in an urban national park (Rock Creek Park), found that dens were occupied by any given individual on average 3.7 times before that individual was located in a new site (exclusive of adult females with young). As many as seven different animals occupied a given den at different times, although most (73%) located dens were associated with only one user. Co-occupancy, or communal denning, occurred among Rock Creek raccoons, most commonly in adult female–adult male or adult male–adult male combinations and never with certainty that individuals occupying the same den site were actually in physical contact.

As many as 23 raccoons have been reported as simultaneously occupying the same den (Mech and

Figure 4.5. Raccoons will use human structures readily. Here, an adult female has used a fireplace chimney as she would a hollow tree, to create a birth den and raise young. Attics and crawl spaces are readily occupied as well. Photo courtesy of Humane Wildlife Services.

Turkowski 1966). Hoffmann and Gottschang (1977) speculated that such co-denning in urban sites was a function of limited availability of appropriate den sites, and more generally Endres and Smith (1993) reported that raccoons used dens of different types based on their availability. O'Donnell and DeNicola (2006) argued for what might be called learned tradition in the observed tendency for raccoons that had used human structures to exhibit a similar preference upon displacement.

BEHAVIOR AND ECOLOGY

Food Habits

Buffon (1828), in one of the first notes concerning raccoon behavior in literature, remarked that they "ate everything," thereby establishing one of the most reliably consistent observations of any concerning these animals. While many studies have been conducted to rigorously identify the foods of raccoons across a variety of habitats, relatively little information is available on urban populations. McComb (1981) identified 22 different foods in the diet of urban raccoons in Connecticut, as opposed to 29 items from agricultural and 35 from forested environments. Manski and Hadidian (1985) found 33 plants from scats and digestive tracts in a preliminary survey of raccoons from an urban national park, with evidence of feeding on anthropogenic food resources from 21% of digestive tracts of road kills and 31% of scats. J. Hadidian (unpublished data) followed this with a systematic collection of 680 scats, mostly from latrine stations, from which nearly two-thirds (64%) of all samples showed evidence of plant foods, followed by vertebrate (17%) and invertebrate (22%) remains. Anthropogenic sources occurred in 14% of samples (e.g., figure 4.6), while unconsolidated, amorphous material that could have been anthropogenic in origin occurred in another 29% of samples. In Toronto, raccoons were observed eating earthworms, commercial pet food, and human refuse (Rosatte 2000). The presence of earthworms, especially in the spring diet of raccoons, may represent a significant food resource that is difficult to identify from scats (Greenwood 1979). Raccoon predation of waterfowl and passerine nests has also been reported from urban areas (Jobin and Picman 1997).

Slate (1985) and Seidensticker et al. (1988) established feeding stations at suburban and rural locations, respectively, that drew most of the radio-collared animals in their studies into a central location. Raccoons are opportunistic foragers, and it is highly likely that short-term and concentrated use of locally abundant food resources occurs as a learned response (Dalgish and Anderson 1979). Future research may reveal the extent to which they focus on localized resources, with intriguing questions about how these resources are initially located and whether raccoons might have a long-term memory for when foods become available seasonally. One adult male raccoon studied during the Rock Creek project significantly moved his center

of activity over the course of two consecutive falls in what might have been an intentional shift to be closer to a grove of ripening persimmon trees, suggesting, if somewhat weakly, that he "remembered" where this resource was and when it would be available for consumption (J. Hadidian, personal observation).

Social Organization

Raccoons are typically described as being "solitary" animals, which may be a consequence of their nocturnal and secretive behavior, rather than any actual behavioral tendencies. Some studies from rural environments provide strong evidence for intolerance and even territorial behavior at low densities (Johnson 1970; Fritzell 1978). However, as investigators are increasingly able to conduct rigorous and longer-term field observations, their social tendencies are being uncovered.

A typical raccoon population consists of adults of both sexes with overlapping home ranges (see figure 4.4), with identifiable male social groups (Gehrt and Fritzell 1998b). Overlap of home ranges may be extensive in urban landscapes, especially where rich patches of anthropogenic resources occur in a spatially predictable pattern (Prange et al. 2004). Although long-term familial bonds have not been described for urban raccoons, they demonstrate mother-offspring and sibling spatial affinities that begin in early summer and continue through autumn and winter, often until the next mating season (Gehrt 2003; J. Hadidian and S. P. D. Riley, unpublished data). In some cases, family members have reestablished bonds following the mating season (Gehrt and Fritzell 1998a). Future research is likely to demonstrate matriarchal associations in which related females share overlapping home ranges, perhaps with definable social hierarchies.

Relatively lengthy mother-offspring bonds may contribute to the raccoon's ability to adjust to urban life. For example, in an urban park in Chicago, the altered foraging behavior of mother raccoons, in which they focused their foraging time at garbage receptacles (Bozek et al. 2007), may have been transferred to their offspring since they followed their mothers closely during the peak picnicking season (S. D. Gehrt and S. Prange, personal observation). To what extent young raccoons are taught urban ways by their mothers, such as searching for and using anthropogenic food or dens (i.e., learned tradition; Gehrt 2004; O'Donnell and DeNicola 2006), warrants further exploration.

Figure 4.6. Evidence of raccoons feeding on trash. Where other species (crows, dogs) might scatter the trash around, the raccoon that visited here reached into the bag and felt for the food items it wanted to remove. Obviously, the solution to this "raccoon problem" is to redefine it as a "trash problem" and provide proper storage facilities. Photo courtesy of the Humane Society of the United States.

Whatever the specifics of raccoon social behavior, the general lack of territoriality and a certain level of social tolerance could contribute to the impressive densities reported for some urban populations. Interactions in which raccoons aggregate appear largely to be restricted to ritualized displays and displacement, indicating a recognized social order (Sharp and Sharp 1956; Slate 1985). However, Totton et al. (2002) reported a relatively high level of direct aggression (biting) at a rural garbage dump, which suggests that this is not always the case. It is ironic that the social system of raccoons, a "solitary" animal, contributes to their ability to attain high densities and exploit clumped resources, whereas territoriality probably limits the densities for species considered more social, such as coyotes.

Interactions with Wildlife

Both direct and indirect interactions with other wildlife species are reported for raccoons. Directly, raccoons may prey on certain species of wildlife (Hartman et al. 1997; Jobin and Picman 1997; Dijak and Thompson 2000), and indirectly they may serve as reservoirs and vectors for diseases (Broadfoot et al. 2001; Rosatte et al. 2006) or as ecological competitors (e.g., opossums; Ginger et al. 2003). Similarly, raccoons may be affected by predation, competition, and interaction with other wildlife species.

Crooks and Soulé (1999) reported that raccoons were one species implicated in the phenomenon of "mesopredator release," in which the removal of a top predator (e.g., coyotes) would allow the "release" of mesopredators, such as raccoons and domestic cats, with consequent impact on prey species (e.g., songbirds) on which mesopredators may have had a limiting effect. However, for raccoons specifically, the effect of coyote presence was not significant. Moreover, Gehrt and Prange (2007) found no raccoon mortality attributable to coyotes, and no avoidance of coyotes or signs of coyote activity, in tests of this hypothesis in which populations of both species overlapped extensively (see chapter 14). Uncertainty may also exist over the ecological impact of raccoons on endangered species, such as nesting sea turtles. Although raccoon predation influences the nesting success of sea turtles (S. Anderson 1981; Ratnaswamy et al. 1997), raccoons may also act to suppress populations of other potential nest predators, such as ghost crabs (Ratnaswamy and Warren 1998; Pennisi 2006), thereby conferring some positive advantage to the survival of turtle hatchlings. These findings suggest complex interrelationships between raccoons and potential prey species, which is not surprising given that raccoons are more opportunistic than obligate predators.

HUMAN-RACCOON RELATIONSHIPS

Raccoons can have complex relationships with humans and their pets. Raccoons are potential predators of domestic pets, including small cats and rabbits, and may be attacked and injured or killed by dogs (Schinner 1969; Hadidian and Riley 1996). Raccoons are easily attracted to feeding stations (Sharp and Sharp 1956; Slate 1985), and they can become regular visitors to backyards where they "entertain" human audiences (Gander 1986). D. Manski (personal communication) knew of two major feeding stations in his study area in Rock Creek Park, Washington, D.C., whose consistently available food resources almost certainly influenced the movement and activity patterns of his radio-collared animals. The extent to which deliberate feeding of raccoons takes place in urban and suburban neighborhoods remains another area of potentially fruitful study.

Raccoons seem to appeal to people at a number of levels. Raccoon young are popular pets (e.g., North 1963), although probably much less so now than in the past. Orphaned and injured raccoons are often treated by wildlife rehabilitators (Ludwig and Mikolajczak 1985). "Ranger Rick," the iconic mascot for the children's educational program for the National Wildlife Federation, was derived from a popular storybook first published almost a half-century ago (Morris 1959) and now is the image that anchors a periodical magazine that reaches many thousands of schoolchildren. Dozens of popular books are based on or around raccoons, ranging from complete natural histories (MacClintock 1981; Zeveloff 2002), to entertaining stories about raccoons as pets (North 1963, 1966), to children's books (Burgess 1942; Moore 1963). This fascination with raccoons extends beyond the United States. The current invasion of raccoons in Japan is partly the result of a popular animated cartoon in the late 1970s, which prompted the importation of many raccoons and subsequent breeding for the pet market (Ikeda et al. 2004). In Germany, where raccoons were released in the 1930s, they are relatively new residents in a number of cities and towns. Hohmann et al. (2001) interviewed residents of one village where raccoons were present and reported that 100% were aware of their presence, and 76% had raccoons using their properties. However,

89% did not consider raccoons a nuisance, even though 33% believed raccoons raided garbage cans, and 52% attributed losses from fruit trees to raccoons.

Such articles as "Those Dirty Raccoons" (McCombie 1999) attest to the highly negative sentiments that these animals can sometimes evoke. Because raccoons make ready use of human structures for den sites and often get into poorly managed trash (figure 4.5; Daag 1970), they are considered a top concern among urban wildlife species for causing conflicts with humans (De Almeida 1987). Gehrt (2003) estimated the economic costs of raccoons in the greater Chicago metropolitan area to be more than $1 million in 1999. If raccoons are estimated to cause approximately 60% of all wildlife damage (De Almeida 1987), then their total economic impact on North Americans would be significant indeed.

Zoonoses and Public Health Threats

Raccoons are susceptible to a number of diseases that have public health consequences for humans, and they can be affected themselves by diseases originating from human pets. Among the primary concerns for humans or domestic animals are rabies, raccoon roundworm (*Baylisascaris procyonis*), and leptospirosis (Kazacos 1983; Jenkins et al. 1988; Rosatte 2000; Rosatte et al. 2006, 2007c). Less studied but potentially important interrelationships between raccoons, people, and other diseases exist as well. For example, in an urban area of Virginia, antibodies to *Toxoplasma gondii* were found in raccoons (Hancock et al. 2005), and raccoons are included among the blood hosts of a mosquito (*Aedes albopictus*) that could be a potential vector of arboviruses (Niebylski et al. 1994).

Raccoon roundworm is an important zoonotic parasite, which has been classified as an emerging infectious disease, having led to greater human mortality than rabies (Kazacos 2001). The disease can be a threat to public health and have important ecological effects. Raccoons are host to the adult form of the worm, which, to them, is relatively benign. However, the worm is capable of producing tremendous numbers of eggs that are shed in the feces and that, once ingested by an intermediate host, can lead to catastrophic neurologic disorder and even death. A wide range of avian and mammalian species has been documented as potentially affected by raccoon roundworm, including humans and most pets (Kazacos 1983).

Prevalence of raccoon roundworm is related to host density, at least in rural areas, and can be as high as 80% in some populations (Kazacos 1983). Given the high population densities reported for raccoons in urban areas, it would be reasonable to assume that prevalence of raccoon roundworm would also be high. However, estimates for raccoons in the Chicago area revealed a lower prevalence of infection when compared with surrounding rural areas (Page et al. 2005), and Jacobsen et al. (1982) found about equal prevalence in urban (27%) and rural (31%) raccoons. The reduced prevalence in raccoons from the Chicago area may reflect differences in intermediate host dynamics or changes in foraging behavior in raccoons (Bozek et al. 2007).

Of all raccoon diseases, rabies has undoubtedly had the greatest social and economic impact on humans. A raccoon rabies outbreak spread throughout the Mid-Atlantic and northeastern United States in the 1980s and 1990s and now extends into parts of the Midwestern United States and southern Canada (Jenkins et al. 1988; Rosatte et al. 2001, 2006, 2007c). Attempts to control the spread of this disease within raccoon populations have focused on large-scale oral baiting programs (Slate et al. 2005; Rosatte et al. 2007c), as well as combined strategies of population reduction and immunization (Rosatte et al. 1997, 2001, 2007). Control using vaccine baits in urban areas will be challenging because of fragmented landscapes as well as high human, pet, and wildlife population densities. Rosatte and Lawson (2001) used extremely high bait densities (200 and 400/km²) in Toronto to achieve raccoon bait acceptance levels of 74% and 82%, respectively, in an area where raccoon density averaged about 20 individuals/km². The number of baits required to reach a significant portion of the raccoons in urban landscapes was about 2.5–5 times that of rural landscapes.

It will be critical to control rabies in urban landscapes as quickly as possible because of high raccoon population densities, which increase the ability of the disease to spread rapidly in cities. In addition, the mobility of raccoons complicates the control of this disease. For example, in eastern Ontario, a rabid raccoon was documented moving 1.7 km in 24 h, and during 1999–2002, rabid raccoons were documented moving 1.7 to 4.1 km (Rosatte et al. 2005, 2006). If rabies is allowed to become enzootic in cities, it may be difficult to control with vaccine baits, as vaccination does not work on animals incubating the disease or clinically active with it (Rosatte et al. 2007d). In Washington, D.C., there was evidence of rabies incidence cycling in the raccoon population 9 years after the initial epizootic, and the disease was potentially continuing to affect raccoon density (Riley et al. 1998). Where it has not yet reached urban centers, Rosatte et al. (1997, 2001, 2007c) have argued that the first line of response may have to be population reduction.

Disease can also be transmitted from domestic pets to raccoons, often with serious population-wide consequences. Canine distemper has been found to be prevalent in both suburban and urban raccoons (Cranfield et al. 1984; Roscoe 1993; Schubert et al. 1998). Of 1,079 raccoons submitted for testing from Ontario, Quebec, and New Brunswick during 1999–2004, 30% were positive for canine distemper (Rosatte et al. 2007d). Schubert et al. (1988) experimentally vaccinated their study population and found they could produce a reduced prevalence of the disease while not necessarily leading to an increased abundance of raccoons, suggesting that much remains to be learned about the actual demographic effects of this disease on raccoons.

MANAGEMENT

Dense raccoon populations living in proximity to humans are frequently cited as potential causes of conflicts, ranging from the transmission of parasites and disease to invading and occupying attics and chimneys (De Almeida 1987; Gates et al. 2006; O'Donnell and DeNicola 2006). Responsibility for the management of urban raccoons is currently fragmented across municipal, state, provincial, and federal agencies (Hadidian et al. 2001) and spans a continuum from localized control of problems caused by individual raccoons to population management at a landscape level. Activities involving control can range from ad hoc responses of homeowners who might trap or kill individual raccoons to strategic, coordinated, multiagency programs, such as oral rabies baiting, that operate at a geographic scale. The state and federal agencies responsible for wildlife protection and conservation may or may not take a direct role in any aspect of management concerning urban raccoons or other urban wildlife (Adams et al. 2006).

Raccoons may be trapped as "nuisance" animals throughout North America, despite being protected under other circumstances by laws regulating the taking of fur-bearing animals. Typically, state wildlife agencies do not report data reflecting the magnitude of such control activities (Barnes 1997; Hadidian et al. 2001). Where data exist, it suggests that impacts can be great. Rosatte (2000), for example, reported that "thousands" of raccoons were euthanized in Toronto

each year because of nuisance situations, with 21.5% of the population euthanized on average each year between 1987 and 1993 in the Scarbourgh suburb alone. Elsewhere, Bluett et al. (2003) reported on a nine-year period (1992–2000) in Illinois during which approximately 128,000 raccoons were killed by wildlife control operators, with another 53,000 released or rehabilitated. Over a 12-year period (1991–2002) in Michigan, more than 79,000 complaints about raccoons led to control actions by private enterprises, with the result that approximately 70,000 raccoons were trapped and released, 43,000 were killed, and another 1,300 sold (Michigan Department of Natural Resources, unpublished data). No estimates exist of the numbers of raccoons trapped by the public, but it is probably as great, or greater, than trapping conducted by other interests. Significantly, as yet virtually nothing is known about the ecological or demographic consequences of control activities, leaving as speculation whether these solve problems, mitigate or exacerbate "nuisance" activities, control or amplify the potential transmission of disease, or stabilize or destabilize wildlife community interactions and relationships.

In countries in which raccoons are exotic, control programs may be initiated to reduce raccoon populations and thereby the ecological damage they may cause to native species. In Japan, there is concern that raccoons may negatively impact some species through predation and others through competition. In an urban park in Hokkaido Prefecture, raccoons apparently displaced native raccoon dogs and caused herons to abandon a nesting colony because of nest predation (Ikeda et al. 2004). Before 2004, raccoon control in the country was unstructured and conducted at the local level in response to nuisance conflicts. However, because of the concern regarding alien mammals, the Invasive Alien Species Act was passed in 2004 with a goal toward a more systematic national effort to eradicate raccoons. In Poland, the raccoon was officially declared a game animal in 2004, but this has apparently not been sufficient to stem the expansion of the animal, especially in urban areas (Bartoszewicz et al. 2008).

CONCLUSION AND RESEARCH NEEDS

The study of urban raccoons should be increasingly emphasized for a number of reasons. Raccoons can serve as environmental monitors (Valentine et al. 1988), opening the door to their use as surrogates in a range of assessments involving anthropogenic influences. Given that environmental impacts varying from pollution to habitat destruction can be amplified and accelerated in cities (Gilbert 1989), urban raccoons might make good candidates for studying phenomena that could take many times longer to document in more remote and less disturbed ecosystems. Further study is needed to better understand the role urban raccoons play regarding biodiversity conservation, disease transmission, educational and social values, and direct conflicts with humans. For those who believe that urban ecosystems are worthy of study in their own right, the raccoon is an important component of the urban fauna, no doubt playing a prominent role in the structuring of urban wildlife communities.

If some populations of urban raccoons are only just becoming established, such as in some parts of Canada or Europe, the opportunity to study them as colonizers of urban habitats may exist and should be seized on. Elsewhere, opportunities may still exist to establish baseline information on raccoon populations that might be in the path of anticipated change. The Mid-Atlantic raccoon rabies outbreak, for example, spread throughout the northeastern United States without baseline (pre-outbreak) population data being collected. Such opportunities should not be overlooked in the future.

The study of urban raccoons is still in its infancy. To date, research illustrates what has long been a common assumption: raccoons fare quite well by living in proximity to humans and perhaps better when associated with human environments than not. The variability evidenced through studies of these animals in different habitats argues that a good deal more research is needed before many generalizations can be made. As that research is conducted, there can be little doubt that many new and original insights into the behavior and ecology of these fascinating animals will be forthcoming.

5

Kit Foxes (*Vulpes macrotis*)

BRIAN L. CYPHER

K IT FOXES are among the smallest of North American canids. They are delicate in appearance with slender bodies, large ears, and relatively short tan to grizzled gray fur. Full-grown adult kit foxes are about the size of a house cat. Adults are about 60–84 cm from nose tip to tail tip, and males on average weigh 2.2 kg while females weigh 1.9 kg (McGrew 1979).

Kit foxes occur primarily in desert and semiarid habitats in the southwestern United States and northwestern Mexico. These habitats include arid shrub and shrub-grassland communities (Cypher 2003). Historically, as many as eight subspecies have been identified (Hall 1981), although some analyses have only supported two (Mercure et al. 1993). One of these subspecies, the San Joaquin kit fox, occurs in a limited range in central California and this population is disjunct from all other kit foxes. Because of extensive habitat degradation and loss, this subspecies has been listed as federally Endangered and California Threatened (U.S. Fish and Wildlife Service 1998).

In the San Joaquin Valley of California, kit foxes occur in some urban areas. This is somewhat ironic given that urban development is one of the primary causes of habitat loss that led to federal and state protections for this taxon. Kit foxes are commonly observed in Bakersfield, Taft, and Coalinga, California (figure 5.1). Taft and Coalinga (human populations of about 10,000 each) are small cities and are bordered on several sides by good kit fox habitat. There likely is considerable interchange between urban and adjacent nonurban areas.

The kit fox population inhabiting Bakersfield is different, however. Ba-

Figure 5.1. Cities with populations of endangered San Joaquin kit foxes in the San Joaquin Valley of California.

kersfield is a large city with a human population of just under three hundred thousand and is growing rapidly. Based on trapping efforts and casual observations, the kit fox population has been estimated at several hundred individuals (Cypher et al. 2003). Only approximately 25% to 30% of the city abuts natural habitat. The remainder is bordered by annual crops such as alfalfa, cotton, and carrots, which provide little habitat value for kit foxes (Warrick et al. 2007). Thus, a relatively small proportion of the kit fox population uses both urban and nonurban habitats. Furthermore, there is little natural habitat within the city. Therefore, most kit foxes in Bakersfield use exclusively anthropogenic habitats and reside exclusively within the urbanized area.

Considerable demographic and ecological information has been collected on the kit foxes in Bakersfield to assess the vigor and viability of the population and to determine whether it can contribute to range-wide recovery and conservation efforts. The information presented here on urban kit foxes is from the Bakersfield population.

COMPARISONS BETWEEN URBAN AND NONURBAN KIT FOXES

Survival

Larger predators are the main cause of mortality for kit foxes in natural habitats (table 5.1). Coyotes are the primary predator, but bobcats, red foxes, badgers, and golden eagles also kill kit foxes (McGrew 1979; Cypher 2003). Because of the plethora and diversity of human activities in urban environments, the potential hazards for kit foxes are numerous. Not surprisingly, kit foxes in urban areas die from a diversity of causes. The primary cause of mortality among urban foxes is vehicles (table 5.1). Other important sources include predation, poisons, and entombment (Bjurlin et al. 2005). Predators include domestic dogs, coyotes, and bobcats. Dogs occur throughout the city, and while most dogs that kill foxes are free ranging, at least one fox was killed after it repeatedly entered a fenced yard in which a dog resided. Coyotes and bobcats occur almost exclusively on the edges of the urban environment and along the Kern River corridor, and foxes killed by these predators were all in these locations.

Four kit foxes are suspected of having died from exposure to some toxicant. Substances potentially toxic to foxes are abundant in urban areas and include vehicle antifreeze, motor oil, and rodenticides. On the basis of laboratory tests, at least two radio-collared foxes are suspected of dying from exposure to commercially available rodenticides commonly used to control commensal rodents (e.g., rats and mice). It is unknown whether the fox exposures were primary (e.g., foxes consuming baits) or secondary (e.g., foxes consuming poisoned rodents). Also unknown is the number of foxes impaired because of rodenticide exposure, causing them to succumb to other mortality agents (e.g., vehicles, predators). Of 24 dead urban foxes tested, 23 exhibited some level of rodenticide exposure. Rodenticides have been identified as a significant mortality factor for urban carnivores elsewhere (Riley et al. 2007).

Other causes of urban kit fox mortality are relatively infrequent. Foxes occasionally have been accidentally entombed when they have occupied dens on construction sites. On a number of occasions, foxes have moved onto construction sites where loosened soil is easy to excavate and food is commonly dropped by workers. On at least one occasion, a fox was entombed by a spontaneous collapse of an earthen den. One fox drowned after apparently climbing into a large fountain and was not able to get out. In an odd and very urban situation, foxes have become entangled in sports netting—usually soccer goal nets—but two foxes also were caught in the netting of baseball batting cages. Some of these animals were discovered in time and rescued, but at least five died in the nets.

Not unexpectedly, some foxes have been maliciously

Table 5.1. Causes of mortality among urban and nonurban populations of San Joaquin kit foxes

Cause of mortality	Population			
	Bakersfield[a] (urban) $n = 229$	Lokern[b] (nonurban) $n = 63$	Elk Hills[c] (nonurban) $n = 341$	Carrizo[d] (nonurban) $n = 41$
Vehicle	27	1	20	1
Predator	17	19	129	17
Entombment	4	—	1	1
Poison	4	—	—	—
Other	4[e]	—	2[f]	—
Undetermined	22	5	73	3

[a] CSUS Endangered Species Recovery Program, unpublished data.

[b] CSUS Endangered Species Recovery Program, unpublished data.

[c] Cypher et al. 2000.

[d] Ralls and White 1995.

[e] Other causes included drowning, gunshot, entanglement, and complications during parturition.

[f] Other causes included drowning and gunshot.

(and illegally) killed by humans. One fox died after being shot with a pellet gun. Six foxes (four pups and their two parents) died after they were entombed in their den by gardeners under direction from their supervisor. Because these instances involved an endangered species, fines were levied against the guilty parties.

Adult survival probabilities for urban foxes in Bakersfield were comparable to and actually slightly higher than probabilities for nonurban foxes (table 5.2). In nonurban kit fox populations, survival rates of juveniles (<1 year) vary from 0.14 to 0.21 (Ralls and White 1995; Cypher et al. 2000). Although the data for urban juveniles have not been analyzed yet, survival probabilities are likely comparable to those of nonurban foxes, based on casual observations.

Reproduction and Social Ecology

Kit foxes apparently have no problem reproducing in urban environments. Litters of kit fox pups have been observed in Taft and Coalinga as well as in Bakersfield, where a number of kit fox litters are reported each year from throughout the city. Natal dens have been located in city sumps, golf courses, vacant lots, railroad rights of way, canal banks, and under portable classrooms at schools.

Reproduction among urban kit foxes appears robust. Annual reproductive success among urban foxes is markedly higher than among nonurban foxes (table 5.3). As with most species, reproductive success is strongly influenced by food availability. In nonurban

environments, food availability fluctuates considerably, depending on annual environmental conditions, particularly precipitation (White and Garrott 1997; Cypher et al. 2000). In urban environments, food seems to be consistently abundant. Because of plentiful water (e.g., landscaping irrigation, canals), prey is abundant and availability does not appear to fluctuate significantly. Also, anthropogenic foods are consistently abundant. Interestingly, mean litter size is virtually identical to estimates for nonurban fox populations (table 5.3).

As with nonurban foxes, relatively little information is available on dispersal among urban kit foxes. Movements of over 8 km have been recorded in Bakersfield. This is not remarkable, as mean dispersal distance among nonurban kit foxes is 7.8 km (Koopman, Cypher, and Scrivner 2000). In Bakersfield, dispersal potential may be somewhat restricted. Survival rates among urban kit foxes are relatively high. As a result, most suitable open space is generally occupied by foxes. Because of the high survival rate of young and the restricted dispersal potential, a greater incidence of philopatric young are observed within parental home ranges.

Kit fox social units consist primarily of a mated pair and their young of the year. In addition, young from the previous year (typically females) may delay dispersal and continue to occupy their natal ranges, and frequently, these individuals will assist the parents in raising the current litter of pups (Koopman et al. 2000; Ralls et al. 2001). Kit foxes generally mate for life

Table 5.2. Annual survival probability estimates of adult San Joaquin kit foxes from urban and nonurban populations

Location	Urban or nonurban	*n*	Survival probability (range)
Bakersfield (1997–2004)[a]	Urban	142	0.70 (0.48–0.95)
Lokern (2001–2004)[b]	Nonurban	44	0.64 (NA)
Elk Hills (1980–1995)[c]	Nonurban	341	0.44 (0.20–0.81)
Carrizo Plain (1989–1991)[d]	Nonurban	33	0.61 (0.50–0.74)

[a] B. Cypher, CSUS Endangered Species Recovery Program, unpublished data.

[b] B. Cypher, CSUS Endangered Species Recovery Program, unpublished data.

[c] Cypher et al. 2000.

[d] Ralls and White 1995.

Table 5.3. Reproductive rates and mean litter sizes among urban and nonurban populations of San Joaquin kit foxes

Reproductive parameter	Population (urban or nonurban)			
	Bakersfield[a] (urban)	Lokern[b] (nonurban)	Elk Hills[c] (nonurban)	Carrizo[d] (nonurban)
No. females monitored	52	24	126	19
% reproductive success	78.8	54.2	61.1	21.1
(range of annual values)	(66.7–100)	(50.0–80.0)	(20.0–100)	(0–57.1)
No. litters	71	23	97	4
Mean litter size	3.8	3.8	3.8	2.0
(range)	(1.9)	(2–9)	(1–6)	(1–3)

[a] B. Cypher, CSUS Endangered Species Recovery Program, unpublished data.

[b] B. Cypher, CSUS Endangered Species Recovery Program, unpublished data.

[c] Cypher et al. 2000.

[d] White and Ralls 1993.

(Grinnell et al. 1937; Egoscue 1956; Ralls et al. 2007). Mating occurs in late November to late December, and young are born in earthen dens from late January to early March after a gestation of 49–55 days. Young appear aboveground at about 4 weeks of age and are weaned at about 8 weeks. Parents will continue to provision pups with food into June. Some pups then begin dispersing in June (Egoscue 1956, 1962; Zoellick et al. 1987; Koopman et al. 2000). In the wild, kit foxes live an average of about 2–3 years (Spiegel 1996), with some animals surviving 7 years (Egoscue 1975; Cypher et al. 2000).

Social ecology is similar among urban and nonurban kit foxes. However, at least one aspect of kit fox social ecology appears to differ between urban and nonurban populations. In urban environments, the incidence of "helper" foxes appears to be higher. These individuals typically are young from the previous year, usually females, that delay dispersal and assist with raising the current year's litter of pups. Even more interesting, older females in Bakersfield that have produced mul-tiple litters have been documented to cease reproducing to assist a daughter in raising a litter of pups. It is unclear whether these older females have become barren or whether some other factor (e.g., social status) is suppressing reproduction.

Habitat and Home Range Use

Arid shrub communities and grasslands are the natural habitats preferred by kit foxes. As mentioned previously, very little natural habitat occurs within Bakersfield. Thus, kit foxes in Bakersfield primarily use anthropogenic habitats in which natural ecological processes are nonexistent or significantly altered. Habitats used by kit foxes include undeveloped lands (e.g., vacant lots, fallow crop fields), storm water catchment basins (sumps), industrial areas (e.g., manufacturing facilities, shipping yards), commercial areas (e.g., office and retail facilities), manicured open space (e.g., parks, school campuses, golf courses), linear rights of way (e.g., canals, railroad corridors, power line corridors), and residential areas.

On the basis of 27 radio-collared kit foxes monitored in 1997, kit foxes used undeveloped lands and sumps disproportionately more relative to the availability of these habitats and residential areas disproportionately less. Use of other habitats was proportional to their availability (Frost 2005). Undeveloped lands and sumps experience less routine anthropogenic disturbance than other urban habitats. Conversely, residential areas have consistently high levels of disturbance, and the presence of numerous dogs as well as fences and walls (surrounding individual yards as well as entire neighborhoods) discourage use of these areas by kit foxes. Consequently, urban kit foxes primarily use nonresidential areas but make occasional forays into adjacent residential areas (figure 5.2).

Home range size of urban kit foxes was 1.72 km² based on the 100% minimum convex polygon estimator and 1.2 km² based on the 95% fixed kernel estimator, and core area size was 0.16 km² based on the 50% fixed kernel estimator (Frost 2005). These estimates are markedly smaller than estimates derived for nonurban San Joaquin kit foxes (table 5.4). Several factors might contribute to smaller home ranges among urban kit foxes. Foods are both more abundant and possibly more concentrated in urban environments, and therefore, kit foxes may be able to obtain necessary resources from a smaller area. High densities of foxes in suitable urban habitats may also result in reduced space use. Finally, the juxtaposition of features in urban environments, particularly potential barriers (e.g., large roads, walls, canals), might impede movements, resulting in reduced space use.

Den Use

Kit foxes are obligate den users (Egoscue 1962; Morrell 1972; figure 5.3). Unlike most other North American canids that use dens primarily during the breeding season to bear and rear young, kit foxes use dens every day of the year. Dens are used for daytime resting, avoiding temperature extremes, conserving moisture, avoiding predators, and bearing and rearing young (Cypher 2003). Thus, dens are a critical component of kit fox ecology. Nonurban kit foxes use an average of 11 different dens per year (Koopman et al. 1998), and these dens are scattered throughout the home range of a fox. Urban kit foxes appear to use fewer dens on average (Frost 1995; Endangered Species Recovery Program, unpublished data), possibly because home ranges of kit foxes are markedly smaller in urban environments.

Urban kit foxes construct dens in a variety of urban habitats and use different types of anthropogenic structures for denning. The 27 kit foxes monitored in 1997 used 190 separate dens over an 8-month period (Frost 2005), comprising 159 earthen dens and 31 man-made structures. Although dens were found in all urban habitats, dens were disproportionately more common in sumps and undeveloped areas compared with the availability of these habitats, and disproportionately less in residential and commercial areas. Man-made structures used as dens included culverts, pipes, seatrains, portable buildings, junkyard debris, and Dumpsters.

Food Habits

Nonurban kit foxes prey primarily on nocturnal rodents, particularly heteromyids such as kangaroo rats. They also consume rabbits, ground squirrels, birds, various reptiles, and insects, particularly beetles, grasshoppers, and crickets (McGrew 1979). Similar to kit foxes in nonurban habitats, prey items consumed by urban kit foxes consist primarily of rodents and insects (Cypher and Warrick 1993; Endangered Species Recovery Program, unpublished data). However, the species consumed vary between the two habitats. Pocket gophers and California ground squirrels are the primary rodents consumed in urban environments (table 5.5). Occasionally, rats and house mice are consumed by urban kit foxes. Insects consumed by urban kit foxes consist of beetles, grasshoppers, and cockroaches. Other soft-bodied insects also may be commonly consumed but are difficult to identify in scats. For example, foxes have been observed feeding extensively on moths (R. Whitehair, personal communication).

Not surprisingly, kit foxes in urban environments opportunistically take advantage of various anthropogenic sources of food. These include trash and discarded food (e.g., fast-food leftovers) but also pet food left outdoors and occasional handouts. Kit foxes will routinely visit locations where such foods are consistently available (e.g., Dumpsters, feeding stations for feral cats, residences where food is handed out; figure 5.4). However, kit foxes in Bakersfield rarely appear to depend on these foods. In one food-habit study (Cypher and Warrick 1993), 12.2% of 180 scats contained obvious anthropogenic food items, and only two scats consisted entirely of such items. In a larger effort, anthropogenic food items were found in 23% of 720 scats from adult kit foxes (Endangered Species Recovery Program, unpublished data). Use of

Figure 5.2. Nightly movements and den locations for two adult female kit foxes in Bakersfield, California. *Above*, Female 6008 was monitored for 11 months and primarily used a golf course with occasional forays into surrounding residential areas. *Opposite*, Female 6010 was monitored for an 18-month period and primarily used a college campus and nearby open space and commercial areas.

Legend:
○ Night locations
△ Den locations

0 125 250   500 Feet

Table 5.4. Home range estimates for urban and nonurban San Joaquin kit fox populations

Location	n	Estimator	Mean home range size	Source
Urban populations				
Bakersfield	28	100% MCP	172 ha	Frost 2005
Bakersfield	28	95% FK	120 ha	Frost 2005
Nonurban populations				
Water Bank	13	100% MCP	251 ha	Knapp 1978
Elk Hills	8	100% MCP	434 ha	Koopman 1995
Elk Hills	21	100% MCP	462 ha	Zoellick et al. 2002
Lokern	32	95% FK	591 ha	Nelson et al. 2007
Lokern	14	95% MCP	668 ha	Spiegel et al. 1996
Carrizo Plain	21	100% MCP	1,160 ha	White and Ralls 1993

Note: Estimators: MCP = minimum convex polygon; FK = fixed kernel.

Figure 5.3. Kit fox in an earthen den.

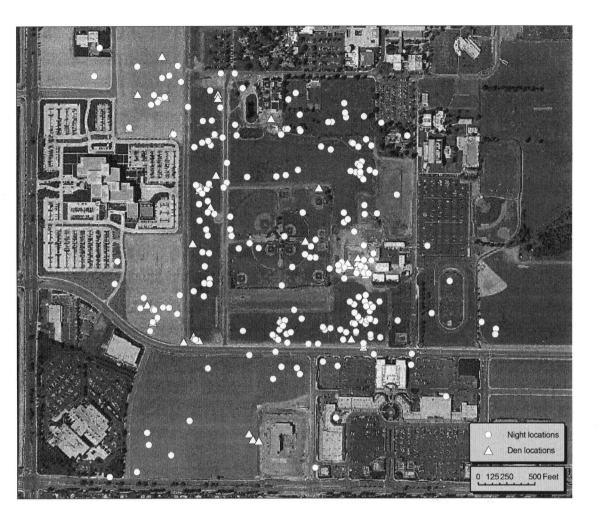

Table 5.5. Food item use by urban and nonurban populations of San Joaquin kit foxes based on fecal analysis

	Percent Occurrence in Diet				
Food item	Bakersfield[a] (urban)	Bakersfield[b] (urban)	Lokern[c] (nonurban)	Elk Hills[d] (nonurban)	Carrizo[e] (nonurban)
Kangaroo rat	—	—	68.0	31.0	8.4
Pocket mouse	—	—	6.2	1.2	20.0
Ground squirrel	10.7	67.8	—	—	6.4
Gopher	7.4	3.9	—	—	2.9
Rabbit	2.2	5.0	3.5	44.1	4.7
Bird	14.0	12.2	3.3	2.5	6.4
Beetles	13.1	22.8	5.6	1.3	17.4
Grasshoppers/crickets	2.6	0.1	23.1	6.4	7.2
Anthropogenic	22.9	12.2	—	—	—
N	720	180	484	4,393	834

[a] CSUS Endangered Species Recovery Program, unpublished data.

[b] Cypher and Warrick 1993.

[c] Cypher et al. 2005.

[d] Cypher et al. 2000.

[e] White et al. 1993, 1996b.

Figure 5.4. Kit fox taking a hard-boiled egg from a pet food bowl in Bakersfield, California. Photo by C. Van Horn Job.

anthropogenic foods is likely underestimated. Unlike natural food items in which insoluble items (e.g., bones, teeth, hair, feathers) facilitate the identification of foods consumed, most anthropogenic foods are more completely digested, and any insoluble remains are rarely recognizable, although food wrappers (e.g., plastic, paper, aluminum foil) are usually obvious. Despite this underestimation, kit foxes still are not likely to depend on anthropogenic foods. For example, in the study by Cypher and Warrick (1993), 178 of 180 (98.9%) scats contained remains of natural prey items.

Interestingly, use of anthropogenic foods apparently can produce physiological differences between urban and nonurban foxes. Compared with nonurban foxes, urban foxes exhibited lower red blood cell counts, hemoglobin, hematocrit, mean corpuscular hemoglobin, mean corpuscular volume, and blood urea nitrogen but higher cholesterol levels (Cypher and Frost 1999). Body mass was also higher for urban foxes, particularly among juveniles. However, these condition indices were measured during a two-year period in which nonurban foxes may have been nutritionally stressed because of low food availability resulting from drought conditions.

INTERACTIONS BETWEEN URBAN KIT FOXES, OTHER SPECIES, AND PEOPLE

In urban environments, kit foxes exhibit interesting interactions with other wildlife species. In addition to consuming California ground squirrels, kit foxes also occasionally usurp and expand squirrel burrows into dens. In their natural habitats, kit foxes commonly occupy and modify the burrows and dens of other species, including badgers, squirrels, and kangaroo rats.

Burrowing owls tend to use the burrows of other species, including those of kit foxes. In Bakersfield, where kit foxes and burrowing owls are abundant, the two species actually may compete for dens. Occasionally, burrowing owls are killed by kit foxes, but owls also have been observed driving kit foxes away from burrows. It is common for the species to alternate uses of a given den. At one den, kit foxes raised a litter of pups; afterward, the den was used by burrowing owls to raise a clutch of young; the following year, the den was used by kit foxes to raise pups again. At least eight dens in Bakersfield have been used by both burrowing owls and kit foxes.

Because of the scarcity of water in arid habitats, striped skunks are rare in natural lands used by kit foxes, but they are abundant in urban environments, and the two species exhibit some interesting interactions. In Bakersfield, skunks are commonly observed in many of the same urban habitats used by kit foxes, particularly areas with irrigated landscaping, such as golf courses, school campuses, and some commercial and residential areas. On the basis of observations, the two species primarily ignore each other. However, there are two scenarios in which the two species are drawn into close contact. At numerous locations throughout Bakersfield, well-intentioned people have established feeding stations for feral cats. In addition to the cats, other species observed feeding at these stations include kit foxes and skunks, as well as red foxes, opossums, and on occasion raccoons. When skunks appear at the feeding stations, all other species, including the kit foxes, appear to defer to the skunks, presumably because of their chemical defenses.

Skunks have also been observed using known kit fox dens. In a project conducted in 2006–7, 12 skunks were captured and radio-collared. At least 38 known kit fox dens also were used by collared skunks. Furthermore, on four occasions, a collared skunk and a collared fox were found occupying the same den simultaneously. Dens and food do not appear to be limiting factors for urban foxes, so shared use of these resources with skunks probably does not adversely affect kit foxes. A more significant implication of shared use of resources is the potential for disease transmission. In particular, skunks are frequently involved in rabies epidemics (Krebs et al. 2003). Indeed, a rabies epidemic among striped skunks was implicated as a possible causal factor in the decline and apparent extirpation of one kit fox population (White et al. 2000).

Nonnative red foxes are rapidly increasing in abundance and distribution in California, including within the range of the San Joaquin kit fox (Lewis et al. 1993). Red foxes are rarely observed in natural habitats used by kit foxes, probably because of the aridity and the abundance of coyotes, which can competitively exclude red foxes (Major and Sherburne 1987; Sargeant et al. 1987; Cypher et al. 2001). However, red foxes are commonly observed and appear to be increasing in urban environments in the San Joaquin Valley where coyotes are less abundant or absent. Red foxes potentially engage in interference and exploitative competition with kit foxes (Ralls and White 1995; Cypher et al. 2001; Clark et al. 2005).

In Bakersfield, red foxes are commonly observed in many of the same habitats used by kit foxes, and the two species have been observed in proximity. Red foxes also have been observed using known kit fox dens. These dens are not available to kit foxes while red foxes are using them. Red foxes also consume many of the same foods as kit foxes, although they appear to rely less on anthropogenic foods compared with kit foxes. However, of 156 kit foxes found dead in Bakersfield, none appears to have been killed by a red fox (Bjurlin et al. 2005). Furthermore, in one instance, kit foxes and red foxes successfully raised litters of pups within ca. 100 m of each other. The kit foxes gave warning barks when the red foxes came near, but no physical aggression was observed. So, the potential effects of red foxes on urban kit foxes are not clear. Red foxes may cause kit foxes to avoid using some areas, but currently they do not appear to be a limiting factor for urban kit foxes. The abundance of food and dens in urban environments may mitigate competition between the two species.

Other species that could potentially affect urban kit foxes include coyotes, bobcats, and raccoons. Coyotes and bobcats have killed some urban kit foxes in Bakersfield (Bjurlin et al. 2005), and raccoons potentially could transmit diseases to foxes. However, these species primarily occur along riparian corridors and urban edges, and therefore are not usually observed in urban habitats used by kit foxes. Thus, they do not appear to be a limiting factor for kit foxes in urban environments.

Kit foxes also interact with domestic animals in urban environments. Feral cats are abundant in urban habitats used by kit foxes. For the most part, the two species appear to avoid and ignore each other. Kit foxes and cats are close in size structurally, but cats generally are two or three times heavier than kit foxes. The two species may eat some of the same foods (e.g., rodents),

and cats have been observed using known kit fox dens. However, as with red foxes, the abundance of food and dens in urban environments may mitigate competition between kit foxes and cats. As previously mentioned, kit foxes sometimes use feeding stations established for feral cats, and when the two species are present simultaneously at a station, kit foxes generally defer to cats. Sometimes, the two may even feed together (figure 5.5).

Domestic dogs may kill kit foxes (Bjurlin et al. 2005). However, most dogs are much larger than kit foxes, and kit foxes likely avoid dogs. How dogs catch and kill foxes is unknown. However, in one instance, a kit fox was killed when it repeatedly crossed under a fence and entered a yard inhabited by a dog.

A suite of carnivores can occur in some urban areas such as in Bakersfield. Although this carnivore abundance is positive from a biodiversity perspective, it also could potentially be cause for concern. A greater diversity of species and associated increases in carnivore density potentially increase the possibility of a disease epidemic. Disease could enter the urban environment through one of the carnivore species or through another species such as bats. In 2007, three rabid bats were found in Bakersfield (*Bakersfield Californian* 2007), although no rabid carnivores were found. An epidemic could significantly affect kit foxes, and if the disease were zoonotic, it could quickly erode public support for maintaining kit foxes (as well as other carnivores) in urban areas.

Kit foxes interact with people in a number of ways. Because they are nocturnal, kit foxes are not as commonly observed by people as more diurnal species (e.g., red foxes). Observations of kit foxes are considered to be a positive experience (Bjurlin and Cypher 2005). In the spring, litters of pups at natal dens can be easy to

Figure 5.5. Kit fox and kitten at a feeding station for feral cats. Photo by C. Wingert.

observe, and accessible locations can attract a number of people who enjoy watching young foxes play.

Urban kit foxes rarely cause nuisance problems. They are virtually silent, and so noise is not an issue. They do not cause property damage and rarely are nuisances (e.g., spilling trash out of receptacles). On one occasion, a business owner complained because dirt excavated by foxes digging a den partially blocked the back entryway to his establishment. Kit foxes have not been reported to attack pets, although they occasionally will consume food left outside for pets. No instances of kit foxes attacking people have ever been reported. Only two instances of kit foxes biting people have been reported. In one instance, an individual was trying to capture a kit fox by hand. In the other instance, a student was attempting to assist a kit fox entangled in a soccer goal net. The most frequent complaint about kit foxes is from golfers who report that foxes run out and steal their golf balls.

SOCIOECOLOGICAL AND ECONOMIC ISSUES

Kit foxes in urban environments appear to pose no physical threat to people. Although a large number of kit foxes reside in Bakersfield close to people, no attacks on people have ever been reported. Kit foxes inhabit school campuses, commercial areas, golf courses, and some residential areas. At one elementary school, administrators only became aware that a pair of foxes was raising young under a portable classroom when they noticed that students were tossing food items over a fence each day.

Urban kit foxes also appear to pose no threat to pets. Indeed, domestic dogs seldom kill kit foxes and thus the foxes mostly avoid dogs. The relationship between kit foxes and cats appears to be one of mutual tolerance and avoidance. Cat hair has been found in 3 of 900 urban fox scats (Cypher and Warrick 1993; CSUS Endangered Species Recovery Program, unpublished data), but it is unknown whether this represents predation or scavenging. Even if predation is suspected, it is extremely rare.

Urban kit foxes rarely cause nuisance problems, such as annoying vocalizations or property damage. The most expensive damage caused by kit foxes that has been reported occurred when one or more foxes excavated a den in the side of a sand bunker at an upscale country club (figure 5.6). Expensive sand apparently is used in these bunkers. The foxes were excluded

Figure 5.6. Kit fox den in a sand bunker on a golf course in Bakersfield, California. Photo by B. Cypher.

from the dens, the dens collapsed, and the bunker was repaired.

A more significant effect involves endangered kit foxes on construction sites. The Metropolitan Bakersfield Habitat Conservation Plan (MBHCP) was adopted in 1994 to permit urban development while also facilitating endangered species conservation (Metropolitan Bakersfield Habitat Conservation Plan 1994). Under the MBHCP, developers in Bakersfield must pay a fee for each acre of land developed, and this fee is used to purchase and manage habitat in nonurban natural areas. If a kit fox den occurs on a site under development, the developer may incur more costs to hire an environmental consultant to monitor, to excavate, and to collapse the den to discourage kit foxes from using the site. However, kit foxes commonly return to sites once construction has begun and may even dig new dens on the site. Foxes may return to a traditional foraging or denning area, or they may be attracted to the sites by food left by workers or disturbed soils in which digging is easy. When foxes reoccupy an active construction site, developers may incur additional costs to hire biologists to exclude foxes again or for project delays.

In general, human residents of Bakersfield favor and enjoy kit foxes (Bjurlin and Cypher 2005). When foxes raise young in easily observed locations, residents visit the location each evening to observe the foxes. People also feed kit foxes (figure 5.4). In some cases, such feeding is opportunistic, such as when workers feed foxes at construction sites. In many cases, foxes are fed regularly, as exemplified by residents bordering golf courses that leave food out for the foxes each evening. In fact, some residents in Bakersfield are quite protective of kit foxes. Actions by these residents include watching over specific kit fox dens, challenging people approaching the dens, reporting incidents of den disturbance or potential threats to the kit foxes, and even confronting researchers attempting to gather information that may contribute to kit fox conservation efforts.

In surveys, some residents expressed concern regarding potential disease transmission by kit foxes to people or domestic animals (Bjurlin and Cypher 2005). The primary zoonotic disease of concern is rabies. However, only two San Joaquin kit foxes have been diagnosed with rabies, and both were from a nonurban population in an area where rabies is endemic in skunks (White et al. 2000). Antibodies to various diseases that could be transmitted to domestic animals have been detected among urban and nonurban kit foxes; such diseases include canine distemper, canine parvovirus, canine adenovirus, leptospirosis, and toxoplasmosis (Cypher 2003). However, kit foxes have never been diagnosed with an active infection of any of these diseases and have never been implicated in transmitting these diseases.

MANAGEMENT AND CONSERVATION

Because urban kit foxes generate little controversy and cause few problems, active management of urban populations is not a critical need for this species. The rare nuisance issues that arise usually are easily addressed by the affected people, California Department of Fish and Game or environmental consultants. As discussed previously, most regulatory issues resulting from the protected status of San Joaquin kit foxes are addressed through guidance provided by the Metropolitan Bakersfield Habitat Conservation Plan (1994).

One management action that may be warranted is the proactive development of a disease response plan. Kit foxes are one species within a complex of carnivores occurring in urban areas, and so a disease such as rabies or distemper could potentially spread rapidly through an urban environment. This could have serious consequences for humans and domestic animals. Such an epidemic also would have conservation consequences. Stakeholders, including city governments, animal control organizations, public health officials, and epidemiologists, should consider preparing and implementing a proactive strategy to prevent a disease epidemic among kit foxes.

Because of their robust demographic and ecological attributes, apparent adaptability, and minimal public opposition, the potential for maintaining viable kit fox populations in urban environments seems con-

siderable. Many factors facilitate successful occupation of urban environments by kit foxes. Kit foxes are foraging generalists, are able to use anthropogenic structures such as dens, have modest space requirements, can use highly altered habitats, are small bodied, which allows them to squeeze through small openings, have activity patterns that are asynchronous with those of humans, and generally are not destructive or annoying. Kit foxes are charismatic. Favorable attributes of urban environments include low abundance of natural kit fox predators, diverse food sources that, along with water, are consistently abundant, and plenty of potential den sites.

Encouraging and maintaining urban populations could contribute significantly to the conservation and recovery of endangered San Joaquin kit foxes. Currently, only three core populations and about a dozen small satellite populations remain (U.S. Fish and Wildlife Service 1998). Urban populations would increase the number of remaining populations. Maintaining more individuals in a greater number of populations also could help conserve genetic diversity. Because urban populations are subject to different environmental influences, they could serve as a "hedge" against catastrophic events in natural habitats, such as prolonged unfavorable environmental conditions (e.g., extended drought) and disease epidemics. Urban kit foxes also potentially could be used as source populations for reintroductions into vacant natural lands. Another benefit is that urban kit foxes serve as excellent ambassadors for their species. Residents of Bakersfield who had observed urban kit foxes expressed a greater appreciation for kit foxes, had greater knowledge of kit foxes, were more supportive of conserving foxes in urban environments, and were more supportive of conserving kit foxes range-wide (Bjurlin and Cypher 2005).

Kit fox conservation in urban environments could be significantly enhanced through a variety of strategies. A basic approach is to minimize adverse impacts to kit foxes. Preconstruction surveys could be conducted on all properties about to be developed, and any kit foxes present could be discouraged from using the site or even be relocated. Any efforts undertaken to conserve existing kit fox dens would be beneficial. Responsible landowners in Bakersfield have done this with great success (figure 5.7). Kit foxes, along with other species, tend to cross roads more frequently at certain locations, particularly where roads cross natural movement corridors (Bjurlin et al. 2005). Road

Figure 5.7. Kit fox at a den in a parking lot in Bakersfield, California. The black fencing, caution tape, and yellow warning sign all were installed to protect this den and the foxes using it. Photo by S. Harrison.

crossing structures (e.g., underpasses, green bridges) could be installed at these locations and speed limits could be reduced. Finally, rodenticide use could be more strictly regulated in urban areas with kit foxes. As discussed previously, rodenticide residues were present in most kit foxes tested in Bakersfield, and at least two foxes died from rodenticide poisoning. Modifications in use of rodenticides (e.g., use of compounds less dangerous to foxes, safer application methods) could benefit kit foxes.

On a landscape scale, urban planning could include development scenarios that create a system of refugia connected by movement corridors. One parsimonious approach could incorporate such a system within greenbelts that double as recreational facilities for people (e.g., bike paths and fitness trails). Such areas could be seeded with artificial dens that would provide cover for kit foxes. Artificial dens also could be easily installed in existing open space, such as utility and railroad rights of way, canal banks, parks, and golf courses. Kit foxes readily use artificial dens, and this has been a successful conservation strategy in Bakersfield. To address the threat of a disease epidemic, a proportion of each urban kit fox population could be vaccinated, particularly against rabies. Vaccination of all foxes would be impractical, but vaccinating at least some segments of the population could ensure that at least a proportion of the population would survive an epidemic. This strategy has been used successfully with Ethiopian wolves (Haydon et al. 2007). To assist landowners, developers, planners, and others in conserving kit foxes, expertise and information could be provided for low or no cost. Finally, outreach and education programs could significantly raise awareness and increase community support for conserving urban kit foxes.

CONCLUSIONS

Many carnivores are not able to occupy urban environments because of factors such as high space requirements and low tolerance of disturbances. For other species, urban environments do not provide optimal habitat conditions, but the species are able to persist in urban areas either continuously or at least intermittently. Finally, some species are not only able to satisfy life history requirements but actually thrive in urban environments and exhibit demographic and ecological attributes that may equal or even exceed those in nonurban environments. The kit fox appears to fall into this last category.

Ecological data suggest that kit foxes have adapted well to urban areas. Indeed, behavioral differences between kit foxes in urban and nonurban habitats have been noted. In recent behavioral studies, foxes in both urban and nonurban habitats were presented with novel objects, including a beachball, new foods, and a simulated predator. Responses to the predators were similar with foxes in both habitats. However, urban kit foxes were far more likely to inspect and try novel foods. Urban kit foxes will occasionally approach people, probably in the hopes of obtaining a food handout. Such behavior is rare among nonurban foxes. Furthermore, demographic attributes for urban kit foxes are robust. Thus, information gathered to date indicates a high potential for kit foxes to maintain viable populations in urban environments.

ACKNOWLEDGMENTS

I would like to thank C. Wingert, C. Van Horn Job, and S. Harrison for permission to use photographs. I also thank S. Phillips and J. Murdoch with assistance in preparing graphics. C. Van Horn Job, C. Wingert, C. Bjurlin, and J. Storlie were the dedicated field biologists who collected many of the data reported in this chapter.

6

Red Foxes (*Vulpes vulpes*)

CARL D. SOULSBURY

PHILIP J. BAKER

GRAZIELLA IOSSA

STEPHEN HARRIS

THE RED fox has the widest geographical range of any wild carnivore. It is found throughout much of the Northern Hemisphere and most of Australia (Saunders et al. 1995). Globally, the species has been introduced and translocated extensively for sport hunting, including restocking wild populations, and to improve the quality of fox furs (Long 2003). Most recently, red foxes have been released in Tasmania, although extensive measures have been implemented to eradicate them to preserve native biodiversity (Dennis 2002; Saunders et al. 2006). The red fox is successful for several reasons: it is a medium-sized generalist predator able to move considerable distances to exploit vacant niches, with no specialist food or habitat requirements and with a high reproductive output (Lloyd 1980). Red foxes are found in a broad range of habitats, from deserts to arctic tundra, and, with the exception of domestic cats and dogs, foxes are the most widespread, and possibly most abundant, urban carnivore.

Historically, urban red foxes were thought to be essentially a British phenomenon (Harris 1977; Kolb 1984, 1985, 1986; Harris and Rayner 1986a, 1986b, 1986c). They were first recorded in London during the 1930s (Teagle 1967), although urban populations were also recorded in Melbourne, Australia, during the 1940s (Marks and Bloomfield 1999a). It is believed that these early British populations arose simply because expanding developments with a new style of housing, built at a density of less than 20 houses / ha so that they had relatively large gardens, enveloped seminatural habitats. Once enclosed by suburban development, the foxes thrived in the new environment, since the gardens provided a diversity of food sources and refugia (Harris and Baker 2001).

In the past 20 years, urban fox populations have been reported in Australia (Coman et al. 1991; Jenkins and Craig 1992; Marks and Bloomfield 1999a), Japan (Tsukada et al. 2000), and North America (Adkins and Stott 1998; Lewis et al. 1999); most are, however, in Europe (Christensen 1985; Schöffel et al. 1991; Willingham et al. 1996; Cignini and Riga 1997; Gloor et al. 2001; Duscher et al. 2005; Reperant et al. 2007; figure 6.1). The timing of colonization of these other cities has varied; for example, foxes colonized many towns and cities in southern England in the 1930s, such as Bristol (Harris and Woollard 1988), although colonization occurred later in cities where foxes were rarer in the surrounding rural areas (Harris and Smith 1987a). Foxes were not recorded in many northern English towns until the 1980s (Wilkinson and Smith 2001). Similarly, Zürich, Switzerland, was not colonized until the 1980s (Gloor et al. 2001). Consequently, it is not entirely clear what factors have driven this process, although, in several instances, colonization has occurred as surrounding rural populations recovered from persecution or disease (Harris and Rayner 1986c; Gloor et al. 2001), possibly suggesting that they may have moved into urban areas as intraspecific competition increased.

PHYSICAL CHARACTERISTICS

The red fox is a relatively small canid with an elongated muzzle, large ears, and a rounded bushy tail. Body size and body mass are plastic across its geographic range; it is substantially larger in Europe (Cavallini 1995; table 6.1). Accordingly, some authorities consider the North American red fox to be a distinct subspecies, *Vulpes vulpes fulva* (Nowak 1991; Larivière and Pasitschniak-Arts

1996). In accordance with Bergmann's rule, body mass also tends to be greater in more northern latitudes (Cavallini 1995); for example, mean mass of males and females in England was 6.7 kg and 5.4 kg, respectively, versus 7.3 kg and 6.2 kg in Scotland (Harris and Lloyd 1991).

Red foxes typically have a reddish brown coat; the belly fur varies in color from white to charcoal gray (figure 6.2). The paws and the back of the ears are typically black, although both can be flecked with white spots. Both males and females may or may not have a white-tipped tail. Several color variants are known, including silver, black, and cross foxes, the latter having a stripe of dark fur stretching along the spine and down across the shoulders. Samson foxes, which lack outer guard hairs, and white foxes have been observed rarely. Red foxes undergo one annual molt.

SOCIAL AND SPATIAL ORGANIZATION

Foxes live in small territorial groups. Territories are exclusive, with little or no overlap in urban areas (Doncaster and Macdonald 1991; White and Harris 1994; figure 6.3). Group size varies from two to ten individuals (Cavallini 1996; Baker and Harris 2004). Territory size varies from <0.2 km² to >20 km² and is smallest in urban areas (Cavillini 1996; Goszczyński 2002). Recorded territory sizes for urban foxes are, however, highly variable: 18–169 ha (Bristol, UK; Soulsbury et al. 2007; figure 6.3), 54–93 ha (Oxford, UK; Doncaster and Macdonald 1991), 420 ha (Orange County, California; Lewis et al. 1993), 259 ha (Virginia; Roundtree 2004), and 379–547 ha (Champaign-Urbana, Illinois; Gosselink et al. 2003). Territory size varies with fluctuations in population density, as was observed in Bristol following a sarcoptic mange epizootic (figure 6.3). Densities in urban areas are typically high; the

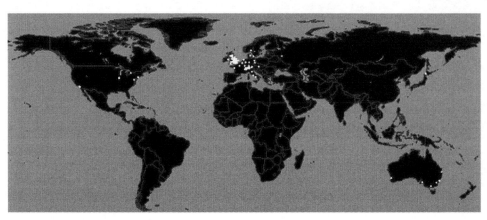

Table 6.1. Adult fox density, body mass, territory size, and habitat preferences in urban areas

	Melbourne, Australia[a]	Toronto, Canada[b]	Zürich, Switzerland[c]	Bristol, UK[d]	Edinburgh, UK[e]	Illinois, USA[f]
Human population	3,592,591	2,503,281	343,157	380,615	448,624	2553–103,913[g]
Adult fox density (adults/km²)	3–16	1.5	11.5	7.8–37.0[h] 0.7–5.5[j]	2–3	3.7–5.3[i]
Body mass (kg)	6.0♂, 5.2♀	5.1♂, ♀4.5	6.4♂, ♀5.7	6.3♂, 5.5♀	—	4.7♂, 3.9♀ [i]
Mean territory size (ha)	45	52	28–40	18[h] 58–170[j]	111	379–547
Habitat preferences[k]	Blackberry/gorse Dense native vegetation Long grass/reeds Paddock/short grass	Floodplain Ravine slopes Low-density residential Vacant residential Medium density residential	Parks/fallow land Allotments Low-density residental Industrial Medium-high-density residential Streets/large buildings	Back gardens Allotments/woodland Grassland/scrub Tarmac Front gardens	Railway lines Cultivated areas[l] Private gardens Wasteland Schools Institutions[m]	Urban grassland Urban undeveloped Urban developed Waterways Grassland Cropland

[a] White et al. 2006.

[b] Adkins and Stott 1998.

[c] Gloor 2002, Contesse et al. 2004, D. Hegglin, personal communication.

[d] Baker et al. 2001a; Newman et al. 2003; Soulsbury et al. 2007.

[e] Kolb 1984, 1985.

[f] Gosselink et al. 2003.

[g] Study was conducted across five urban areas: Champaign-Urbana (pop. 103,913), Rantoul (12,857), Mahomet 4,877, Monticello (5,318), and Villa Grove (2,553).

[h] Pre-mange period.

[i] T. Gosselink, personal communication.

[j] Post-mange period.

[k] Habitats are ranked from most preferred to least preferred.

[l] Cemeteries, nurseries, and allotments.

[m] Grounds of large buildings such as hospitals or colleges.

Figure 6.2. A profile of an urban red fox from England. Photo by Paul Cecil.

Figure 6.1. Opposite, The localities with reported urban fox populations (white dots). One hundred fourteen urban populations are denoted, mostly from the United Kingdom (56) and mainland Europe (40). Urban fox populations are also recorded in North America (10), Australia (6), and Japan (2). Absence of fox populations in some localities may reflect a lack of reports rather than a lack of populations. Some European urban fox locations taken from www.zor.ch/fuchs_frame.html.

population in Bristol peaked at 37 adults/km² before an outbreak of mange (Baker et al. 2001b). Typical urban densities are 2–12 adults/km² (Harris and Rayner 1986a; Marks and Bloomfield 1999a; Contesse et al. 2004; De Blander et al. 2004; Rosatte et al. 2007c; see table 6.1). In comparison, densities in rural areas are typically lower: 0.2–2.7 adult foxes/km² (Larivière and Pasitschniak-Arts 1996; Webbon et al. 2004).

Social groups typically consist of extended families containing a dominant breeding pair and nondispersing offspring from one or more previous breeding attempts. In most studies, nondispersing individuals were typically females (Baker and Harris 2004), although subordinate males were as common as females at high densities in Bristol (Harris and Smith 1987b; Baker et al. 2000). Older females that had previously been the dominant breeding vixen may also remain in the group as nonbreeding subordinates.

REPRODUCTION AND DEVELOPMENT

The reproduction and development of red foxes is similar in urban and nonurban habitats. Foxes are

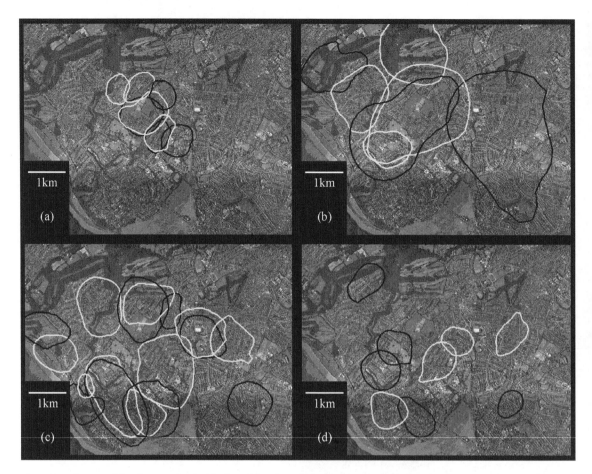

Figure 6.3. The spatial distribution of male (black) and females (white) fox territories and changes in territory size caused by a sarcoptic mange epizootic: pre-mange (a) 1990 (n = 8 territories, $\bar{x}$ = 29 ha, density = 7.8 adults/km²). Following the mange epizootic, territory sizes increased significantly: (b) 1999 (n = 7 territories, $\bar{x}$ = 169 ha, density = 1.1 adults/km²), before recovering slowly: (c) 2003 (n = 17 territories, $\bar{x}$ = 58 ha, density = 4.5 adults/km²), and (d) 2006/7 (n = 9 territories, $\bar{x}$ = 32 ha, density = 12.5 adults/km²).

annually monestrus (produce one litter each year). The estrus cycle depends on photoperiod such that birth dates vary with latitude and are shifted by 6 months in the Southern Hemisphere (Marks 2001; McIlroy et al. 2001); births in the Northern Hemisphere typically occur between February and May (Lloyd 1980; Cavallini and Santini 1995). During the breeding season, it is thought that the dominant male guards the female(s) within the group, but males also make extraterritorial forays in search of extra-pair copulations (Saunders et al. 1993; White et al. 1996a; Baker et al. et al. 2004a): the ability of males to sire cubs outside their own group appears to depend on their size (Iossa et al. 2008). Females are sexually receptive for three to five days a year (Tembrock 1957). Following mating, the male's penis may become trapped in the female's vagina (Asa and Valdespino 1998): this postcopulary

lock may last up to an hour and is thought to serve as a paternity-assurance mechanism.

Within social groups, it was commonly thought that subordinate female reproduction was suppressed by a combination of (1) "social stress" (Hartley et al. 1994), resulting in the complete or partial reabsorption of fetuses, the abandonment of viable litters, and the production of nonviable cubs (Harris 1979); and (2) infanticide by the dominant female (Vergara 2001), although subordinate females are increasingly likely to breed when food is abundant (Storm et al. 1976; von Schantz 1981, 1984a, 1984b; Zabel and Taggart 1989). Consequently, foxes have been classified as socially monogamous (Kleiman 1977); that is, each group produces one litter of cubs each year, sired by the dominant male. However, given the productivity of urban environments, multiple litters are common within urban fox groups (e.g., Macdonald 1979b; Baker et al. 2004a). Furthermore, in the high-density Bristol population, before the outbreak of mange, genetic analyses indicated that males and females would frequently mate with animals outside their own group (Baker et al. 2004a; Iossa et al. 2008). Given that the mating season is the peak period of mortality for adult males (e.g., 25% of 263 adult male foxes recovered dead were killed in January; Harris and Smith 1987b; Baker et al. 2007), evolutionary theory would suggest that males are likely to obtain some level of extragroup paternity to offset the cost of increased mortality during the mating season and that nonmonogamous patterns of mating may be more common than previously supposed.

Gestation lasts approximately 53 days. Cubs weigh approximately 100–150 g at birth and are born onto bare ground; their eyes open after ca. 2 weeks. Litter size tends to increase as population density declines or as population turnover increases. During the first three to four weeks, the cubs remain in the natal den. The female only emerges for short periods, as the cubs have poor thermoregulatory capabilities during the first 2 weeks of life; the male brings food to the den for the female. This is a period of high juvenile mortality, with approximately 15% of cubs perishing before they emerge from the den (Harris and Smith 1987b; Harris and Baker 2001). Litter size at emergence (about four weeks of age) has been recorded as: 3.0–4.0 (Orange County, California: Lewis et al. 1993), 4.0–4.7 (Bristol: Harris and Smith 1987b; Soulsbury et al. 2007), 4.4 (Copenhagen: Wincentz 2004; Melbourne: Marks and Bloomfield 1999a), and 4.8 (London: Harris and Smith 1987b). Comparative figures for rural areas are 3.7–4.9 (Iowa and Illinois: Storm et al. 1976) and 4.2 (Wiltshire, UK: Baker et al. 2006).

Lactation lasts approximately 5 weeks, and weaning occurs gradually, with cubs taking their first solid food at about 3 weeks old (Henry 1986). If more than one female in a group is lactating, cubs may be suckled by a female that is not their mother. Consequently, even very young cubs may survive if their mother dies (Macdonald 1980a). When the pups are about 3 months old, both dominant and subordinate adults may bring food back to the den for the cubs to eat (Macdonald 1979b; Zabel and Taggart 1989; Baker et al. 1998), although this is not always the case (von Schantz 1981, 1984b; Baker et al. 1998). These foods tend to be large, easy-to-carry items, such as birds, so that the adults reduce the number of trips they need to make (Lindström 1994; Lovari and Parigi 1995). Adults will also act as babysitters, watching the cubs and warning them if danger approaches with a characteristic bark (Newton-Fisher et al. 1993).

The benefits of parental and alloparental care in red foxes are equivocal. Early studies surmised that the presence of nonbreeding alloparents (helpers) would act to increase the number of cubs raised by a group by increasing the volume of food delivered to the cubs (Macdonald 1979b). Indeed, in Zabel and Taggart's (1989) study of an island population of red foxes, the presence of an additional adult in a social group significantly increased the number of cubs that survived as a result of an increase in food brought back to the den and protection of the cubs against infanticide by other foxes. It is not known, however, whether such high levels of intraspecific predation are present in nonisland populations. Conversely, increasing numbers of subordinate animals in groups in Bristol did not increase the number of visits to the cubs. Instead, the total number of visits remained approximately constant across groups of different size, but each group member made fewer provisioning trips per night as group size increased (Baker et al. 1998). Whether this helped to reduce the mortality rates of breeding adults (the longevity of dominant animals was substantially greater than that of subordinates; Baker et al. 1998) is unclear.

For the first few months after they emerge, cubs restrict their activity to the area immediately around breeding dens, although they will make movements between dens that may be several hundred meters apart (Robertson et al. 2000). Over time, however, they increasingly range over larger and larger areas; by five or six months old they are using most of their natal territory (Robertson et al. 2000). Throughout this period, the amount of contact pups have with littermates declines. Early on, the entire litter is commonly found together, and this is the period when the cubs develop a social hierarchy (Meyer and Weber 1996); this often involves fights in which cubs are seriously injured or killed. The litter then splits into smaller and smaller groups, until the juveniles are independent.

At approximately 3 months old, the cubs no longer receive food from their parents. At this stage, however, they are not efficient predators and are not readily able to catch small rodents and birds (Henry 1986). This transitional period to independent foraging is marked by a loss of fat reserves and an increased reliance on easy-to-catch foods, such as insects and earthworms (Soulsbury et al. 2008a); earthworms are hunted at the soil surface following periods of rain (Macdonald

1980b). Consequently, the full-grown (≥6 months old) body mass of cohorts of foxes in Bristol was strongly correlated with the total rainfall that fell during the month immediately after their parents stopped feeding them (Soulsbury et al. 2008). Such interannual variation in rainfall can also have far-reaching effects, as adult size affects male, but not female, territory size and reproductive success (Iossa et al. 2008).

Juveniles remain on their natal territory until five to six months of age, after which they may disperse and attempt to breed elsewhere. Both sexes reach sexual maturity at 9 months. At low density, females typically breed in their first year, whereas most males begin breeding at 2 years of age (Soulsbury et al. 2008); at high densities, females may have to wait longer to reproduce (Baker et al. 2004a).

DISPERSAL

Both males and females may disperse, but dispersal is male biased. In London, 85% of males versus 20% of females dispersed (Page 1981), whereas in Bristol, 67% of males versus 32% of females dispersed (Trewhella et al. 1988). The dispersal period is protracted, with most animals leaving their natal range between 6 and 12 months old. Some animals will, however, disperse from their natal group as adults, even if they have bred successfully (Harris and Trewhella 1988). Dispersal is also male biased in rural populations, with dispersal rates similar to those found in urban areas (males: 60%–88%; females: 29%–58%; Storm et al. 1976; Pils and Martin 1978; Lloyd 1980).

In both urban and rural areas, the physical act of dispersing may take several forms (Woollard and Harris 1990). Animals may make simple one-off permanent movements. Alternatively, they may leave their natal range for one or more days, only to return to their parents' range before disappearing again a few days later. When they do move, it is common for dispersers to set up very small ranges either within another fox's territory or at the border(s) between territories (Robertson and Harris 1995a); these small ranges may help them avoid conflict with territory holders. Similarly, dispersing individuals appear to avoid the areas of core activity within the territories of the foxes whose ranges they traverse (C. D. Soulsbury, P. J. Baker, G. Iossa, and S. Harris, unpublished data). Occasionally, dispersing foxes may return to their natal range in the late spring or early summer, after the next litter of cubs is born.

Although dispersing foxes can move considerable distances (Trewhella et al. 1988), most urban foxes do not disperse far, principally because territories are small in urban areas. For example, the average dispersal distance in Bristol was just 3.1 km for males and 1.8 km for females (Trewhella et al. 1988). Consequently, many foxes born in urban areas live their entire lives in such habitats. Where cities are large, there may be significant obstacles to movement between different areas or marked differences in density, such that there can be significant genetic differentiation between subpopulations in the same city (Robinson and Marks 2001; Wandeler et al. 2003). Furthermore, genetic data from Zürich indicated that colonization was a one-time event in that city (Wandeler et al. 2003), but it is not known whether this same pattern has occurred elsewhere. Urban foxes will disperse to rural areas and vice versa (Harris and Trewhella 1988; Wandeler et al. 2003), although immigration into urban areas may be limited (Gosselink 2002). Consequently, the widely purported belief popularized in the national press in Britain that the "town" or "city" fox is different from its rural counterpart is a fallacy.

The factors determining whether an individual chooses to disperse or remain in its natal group appear complex. Competition to remain on the natal site appears key, so primary factors potentially affecting the decision to disperse or to remain appear to include group size and litter size, with animals in larger groups and larger litters tending to have a higher likelihood of dispersing (Harris and Trewhella 1988). In addition, lower status animals are more likely to leave. Harris and White (1992) examined the degree of amicable social interaction received by individual foxes by examining their ear tags: tags were chewed during social grooming and the number of teeth marks left in them indicated the amount of grooming an animal had received. Juvenile and adult males that received the least amount of grooming from their littermates or other adults, respectively, were significantly more likely to disperse, although no such relationships were evident for females (Harris and White 1992). These data suggest, therefore, that those individuals most loosely bonded to a group are the ones most likely to leave and that interactions with other group members play a significant role in maintaining group cohesion (White and Harris 1994; Baker and Harris 2000).

Dispersal may entail high costs with respect to survival, as dispersing individuals have to traverse unfamiliar terrain, cross busy roads, and run the risk of encountering other foxes. Of 85 males and 74 females tagged as cubs in Bristol and subsequently recovered dead as adults, the mean age at recovery of males that had dispersed (121 ± 52 weeks) was significantly lower than that of animals that had not dispersed (142 ± 46 weeks), but this was not the case for females (138 ± 68 vs. 149 ± 77 weeks; Harris and Trewhella 1988). However, with a larger data set, there was actually little difference in the patterns of mortality exhibited by dispersing and nondispersing males and females both during the main dispersal period and subsequently (Soulsbury et al. 2008a). Instead, the major cost associated with dispersal appeared to be the risk associated with missed breeding opportunities in females, whereas both philopatric and dispersing males had an equal probability of breeding (Soulsbury et al. 2008a). Such detailed studies on dispersal costs are absent from rural areas, although Storm et al. (1976) indicated no differences in mortality of dispersing and nondispersing individuals in rural Illinois and Iowa.

SURVIVAL AND MORTALITY

The principal cause of death for foxes living in urban areas is typically collisions with motor vehicles (table 6.2), although other factors such as human control or disease may be more important in different populations or at different times. There is an increased risk associated with larger roads, as these carry a higher volume of traffic and at faster speeds, yet even minor category roads may have a significant influence on movement patterns (Baker et al. 2007). There are peaks in the recovery of juveniles due to collisions with cars at approximately 3 and 7 months old (Harris and Baker 2001). These are associated with an increase in the area ranged around breeding dens (Robertson et al. 2000) and with the onset of the dispersal period, respectively. During the mating season, adult males are also increasingly vulnerable to traffic as they make extraterritorial movements searching for estrous females (Baker et al. 2007).

Reported annual mortality rates for juveniles in urban areas range from 54%–57% in Bristol to 64%–66% in London and 66%–68% in Illinois (Harris and Smith 1987b; Gosselink et al. 2007); comparable figures for adults are 50%, 53%–56% and 61%–74%, respectively. Mortality rates of juveniles, but not adults, were higher in urban versus rural areas in Illinois, principally through sarcoptic mange infection (Gosselink et al. 2007). Within urban areas, fights with domestic dogs can also be a significant cause of death (Harris and Smith 1987b). Young cubs may also be killed by

adult foxes from their own or another group (Vergara 2001) or by their littermates (Harris and Smith 1987b). Sarcoptic mange can be a significant cause of mortality (Gosselink et al. 2007; Soulsbury et al. 2007; see "Diseases and Parasites").

Historically, local authorities in some cities in Britain culled foxes by shooting, trapping, and gassing dens (Teagle 1967) to reduce nuisance complaints. Such control, however, was not successful and was largely abandoned by the 1980s (Harris 1985), although some residents continue to employ private companies to shoot or to trap the foxes on their property. Until the mid-1980s, foxes were also killed for their fur in towns and cities in the United Kingdom (Macdonald and Carr 1981).

DISEASES AND PARASITES

Red foxes harbor diseases and parasitic agents that are potentially transmissible to humans or their pets (Richards et al. 1993, 1995; Samuel et al. 2001; Williams and Barker 2001; Wolfe et al. 2001). The most important of these are rabies, echinococcosis (caused by the parasitic tapeworms *Echinococcus multilocularis* and *E. granulosis*), sarcoptic mange (*Sarcoptes scabiei*), toxocariasis (*Toxocara canis*), French heartworm or canine lungworm (*Angiostrongylus vasorum*), and the canine heartworm (*Dirofilaria immitis*).

Red foxes are the main wildlife vector of rabies in Europe (Macdonald and Voigt 1985), Canada, and Alaska; in the lower 48 states, other species are also important vectors (Rosatte et al. 1992a). Rabies in red foxes and other wildlife has been successfully managed using vaccines distributed in oral baits. For example, the disease has been eradicated from many areas of Europe (Brochier et al. 1991; Flamand et al. 1992). In both Europe and Australia, there have been many studies on bait uptake rates of urban foxes, as these are seen as key in eliminating rabies from urban areas (Smith and Harris 1989; Trewhella et al. 1991; Marks and Bloomfield 1999b). In Europe, baiting campaigns have avoided urban areas, but a baiting campaign eliminated rabies in red foxes in Toronto (Rosatte et al. 2007c). In developed countries, rabies poses little direct threat to humans and can be effectively treated following exposure.

Within urban areas, *E. multilocularis* is spread principally between rodents and carnivores, with foxes as a host in many countries (Jenkins and Craig 1992; Hofer et al. 2000; Tsukada et al. 2000; Deplazes et al. 2004; Duscher et al. 2005; Brochier et al. 2007). Prevalence

Table 6.2. Causes of mortality (%) in urban fox populations

	Bristol, UK (pre-mange)[a]	Bristol, UK (post-mange)[a]	Copenhagen, Denmark[b]	London, UK[c]	Illinois, USA[d]
Vehicle collisions	62.1	31.7	89	11.8	33.3
Disease	17.2	61.0	—	6.9	39.0
Human control	10.3	—	—	78.1	5.9
Fights	8.6	4.9	—	—	—
Predation	—	—	—	—	11.9
Misadventure	1.7	2.4	—	0.8	—
Unknown	—	—	11	2.4	9.9
Samples size (foxes)	80	67	73	445	120

Note: Misadventure includes unusual causes of death, such as electrocution, drowning, and accidental strangulation. UK = United Kingdom.

[a] Harris and Smith 1987b, Soulsbury et al. 2007.

[b] Wincentz 2004.

[c] Harris and Smith 1987b.

[d] Gosselink et al. 2007.

rates in foxes vary between cities, for example, 20%–30% in Geneva and 47%–67% in Zürich (Hofer et al. 2000; Reperant et al. 2007). Comparative data indicate that prevalence rates are lower in urban versus rural areas (Gloor 2002; Fischer et al. 2005), but this may vary with location. However, parasite prevalence in foxes is closely associated with the abundance of susceptible rodent species (Tsukada et al. 2000; Stieger et al. 2002; Hegglin et al. 2007) and vice versa (Saitoh and Takahashi 1998). Echinococcosis, or hydatid disease, is contracted through contact with fecal material, particularly from pets (Kreidl et al 1998; Campos-Bueno et al. 2000); the ingestion of eggs leads to the formation of cysts within the liver, lungs, kidneys, or spleen. In general, the disease is asymptomatic in humans but can be fatal if a cyst ruptures. Following the successful management of rabies in Europe, urban and rural fox populations have increased, leading to increased incidence of human alveolar echinococcosis (Schweiger et al. 2007). *Echinococcus multilocularis* is absent from Britain; *E. granulosis* is absent or very rare (Smith et al. 2003). Baiting with anthelmintics has been shown to reduce parasite prevalence in foxes (Hegglin et al. 2003).

Sarcoptic mange is an ectoparasitic infection that has been reported in foxes in many urban areas (Marks and Bloomfield 1999b; Bartnik 2001; Gloor et al. 2001; Wilkinson and Smith 2001; Gosselink et al. 2007; Soulsbury et al. 2007). It is notable for causing large-scale reductions in the size of both rural and urban fox populations (Lindström et al. 1994; Soulsbury et al. 2007); in Bristol, the population declined by >95% in just 2 years (Baker et al. 2000). In foxes, the disease causes a range of symptoms; the most notable is fur loss, which

occurs when a thick crust of parasite excreta builds up on the skin surface. It is also intensely irritating, and infected foxes will chew their own tails off or mutilate their limbs to relieve the itching. In advanced stages, tissues are catabolized to obtain sufficient nutriment (Newman et al. 2002), and animals will often seek shelter in buildings in cold weather to keep warm. In naive populations, death is almost inevitable following infection, typically within 4–6 months; in enzootic phases, infected animals may live much longer (>12 months), and death is not inevitable, although mange remains a significant cause of mortality (table 6.3). Sarcoptic mange is transmissible to domestic dogs but can be readily treated using acaricides. During the epizootic in Bristol, a high mange prevalence rate and high fox density was associated with a spillover of mange into domestic dogs (Soulsbury et al. 2007). In the enzootic phase, prevalence rates remained high, but the fox population density was significantly lower, and reports of mange in dogs declined to negligible levels. These data suggest that dogs are most at risk where fox densities are particularly high.

Three nematode parasites, *T. canis*, *A. vasorum*, and *D. immitis*, are globally important as they are commonly found in domestic pet populations, which in turn are a source of infection for humans (Holland et al. 1995). There is evidence that the latter two parasites are spreading in Europe (Genchi et al. 2005; Morgan et al. 2005, 2008). Foxes have been identified as a possible infective source (Bolt et al. 1992; Marks and Bloomfield 1998) for all three nematodes, but the relative contribution of foxes to infections in dogs is not clearly understood. For example, *T. canis* has been

Table 6.3. The relative importance of different foods (%) in the diet of urban foxes

Food group	Melbourne, Australia[a]	Brussels. Belgium[b]	Zürich, Switzerland[c]	Bristol, UK[d]	London. UK[e]	Oxford, UK[f]
Mammals	22	13	11[g]	6	16	16[h]
Birds	5	15	5	9	20	8[h]
Earthworms	—	—	—	6	12	27
Insects	18[i]	15[i]	4[i]	10	9	2
Fruit and vegetable matter	—	26	18	5	8	9
Scavenged meat	12[j]	28[k]	21	33	24	37[k]
Other scavenged food items	—	—	28	32	11	—
Method	Percentage composition of feces	Percentage occurrence in stomachs	Percentage by stomach volume	Percentage by stomach volume	Percentage by stomach volume	Percentage dry weight of feces

Notes: Scavenged items are those of anthropogenic origin, i.e., they include provisioning. "Other scavenged food items" included bread and birdseed. As studies use different categorization, it is not possible to present figures that add to 100%.

[a] White et al. 2006.

[b] De Blander et al. 2004.

[c] Contesse et al. 2004.

[d] Saunders et al. 1993.

[e] Harris 1981a.

[f] Doncaster et al. 1990.

[g] Rodents only.

[h] Excludes domestic animals.

[i] Earthworms and invertebrate items pooled.

[j] Defined as bones.

[k] All scavenged items pooled.

found to be morphologically and genetically identical in fox and domestic dog populations (Epe et al. 1999), suggesting that cross infection does occur. Prevalence of *T. canis* in urban foxes is generally high: 17.9% in Brussels (Brochier et al. 2007), 41%–61% in Bromley, United Kingdom (Richards et al. 1995), 44% in Geneva (Reperant et al. 2007), 44%–58% in Bristol (Richards et al. 1995), 47% in Zürich (Hofer et al. 2000), and 81% in Copenhagen (Willingham et al. 1996). In contrast, infection rates for *D. immitis* (6% in Melbourne: Marks and Bloomfield 1998) and *A. vasorum* (36% in Copenhagen; Willingham et al. 1996) appear lower but have been less well studied. There is no clear evidence that the prevalence rates of any of these three parasites are higher in urban than rural populations (Richards et al. 1995; Marks and Bloomfield 1998, Gloor 2002; but see Saeed and Kapel 2006).

DIET AND PREDATION

Foxes are opportunistic solitary omnivores that eat a wide range of foods, including insects, earthworms, fruits, vegetables, rodents, lagomorphs, and birds. They also readily scavenge the carcasses of larger animals

that have been killed: in the outback of Australia, they have even been recorded to consume the remains of people that have died, and in Britain, they often disrupt the burial places of murder victims in both urban and rural localities. Yet in many urban areas, the commonest sources of food are those supplied by humans (table 6.3; figure 6.4; see also Lavin et al. 2003). These anthropogenic food sources include compost heaps; allotment gardens (a small piece of land rented by a person to grow vegetables); food put out deliberately for other animals, including birds; and refuse left in garbage bags and open-topped public garbage bins. The latter brings them into conflict with health officials and homeowners, although rifling through bins occurs a lot less regularly than is generally believed (Harris and Woollard 1988; Contesse et al. 2004; see section "Relationships with Humans") and is easily rectified by ensuring that garbage is placed in a secure receptacle.

In urban areas, a significant proportion of people may feed foxes (Harris 1981a; Lewis et al. 1993; Contesse et al. 2004). For example, immediately before the outbreak of mange in Bristol, approximately 10% of all homeowners were supplying food for foxes (Baker et al. 2000). In part, this high level of feeding was

Figure 6.4. An urban fox eating scraps of chicken provided by a homeowner. Photo by Paul Cecil.

attributable to the presence of several very "tame" animals in the population that would actively take food from people's hands (Baker et al. 2000, 2004b), although such behavior is not recommended. Many urban residents gain much pleasure from feeding foxes, and, on the whole, this does not generate any significant management issues, at least in Britain where foxes do not carry any significant zoonotic diseases. There is, however, a huge dichotomy in attitudes toward foxes by urban dwellers, and this can lead to conflict between neighbors (C. D. Soulsbury, P. J. Baker, G. Iossa, and S. Harris, personal observations).

Foxes will kill domestic poultry and pets within urban areas (Harris 1981a; Lewis et al. 1993), although this is not common. For example, in a survey of residents living in a high-fox-density suburb in Bristol before the outbreak of mange, only 0.7% of households reported losing a pet cat to foxes in the previous 12 months (from a population of 1,225 cats). Similarly, only 8% of 262 households with chickens, rabbits, or guinea pigs had lost an animal to foxes in the previous 12 months (Harris 1981a). Following the outbreak of mange, domestic cat remains were identified in only 2.4% of scats (Ansell 2004); Lewis et al. (1993) also found cat fur in just 2.7% of scats in Orange County, California. However, such analyses cannot discriminate between the active predation of cats rather than scavenging carcasses. As with the other nuisance problems, this issue can be easily rectified by ensuring that vulnerable animals are housed securely and not let out unattended, particularly at night when fox activity is greatest (Doncaster and Macdonald 1997).

HABITAT ASSOCIATIONS AND DISTRIBUTION WITHIN URBAN AREAS

Foxes require safe areas to rest during the daytime and dens in which to breed. Daytime resting sites are often located away from humans, domestic dogs, and other possible predators, and where suitable cover is available (Gosselink et al. 2003; Marks and Bloomfield 2006). Foxes are not restricted to terrestrial resting sites and may climb onto sheds or buildings to sleep (figure 6.5). In Britain, there is a strong association between fox density and the availability of medium-density housing (Harris 1981b; Harris and Rayner 1986a, 1986b, 1986c; Harris and Smith 1987a). This housing is characterized by moderately sized enclosed, secure gardens that provide a wide range of foods (Ansell 2004) and where free-roaming dogs are rare or absent. Consequently, these private residential gardens are highly favored as both daytime rest sites and during nocturnal movements (Saunders et al. 1997; Newman et al. 2003; table 6.1), although all habitats are used (figure 6.3). In other countries, medium-density housing is less common, and foxes appear to use habitats away from human habitation, such as in remnant natural vegetation or urban parkland/grassland (Lewis et al. 1993; Adkins and Stott 1998; Gloor 2002; Roundtree 2004; White et al. 2006). The planned future construction of large numbers of high-density housing units with small gardens in Britain is, therefore, likely to be unfavorable for foxes (Baker and Harris 2007).

Breeding dens are often in the space beneath buildings or under garden sheds (London: 47%; Melbourne: 44%) or in banks of earth (London: 30%; Melbourne: 31%; Harris 1977; Marks and Bloomfield 2006; figure 6.6). Most dens are in unoccupied structures, areas without public access, or areas with restricted access such as in private residential gardens, schools, or industrial properties (Harris 1981b; Gosselink 1999; Marks and Bloomfield 2006). Foxes will, however, tolerate high levels of disturbance in some instances (Harris and Baker 2001). For example, Gosselink (1999) recorded dens in both cemeteries and on golf courses where human activity was high. Such habitats are also frequently used in Britain (Teagle 1967). The choice of breeding den sites may also be influenced by fox density. For example, before the outbreak of sarcoptic mange, when fox density was high, most dens in Bristol were located under sheds; following the mange epizootic, dens were most commonly recorded in occupied and unoccupied badger setts (Newman et al. 2003). This may suggest

Figure 6.5 A fox asleep on a rooftop. Photo by Paul Cecil.

Figure 6.6. An urban fox den in a bank of earth in Bristol, United Kingdom. Photo by S. Harris.

that foxes prefer more natural den sites when they are available.

The distribution of red foxes can be negatively affected by the distribution of predators, such as free-roaming dogs (Harris 1981b) and coyotes (Gosselink et al. 2003). During the night, the effect of pet dogs on fox activity is likely to be limited. Although foxes avoid gardens where a dog is running freely, they will readily forage in a neighboring garden even when the dog is visible through a fence. Furthermore, most pet dogs are kept indoors after midnight, so that even dog-occupied gardens are used for foraging in the latter half of the night. Human activity also appears to have only a limited effect on fox movements. For example, foxes will forage in gardens if people are relatively quiet and nonthreatening, but most will flee if approached too closely.

RELATIONSHIPS WITH HUMANS

Human perceptions of, and attitudes toward, wildlife are complex and dynamic (Butler et al. 2003). Experiences with urban wildlife can be negative (e.g., disease transmission, damage to physical structures, food supplies, and ornamental vegetation, traffic accidents, attacks on companion animals, and general nuisance) or positive (e.g., observing animals in an environment where wild animals are scarce). In surveys aimed at quantifying people's attitudes and perceptions toward wildlife, most indicate that urban residents value the wildlife in their environment. For example, only 8.5% of respondents in Guelph, Canada, stated that they found urban mammals a nuisance (Gilbert 1982), and 60% of respondents in Munich, Germany, stated that they were pleased to see foxes (König 2008).

The activities of urban foxes can, however, bring them into conflict with householders. For example, complaints to local authorities in the United Kingdom tend to increase during January and February because of the noise associated with mating. In summer, many complaints relate to damage caused by exuberant cubs damaging flowerbeds during play and because of the food debris that accumulates at den sites (Harris 1985). Urban foxes are also frequently portrayed as scavenging from dustbins (Vesey-Fitzgerald 1965), but this actually occurs infrequently: a survey of Bristol residents indicated that 81% had never had their garbage rifled and 16% stated that it occurred only occasionally (Harris 1981a; see also Contesse et al. 2004). Current incident rates are likely to be substantially lower as many U.K. residents now use larger animal-proof garbage bins.

An examination of the number of complaints per head of population indicates that foxes are not a significant problem in Britain, despite vociferous protests by some people to reinstate widespread lethal control. For example, Harris (1985) reported an average of 400 complaints from a population of 297,500 individuals in one area of London. In Scotland, only the cities of Edinburgh (population 430,000) and Glasgow (population 629,500) experience >50 complaints per year; the majority (56%) of urban areas receive <25 complaints per year (Campbell and Hartley 2007).

Attitudes toward urban foxes in Europe are more variable, principally because foxes in these countries transmit diseases that pose a risk to human health. For example, in Hunziker et al.'s (2001) survey of Zürich residents, most replied that they had a positive (36%) or ambivalent (56%) attitude toward foxes, whereas a second similar study reported that 40% of residents did not like foxes (Bontadina et al. 2001); attitudes varied with age and social background and were likely to be negative if respondents had suffered fox-related damage (Bontadina et al. 2001). Similarly, 74% of residents in Munich thought foxes were a danger because they transmit diseases, and 65% believed the fox population should be reduced dramatically, though 60% of people were pleased to see a fox, indicating that urban foxes were still viewed positively (König 2008). In North America, foxes are tolerated in urban areas more than coyotes (Gosselink et al. 2007).

The other major nuisance problem posed by foxes is the deposition of feces in residential gardens. Urine and feces are used by carnivores for communication (Macdonald 1979c, 1980c; Henry 1980) and are placed throughout the territory, often in prominent positions. In urban areas, this may include on items left out overnight, such as children's toys. Apart from the annoyance associated with cleaning up pungent feces, there is a potential risk of zoonotic infection, although this risk appears extremely low (Richards et al. 1993); a much greater volume of feces is deposited by domestic cats (Dabritz et al. 2006). However, problems with fouling can arise where a single householder provides a large volume of food for foxes; this may lead foxes to confine a substantial part of their activity to a few gardens, thereby markedly increasing the local density of their feces.

MANAGEMENT STRATEGIES

Historically, the management of urban foxes in Britain was attempted using culling to reduce abundance (Harris 1977), but this proved universally unsuccessful. For example, foxes in London were controlled for >25 years, yet it did not stop them from spreading across the city, nor did it lower the number of complaints received (Harris 1985). Furthermore, a comparison between London (with fox control) and Bristol (no organized control) indicated that the density of breeding groups was the same, but groups in London were smaller because of the higher mortality rate (Harris and Smith 1987b). Consequently, by 1980, only 59 of 401 authorities in England and Wales undertook some form of fox control (Harris 1985). At present, no organized control occurs in any town or city in Britain, but individual householders may contract pest control agents to kill foxes causing problems in their gardens. In Europe, there appears to be no consistent pattern of control, although it has occurred in some cities (Zürich; Hofer et al. 2000).

Most of the problems caused by urban foxes are relatively trivial and do not warrant lethal control measures, which are likely to be unsuccessful anyway as any animal removed is likely to be rapidly replaced. Preventive nonlethal measures are readily available for most nuisance problems caused by foxes. For example, rifling through garbage bins can be prevented by ensuring rubbish is placed in secure containers, and predation on pets / fowl will not occur if these are housed in fox-proof enclosures. Commercial chemical deterrents are also available that aim to discourage foxes from urinating or defecating in gardens, although whether they do so is equivocal. Foxes breeding under garden sheds can be easily persuaded to move by low-level disturbance, such as placing chemical deterrents at entrance holes or lightly blocking holes with materials such as hay; this will cause the female to move the cubs to another den site, which is something that they do naturally during cub rearing. Similarly, removing patches of nonnative vegetation, which are preferred habitats for resting foxes in Australia (Marks and Bloomfield 2006; White et al. 2006), has been suggested as one possible means for reducing fox density (White 1999; Marks and Bloomfield 2006).

Because of the proximity of foxes and humans in urban areas, the management of rabies and echinococcosis is of particular concern. Effective rabies management requires that the density of susceptible individuals is reduced below a critical threshold, either using oral vaccines or by culling using poisoned baits. Given the high density of foxes in urban areas, computer simulation models have indicated that 70%–90% of foxes would need to consume baits for such approaches to be successful (Harris et al. 1988; Smith and Harris 1989, 1991; White et al. 1995; Smith and Wilkinson 2003). Bait uptake studies have indicated that such levels are achievable in some (Rosatte et al. 1992b; Marks and Bloomfield 1999b) but not all circumstances (Trewhella et al. 1991; Baker et al. 2001a). A long-term baiting program in Toronto appears to have eliminated rabies from the city's foxes (Rosatte et al. 2007c). Antihelmintic baiting programs also appear effective in controlling *E. multilocularis* (Tsukada et al. 2002; Hegglin et al. 2003).

Urban foxes are also readily caught in baited box or cage traps (Baker et al. 2001a). These are often used to catch sick or injured foxes by animal welfare organizations that euthanize or treat and release the individual, and pest control agents who shoot or release the animal elsewhere at the householder's request; the ethics of the latter approach are questionable, as translocated animals are likely to experience a substantially reduced life expectancy (Robertson and Harris 1995a, 1995b). Given the relative unimportance of the problems caused by foxes in Britain (the only issues imposing a financial burden are the costs associated with vehicle collisions, including the risk of injury to the driver, and the treatment of dogs for sarcoptic mange), economic analyses have suggested that the use of cage trapping to control fox populations would only be viable if densities approached 50–80 foxes / km² (minimum-maximum estimated costs) in spring and summer (White et al. 2003); such densities are very unlikely. The cost-benefit of fox control, however, would be markedly different in the presence of rabies or echinococcosis.

CONCLUSION

The red fox is the most widely distributed and probably most abundant wild carnivore found in urban areas of Australia, Europe, Japan, and North America. Urban fox ecology is similar to foxes in rural areas, with comparable litter sizes and mortality and dispersal rates, though home ranges are smaller and densities are higher than usually found in rural areas. In urban environments, mortality is principally caused by road traffic accidents and disease. Diet is heavily reliant on anthropogenic food sources (scavenging and provisioning), but vegetable matter, invertebrates, and vertebrates are also important; the relative importance of dietary components appears to vary with locality. Urban foxes are potentially important vectors of diseases such as echinococcosis and rabies, but in many instances, the risk to humans or pets is not fully quantified. Human-fox conflict is generally limited, with foxes posing no physical threat to humans and causing little damage to property and pets. Complaints are relatively few and, in Britain and other countries, urban foxes are generally viewed positively. In Britain, urban fox control was found to have limited impacts on populations and was uneconomical, so council-run fox control largely ceased in the 1980s. An economic analysis suggests that control would be only viable at fox densities higher than recorded in urban areas. In other cities, urban foxes have been included in management strategies to control rabies and *E. multilocularis*.

Red foxes are an important part of the urban fauna and provide urban residents with considerable pleasure. Red foxes are probably the most intensively studied wild urban carnivore, and our knowledge of their basic ecology and interaction with urban environments is greater than for any other species of wild carnivore. However, most research has been on urban foxes in Europe and Australia, and there is a significant need for more research on populations in North America and Japan. In addition, future work must examine the management of urban populations, particularly with respect to zoonotic diseases such as *E. multilocularis*.

ACKNOWLEDGMENTS

We thank Daniel Hegglin and Clive Marks for providing unpublished data and Sandra Gloor for making her Ph.D. thesis available. Over the years our research has been funded by the Dulverton Trust; the International Fund for Animal Welfare; the Leverhulme Trust; the Ministry of Agriculture, Fisheries, and Food; the Natural Environment Research Council; the Nature Conservancy Council; the Royal Society; the Royal Society for the Prevention of Cruelty to Animals; the Science and Engineering Research Council; and the Wellcome Trust. Finally, we are particularly grateful to all the members of the public who have helped with our studies.

Arctic fox in urban setting. Photos by Paula A. White

Arctic Foxes (*Alopex lagopus*) and Human Settlements

Paula A. White

The arctic fox is a small (<10 kg) canid with a circumpolar distribution in the arctic and alpine tundra zones throughout North America, Eurasia, and Scandinavia (Hall and Kelson 1959; Macpherson 1969; Nowak and Paradiso 1983). Throughout their range, arctic foxes occur in habitats that are largely unoccupied by humans. However, arctic foxes alter their behavior patterns and ecology in response to human settlements. Contact between arctic foxes and humans occurs most frequently in outlying native villages such as the towns of St. Paul and St. George on the Pribilof Islands (see figure), remote outposts such as oil-drilling or construction camps, and modern or permanent settlements such as Dutch Harbor on Unalaska Island and the United States Air Force (USAF) Eareckson Air Station on Shemya Island.

Food resources typically serve as the initial attractant encouraging arctic foxes to forage in urban areas. Such resources consist of garbage, pet food left outdoors, deliberate feeding, and remains of harvested animals and fish. These sources can result in concentrations of foxes that in turn escalate territoriality and aggressive encounters among foxes, thereby negatively affecting fox health. Urban arctic foxes on St. Paul Island showed more variability in body weight and had a higher incidence of fresh scarring than nonurban foxes (P. A. White, unpublished data). Access to artificial foods also increases pup survivorship, and urban arctic foxes are more likely than nonurban foxes to breed as yearlings (White 1992). Pups reared in an urban environment tend to remain where artificial food and shelter are abundant, further increasing fox density. If they are subsequently forced to disperse as juveniles, these urban-raised pups may be less skilled at finding natural prey and therefore may experience higher mortality compared with pups reared in a nonurban setting.

Conflicts arise between humans and arctic foxes because of dens in inconvenient locations, aggression and territoriality during the breeding season, scent-marking, and foraging activities. Foxes will dig dens beneath sheds, houses, and cement building pads or use artificial structures such as culverts and pipes. Breeding adults as well as nonbreeding adult "helpers" will actively defend dens during the breeding season and will chase other foxes as well as humans who venture too close. Foxes have been observed chasing and nipping at human passersby.

Once arctic foxes establish residence in or near human settlements, they actively mark and defend their new food patch. On St. Paul Island, urban arctic foxes often jump into the beds of parked pickup trucks and will even enter a house in search of food if the door is left open. During such forays, foxes snatch and run off with food and sometimes nonfood items. Foxes commonly scent-mark during these excursions, adding to the dissatisfaction of local human residents.

Urban arctic foxes are subject to direct encounters with domestic animals and subsequent exposure to zoonotic disease. On St. Paul Island, urban foxes were found to have titers showing past exposure to a feline virus (P. A. White, unpublished data). Foxes that exist in or forage near human settlements also encounter artificially high bacterial loads from human garbage. Arctic foxes living at the city dump were observed scavenging human excrement from discarded baby diapers. This is not only detrimental to fox health but also increases concerns of disease-carrying foxes as threats to human safety. Arctic foxes also are susceptible to rabies, and the proximity of urban foxes to both humans and domestic pets may pose a significant risk to human health during years of rabies outbreaks (Crandell 1975; Garrott and Eberhardt 1987).

Several control measures have been attempted to alleviate problems associated with urban arctic foxes. Short-distance relocation has been found to be ineffective. Foxes on St. Paul Island that were moved to the other end of the island returned to the town by early the following day, a straight-line distance of 19 km. Lethal control is effective in the removal of individual foxes, but if the source of attraction (e.g., food) is not eliminated, other foxes quickly move into the vacated territories.

To combat food availability, the city of St. Paul and the island's Stewardship Program instigated a garbage-control program that entailed two aspects. The first was a household garbage containment program. Historically, townspeople used open-topped, 55-gallon drums outside of their houses as refuse containers. Open drums presented arctic foxes with easy access to garbage. Residents can request simple enclosure system for garbage cans provided free of charge, and reduced access to household garbage has greatly reduced the number of animals residing in town. The second aspect of garbage control was the installation of an incinerator at the city dump. Before initiation of the program to burn and cover garbage, arctic foxes and abandoned house cats lived at the dump, subsisting on refuse for food (P. A. White, unpublished data). Now that garbage processing is more efficient, the presence of foxes and cats at the city dump is rare.

Under certain circumstances, the human/fox association may be beneficial for humans, thereby generating favorable attitudes toward foxes. On the Pribilof Islands, the artificial breakwaters built to increase the islands' ability to service larger fishing vessels have rendered both St. Paul and St. George Islands vulnerable to rat introductions. Foxes inhabiting the breakwaters are recognized by United States Fish and Wildlife Service as a first-line defense against possible rat invasion. On Shemya Island in the Aleutian chain, an introduced population of arctic foxes is purposely maintained to discourage Aleutian cackling geese and gulls from roosting or nesting on the U.S. Air Force runway. The resident foxes subsist mainly by beachcombing and gleaning invertebrate prey from the intertidal zone. As a result of the minimal natural resources and very small surface area of the island, arctic foxes crisscross the entire surface on a regular basis, thereby effectively discouraging large birds from loitering on the runway.

As human presence and urbanization expand in arctic regions, the number of arctic foxes encountering urban areas likely will increase. With proper management as well as public outreach, conflicts between humans and arctic foxes can be minimized, thereby facilitating coexistence.

7

Coyotes (*Canis latrans*)

STANLEY D. GEHRT

SETH P. D. RILEY

THE COYOTE has long been considered an icon of wilderness and has elicited strong emotions in people from the beginning of pioneer settlement of the West. These emotions have contributed to real or imagined conflicts, with resulting intensive efforts to reduce or to exterminate coyotes from the landscape. Despite intensive control campaigns carried out over decades, coyotes are unique among the order Carnivora because they have essentially doubled their historic range across the North American continent (Laliberte and Ripple 2004). Increasingly, coyotes are extending their success to the last remaining vacancies in their range, the metropolitan landscapes. The icon of wilderness has quickly become a denizen of the city. Consequently, emotions toward coyotes that used to be confined to rural areas have emerged in cities and suburban neighborhoods across much of the United States and Canada. These emotions run the gamut from fear to fondness and from respect to loathing, with intensities possibly as great as those previously experienced in the western United States. The result is the emergence of the coyote as arguably the most controversial carnivore species in North American metropolitan areas (Gehrt 2007).

Although the coyote is one of the most studied canids in North America (Bekoff and Gese 2003), relatively little attention has been directed toward coyotes in urbanized landscapes, especially those areas with the highest levels of urban development. This paucity of urban research more likely reflects the relatively recent appearance of coyotes in many metropolitan areas joined with the incredible challenges facing researchers attempting to capture and track coyotes in public areas (Gehrt 2007). Unfortunately, the limited research has left an information void to be filled by popular media or professional papers focused exclusively on conflicts or threats that coyotes pose for people or pets.

SCOPE OF REVIEW

In this chapter, we survey published literature, theses, and our own unpublished data to summarize the current state of knowledge regarding the relationships between coyotes and the many ecological layers of urbanization. Before we do so, a caveat is in order, as it is for most of the chapters in this volume. In addition to the usual caution necessary for comparisons among studies that differ in sample sizes, in techniques for data collection, and in analytical approach, urban studies have that added complication associated with different interpretations of the terms *urban*, *suburban*, and *exurban* (i.e., the different levels of urbanization associated with different sizes and densities of metropolitan areas; table 7.1). Unfortunately, important features of the landscape within the actual study areas, such as human population density, types of development, and road densities, are not consistently reported across publications. Most notably in the case of coyote research, inspections of study area maps suggest that there are considerable differences among studies in the location of study areas relative to cities and truly urban landscapes. Most studies have occurred at the edge of urban development with large areas of nonurban habitat available to coyotes, and with relatively few animals residing within the urban matrix. Consequently, we often refer to *urban coyotes* in this chapter, but that in reality there was considerable variation in the amount of urbanization and associated human activities to which the coyotes within and among the various studies were exposed.

We frequently refer to two study areas in this chapter: the Cook County Study Area (CCSA; figure 7.1, *above*), which is associated with the Chicago metropolitan area in Illinois, and the Ventura County Study Area (VCSA; figure 7.1, *opposite*), which is associated with Los Angeles metropolitan area. These areas are the locations of our own, long-term studies of urban coyotes and illustrate the variation in levels of development typical of urban coyote studies.

Another source of variation across studies is the history of residence for coyotes in different metropolitan areas. Whereas coyotes are apparently recent inhabitants of many Midwestern and eastern U.S. cities, coyotes may have always been a component of the urban fauna of some southwestern metropolitan areas. For example, it appears that coyotes may never have been totally excluded from the Los Angeles area during its development (Gill 1970). It is possible that coyotes may respond differently to human activities or anthropogenic resources if they are the product of multiple generations of urban life versus those with a limited history of living with people.

DESCRIPTION

The coyote is a member of the family Canidae (including wolves, foxes, and the domestic dog) and closely resembles some dog breeds with their upright ears, long

Table 7.1. Location, associated human density, and sample size for select radiotelemetry studies of urban and suburban coyotes

Metropolitan area	State	Human density (people / km²)	No. radiocollared	Source
Tucson	Arizona	950	19	Grinder and Krausman 2001a, 2001b
Los Angeles (VCSA)	California	534	110	Riley et al. 2003, 2006; NPS, unpublished data
Barnstable County (Cape Cod)	Massachusetts	203	11	Way et al. 2002, 2004
Albany	New York	217[a]	21	Bogan 2004
Chicago (CCSA)	Illinois	5,602	150	Morey 2004; Gehrt 2004b; S. D. Gehrt, unpublished data
Banff	Alberta	—	11	Gibeau 1998

Note: Data illustrate the variation in urbanization and sampling intensity among studies. CCSA = Cook County Study Area; VCSA = Ventura County Study Area.

[a]Human density is estimated by U.S. Census Bureau data (2004) for the primary county in the metropolitan area.

snout, and long, bushy tail (figure 7.2). Coyotes vary in size across their range, with a body length of approximately 1.5 m and mean weights ranging from 10 kg to 15 kg across locations. Adult males are slightly larger than females but with considerable overlap. The current range of coyotes is extensive; they are geographically distributed from the Atlantic to the Pacific Coast and from northern Alaska south to Central America (Bekoff and Gese 2003). Larger carnivores (e.g., mountain lions, black bears) have been reported in metropolitan areas, but the coyote is the largest wild carnivore in most metropolitan areas across its range.

Coyotes in the northeastern United States and eastern Canada apparently differ genetically, morphologically, and possibly behaviorally from more western and southern coyotes (Hilton 1978; Parker 1995) because of hybridization with wolves during a northeastern expansion of the coyote's range (Wilson et al. 2004). However, results from Cape Cod, Massachusetts (Way et al. 2002; 2004), indicate eastern coyotes respond to urbanization in similar ways to other coyotes, so we do not discriminate between eastern coyotes and coyotes elsewhere. Nevertheless, as new research develops, it will be interesting to see whether differences exist between eastern and western coyotes in their abilities to exploit urban landscapes and in their relationship to the public.

The similarity to domestic dogs presents interesting complications for the human-coyote relationship in urban areas, where a public that is largely unfamiliar with wildlife often confuses coyotes with dogs running loose and vice versa. A common question from the public and researchers alike is whether the constant proximity of coyotes and dogs that inevitably occurs in urban areas leads to hybridization. Although rare, coyotes are certainly capable of hybridizing with dogs (Mengel 1971; Nowak 1978; Adams et al. 2003), but there is little indication that coyotes living in urban areas are more genetically intermingled with dogs than has occurred historically in nonurban areas.

URBAN ECOLOGY

Social Organization

Few urban studies have addressed the social system of coyotes, and none have provided insight into how urban landscapes may influence social behavior. Coyotes exhibit intraspecific variation in social organization (Bekoff and Gese 2003), but the typical population comprises resident groups, or packs, maintaining exclusive territories across the landscape, with solitary individuals of both sexes inhabiting large home ranges that overlap one another and the smaller group territories (Bekoff and Wells 1980; Gese and Ruff 1997, 1998; Gese 2001). This system appears to hold for coyotes in urban landscapes (Way et al. 2002; Gehrt 2004b), although group sizes and internal pack dynamics may vary across sites. Packs are formed for territory defense (Bekoff and Wells 1980; Bowen 1982), but pack size is affected by food abundance, mortality rates, and population density (Andelt 1985; Bekoff and Gese 2003). In a mixed agricultural and urban landscape in Indiana, Atwood (2006) determined that coyote group size was greater in territories with aggregated habitat patches, in contrast to those with dispersed habitat patches. Thus, it could be expected that group size may differ substantially across a metropolitan landscape with widely varying habitats and patterns of fragmentation.

Packs of rural and urban coyotes are typically composed of an alpha pair (adult male and female) that breed and are dominant over one or more associates that help defend territories and provision the young of the alpha pair. Pack size on Cape Cod was typically three to four adults based on visual sightings (Way et al. 2002), with one example of a group of six adults (Way 2003). In the CCSA, pack size was typically four to six coyotes, not including the pups of the year (Gehrt 2004b, 2006), although these numbers may be conservative because they are based on visual sightings of the group. Genetic analysis of 241 individuals during 6 years in the CCSA population (including 19 litters with 96 pups), revealed that: (1) all but one pup was the product of a single alpha pair, (2) the male and female of alpha pairs ($n = 7$) were not closely related, and (3) pack members were usually (>85% of associate pairs, $n = 116$) closely related to one another and to the alpha pair (probably as offspring from previous litters; Hennessy 2007). Coyotes born into a pack either stayed as an associate in the pack; left to become a local, solitary nomad; or dispersed from the area altogether. We documented at least two instances of territorial inheritance, in which an offspring assumes alpha status following the loss of the same-sex alpha member. Nomads replaced residents that were killed or removed from territories or established new territories with another nomad of the opposite sex. Occasionally, older individuals would leave territories and become nomads, including previous alpha adults upon the death of their mate.

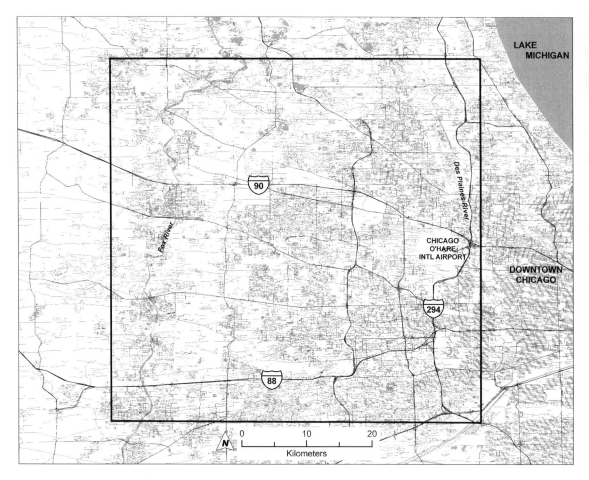

Figure 7.2. A coyote from the Chicago metropolitan area. Max McGraw Wildlife Foundation.

Population Biology

Demographic characteristics can provide important insight into the adaptability of coyotes to urban landscapes; however, few studies have had the sample size to provide estimates of age, sex structure, or reproductive rates. If metropolitan areas are challenging

Figure 7.1. Study areas in Chicago area (*above*), the Cook County Study Area (CCSA), and Southern California (*opposite*), the Ventura County Study Area (VCSA). The VCSA is further divided into study areas A (dashed line), B (dotted line), and C (solid line), representing specific studies (e.g., Fedriani et al. 2001 in study area A; Riley et al. 2003 and Tigas et al. 2002 in study area B and more recent research in increasingly urban and fragmented parts of the landscape in study area C).

to coyotes, then one could predict that demographic characteristics of the urban population might resemble that for exploited rural populations, with a younger age structure, possible bias toward females, larger litter sizes, greater percentage of young females reproducing, and smaller pack sizes (Gier 1968; Knowlton 1972; Berg and Chesness 1978; Andelt 1985; Knowlton et al. 1999; Bekoff and Gese 2003). The converse would be predicted if urban populations resemble nonurban populations protected from exploitation.

Density

Urban coyote populations are often thought to exist at higher densities than those in nonurban areas because of increased food resources, reduced mortality

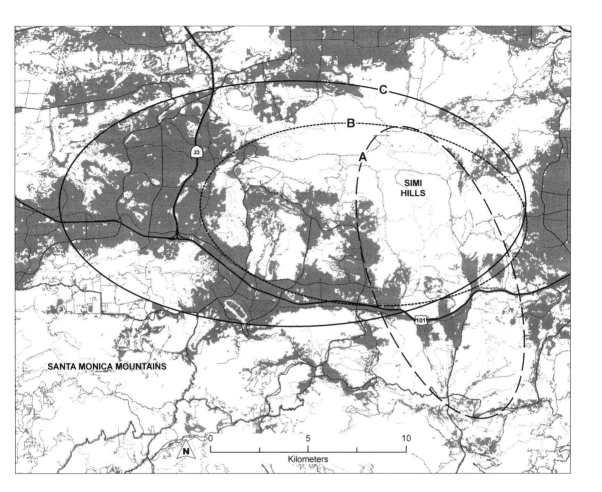

from trapping and hunting, or both. Fedriani et al. (2001) estimated coyote densities using scat genotyping (Kohn et al. 1999) and capture rates in three sites within study area A in the VCSA (see figure 7.1, *opposite*) in Southern California. They estimated densities on the sites at 2.4–3.0, 1.6–2.0, and 0.3–0.4/km², and found that density was significantly higher in the most urban-adjacent of the three locations, although this site was still largely composed of natural habitat that bordered on development. The authors speculated that increased food resources in more urban areas, such as trash and domestic fruit, may contribute to higher coyote densities, as the highest percentage (14%–24%) of anthropogenic food items in the coyote diet was at the most urban site. Concurrent and more recent work involved extensive radio-tracking of coyotes in these same areas and in fragments of habitat to the west (figure 7.1, *opposite*: study areas B and C; Riley et al. 2003, National Park Service unpublished data). Minimum density estimates based on numbers of radio-collared animals and the area they used were 0.21 coyotes/km² in the fragmented areas and 0.53 coyotes/km² in

the contiguous natural areas adjacent to urbanization (figure 7.1, *opposite*; study area A, including the site of highest density from the Fedriani et al. [2001] study). These density estimates are very conservative because generally only adults or yearlings were collared, and not all adult coyotes in the area were followed.

In the CCSA, density estimates based on radio-collared animals and observations (during radio-tracking or on helicopter flights) of uncollared animals from the same packs were relatively high, especially in forest parks (figure 7.1, *above*), ranging from 2 to 6 coyotes/km². In their review, Bekoff and Gese (2003) report coyote densities from 12 different non-urban studies and from various times of year, ranging from 0.1 to 0.9 coyotes/km² with one fall estimate of 1.5–2.3 coyotes/km² (Knowlton 1972). Density of carnivores, including coyotes, is difficult to measure, and methods vary significantly between studies. More estimates from urban areas are also needed. However, the minimum estimates from local areas in the CCSA are as high, or higher, than any other reported estimates we are aware of, as are the higher estimates from the

Fedriani et al. (2001) study, based on scat genetics (Kohn et al. 1999).

Survival

Radiotelemetry studies have provided annual survival estimates for some metropolitan areas. Annual survival estimates were similar for coyotes in the Tucson and Los Angeles areas, ranging from 0.71 to 0.74 across studies (Grinder and Krausman 2001a; Riley et al. 2003). In the CCSA, annual survival estimates ranged from 0.53 to 0.68 (mean = 0.62) during a six-year span (2000–2005) for all sex-age groups combined (S. D. Gehrt, unpublished data). Annual survival rates (and 95% confidence intervals) were not significantly different among sex-age groups: adult female 0.70 (0.57–0.83), adult male 0.59 (0.48–0.70), subadult female 0.58 (0.43–0.73), subadult male 0.63 (0.48–0.78), juvenile female 0.61 (0.45–0.77), and juvenile male 0.61 (0.42–0.81). These relatively high survival estimates for urban studies are in stark contrast to an annual survival estimate of 0.20 for coyotes in the Albany, New York, area (Bogan 2004). With the exception of the Albany study, the overall pattern of survival estimates for urban populations are similar to estimates for unexploited, or protected, rural coyote populations and considerably higher than those for coyote populations hunted or trapped (Bekoff and Gese 2003), indicating that urban landscapes may be favorable for coyotes.

The refuge effect of large metropolitan areas may be particularly noticeable in the Midwestern United States, where the landscape is dominated by row crops that undergo an annual fall harvest, and coyotes are allowed to be taken by hunters and trappers without limits. In this case, the landscape undergoes a dramatic transformation where the harvest of crops removes cover over large areas at roughly the same time the demand for coyote fur is highest. In rural Illinois, annual survival of coyotes dropped during the fall harvest season, particularly for the pup age class with an annual survival of only 13% (Van Deelen and Gosselink 2006). By comparison, annual survival of the pup age class in the CCSA from 2000 to 2006 was 61% for both sexes, with no seasonal variation, which is nearly five times higher than the estimate reported by Van Deelen and Gosselink (2006). Population modeling of rural populations has indicated that pup survival is an important determinant of population growth, in which a pup survival rate of 33% is necessary for stable populations, given a 60% annual survival rate for adults (Knowlton 1972; Nellis and Keith 1976).

Within metropolitan areas, collision with vehicles is a common cause of mortality for coyotes, and this seems to be a cost of living in urban areas. Mortality resulting from vehicles represented 35% to 50% of mortalities in Albany, Tucson, and Los Angeles, whereas it represented 62% (n = 68) of the deaths in the CCSA. The relatively high vehicle mortality rate for the CCSA likely reflects the study area occurring in a landscape dominated by roads, compared with other studies (figure 7.1, *above*). Interestingly, the Albany study that reported a low annual survival rate also reported a relatively high rate (43%) of hunting mortality (Bogan 2004), which is not a common activity in urban areas and may be a result of a substantial rural component to their study area.

Death from anticoagulant rodenticide poisoning was also a significant source of mortality in coyotes in the VCSA (Riley et al. 2003; National Park Service, unpublished data). Over the 9 years of the study, 12 of 45 (27%) mortalities of known cause were from anticoagulant poisoning, versus 23 from vehicles (51%), although some of the undetermined mortalities (15 of 60 total deaths) were likely antiocoagulant related as well. Death from anticoagulant toxicity is characterized by extensive internal bleeding, often in the thoracic, peritoneal, and pericardial cavities, with no other signs of trauma or injury, coupled with anticoagulant rodenticide residues in the liver. Overall, 20 of 24 coyotes (83%) tested in the Santa Monica Mountains area were positive for anticoagulants, with 9 of the 20 exposed animals (45%) positive for multiple (2–4) compounds. Although there are cases of urban residents using anticoagulants to attempt to poison coyotes intentionally, the evidence (e.g., frequent exposure to multiple compounds) indicates that the majority of exposures are secondary, through eating poisoned rodents. Interestingly, coyotes in the CCSA have not suffered any known mortality from anticoagulant rodenticides. Mortality from anticoagulants was also documented in the Cape Cod study (Way et al. 2006). The pattern of rodenticide use, especially outdoor use, may vary in different regions and landscapes.

Disease

Coyotes are exposed to a variety of transmissible diseases in urban areas, although it is unknown whether living in urban landscapes alters the dynamics of these diseases. Urban coyotes may be exposed to canine parvovirus, canine distemper, canine herpesvirus, canine adenovirus, and *Leptospira interrogans* (Grinder and

Krausman 2001a; National Park Service, unpublished data), although extensive mortality from these pathogens has not been reported. Although coyote-strain rabies is restricted to Mexico and south Texas, raccoon-strain rabies has occasionally spilled over into coyotes. A coyote attack on a man in an urban park in Cleveland, Ohio, was the result of this type of spillover (Ohio Department of Public Health, unpublished data). Sarcoptic mange can be present in urban populations (Grinder and Krausman 2001a), and 10% of mortalities in CCSA were due to mange.

Coyotes may serve as host reservoirs for heartworm (*Dirofilaris immitis*), although their importance may vary among cities. In the CCSA, serology tests ($n = 85$) yielded an overall seroprevalence of 27%, with 30% of males and 24% of females testing positive for antibodies to heartworm (S. D. Gehrt, unpublished data). However, infection rate was higher for carcasses collected during the same period, for which 41% ($n = 62$) were infected. Interestingly, the infection rate of coyote carcasses from CCSA was considerably higher than the 4.6% of coyotes infected with heartworm from rural areas in northern Illinois (Nelson et al. 2003), suggesting heartworm dynamics may shift with urbanization.

Home Range

Most radiotelemetry studies of urban coyotes have focused on resident, territorial individuals. The trend across studies is relatively small home ranges when compared with rural studies, but with a range in mean home range estimates from 3 to 36 km² (table 7.2). Published home range estimates of resident coyotes from nonurban areas in the review by Bekoff and Gese (2003) ranged from 3 to 42 km², with a mean of 17.5 km², suggesting a trend for smaller home ranges in urban landscapes: the mean of the studies in table 7.2 is 13.4 km², and for those with $n \geq 10$, the mean is 8.2 km². Similarly, Atwood et al. (2004) reported a negative relationship between home range size and anthropogenic development in Indiana. However, the ranges of both urban and rural data overlap considerably, with some rural home range estimates as small as those reported for urban populations. At the landscape level, small home ranges can indicate high population densities (Andelt 1985; Fedriani et al. 2001) in either urban or rural areas.

In contrast to the trend in home range size between urban and rural studies, at the local scale within metropolitan areas, coyote home ranges may become larger when they encompass larger amounts of urban land use. Within the CCSA, there was a positive relationship ($r^2 = 0.40$, $P < 0.001$) between the amount of developed property in the home range and home range size, but there is also considerable variation in home range size not explained by that relationship (Gehrt 2007). The relatively large home ranges of coyotes in the highly

Table 7.2. Annual home range sizes (km²) from radiotelemetry studies of coyotes in urban areas

Metropolitan area	Resident		Transient		Source
	N	Home range	N	Home range	
Tucson, AZ	13	13			Grinder and Krausman 2001a, 2001b
	7	23	1	62	Grubbs and Krausman 2009
Lincoln, NB	1	7		—	Andelt and Mahan 1980
Lower Fraser Valley, BC	13	11			Atkinson and Shackleton 1991
Los Angeles, CA	53	5	8	50	Riley et al. 2003; NPS, unpublished data
Barnstable County (Cape Cod), MA	5	36	1	115	Way et al. 2002[a]
Albany, NY	17	7			Bogan 2004[b]
Chicago, IL	118	5	40	27	S. D. Gehrt, unpublished data

Note: All estimates are minimum convex polygons (MCP), unless otherwise noted. NPS = National Park Service.

[a] 95% modified MCP.

[b] 95% Adaptive kernel.

developed landscape on Cape Cod, Massachusetts, is consistent with this pattern, although this may be a result of the larger size and behavioral differences in eastern coyotes (Way et al. 2001). The mean reported by Grubbs and Krausman (2009) is primarily for members of one pack residing in an urban park in central Tucson and may be the result of urban development, a small sample size, or the use of Global Positioning System data. In the VCSA, Riley et al. (2003) reported a positive relationship between home range size and the amount of nonnatural land (developed plus "altered" areas, such as golf courses, other landscaped areas, low-density residential areas, etc.) in the home range, a significant relationship between home range size and development for adult males, and a positive but nonsignificant relationship between development and home range size for all coyotes. Overall, these results suggest that while coyotes are capable of living within the urban matrix, it may come as a cost in needing a larger home range to meet energetic requirements (Riley et al. 2003) or may reflect an avoidance of developed habitat.

Compared with resident, territorial coyotes, little information is available for nomadic, solitary individuals despite their potential importance for population dynamics. Under the best of circumstances, it can be difficult to radio-track coyotes in urban landscapes, and this difficulty can be exacerbated by the large-scale movements and dispersal associated with nomads. The few reports of solitary coyotes suggest that, like rural populations, the typical home range size of solitary, nomad coyotes is quite large (90–100 km²; Grinder and Krausman 2001b; Way et al. 2002). Mean home range size for 40 nomad coyotes in the CCSA was 27 km², with a maximum size of 101 km². In this study, nomads included both sexes (23 females, 17 males) and all age classes (juvenile, subadult, and adult). Similarly, the average home range size of transient coyotes in the Los Angeles area (50 km², $n = 8$) was actually an order of magnitude larger than for resident animals (table 7.2), transients of both sexes were identified, and transient home ranges were as large as 100 km².

Response to Roads

In the VCSA in Southern California, coyotes were documented crossing roads, both freeways and secondary roads, and dispersing between habitat fragments (Tigas et al. 2002) and from fragmented urban areas to large contiguous natural areas to the north. While 58 of 110 radio-collared coyotes (52%) crossed major secondary

roads in the area, only 5 of 110 (4.5%) crossed U.S. 101, the major road separating the Santa Monica Mountains from the Simi Hills (figure 7.1, *opposite*; Riley et al. 2006). This freeway, a ten- to twelve-lane road with 150,000 vehicles / day and a cement median barrier, was also a home range boundary for coyotes that lived near it, and no 95% Minimum Convex Polygon home ranges overlapped it. The freeway was enough of a barrier to movement that it also potentially reduced gene flow for coyotes, as genetic differentiation between coyote subpopulations was significantly greater across the freeway than over similar or greater distances along it (Riley et al. 2006).

In contrast, coyotes in the CCSA often crossed both secondary roads and freeways, including roads with traffic volumes, exceeding 100,000 vehicles / day. They also crossed major rivers (e.g., Fox and Chicago rivers), over bridges, over the ice, or by swimming. Coyote use of bridges was also observed on Cape Cod (Way 2002).

Activity Patterns

A consistent observation among virtually all urban coyote studies is an increase in nocturnal activity with an increase in the level of development or human activity within the home range (Atkinson and Shackleton 1991; Quinn 1997a; Gibeau 1998; Grinder and Krausman 2001b; Tigas et al. 2002; Riley et al. 2003; Grubbs and Krausman 2009). Conversely, then, radio-collared coyotes typically reduce their activity during the day because of living close to people. Using camera traps, George and Crooks (2006) also determined that coyotes avoided areas and periods of high human activity within a large urban park. Exceptions to these trends may involve coyotes habituated to human activities (Gehrt 2006).

At least two benefits to coyotes may result from nocturnal behavior (Gehrt 2007). Coyotes can more easily avoid humans during nighttime because people may have more difficulty observing them and less human activity occurs at night. Second, traffic volumes are usually lower during nocturnal hours, and this may allow coyotes to cross roads more easily. Given that a major mortality factor is often collisions with vehicles, a shift to nocturnal activity may be particularly important to survival in urban landscapes.

Landscape Use and Selection

An important issue regarding coyotes in metropolitan areas is the extent to which coyotes avoid, or are attracted to, developed areas. It is obvious that coyotes have flexible habits that they are not as negatively impacted by urbanization as are perhaps other carnivore species. Using track stations, Crooks (2002) determined that coyotes are capable of occurring in relatively small habitat fragments separated by relatively long distances. In that study, probability of occurrence was at least 50% for fragments >1 ha in size, and 50% for an isolation distance of 883 m. This implies that coyotes are more able to exploit habitat fragments within the urban landscape than other large carnivores.

Although coyotes may maintain territories within the urban matrix, there is a trend among studies for coyotes to center territories within natural habitat patches or to use these patches more frequently than residential areas within the home range (Quinn 1997a; Riley et al. 2003; Gehrt et al. 2009). Studies have also reported avoidance of developed areas in the home range (Quinn 1997a; Gehrt et al. 2009), or at most use in proportion to their availability (Gibeau 1998; Grinder and Krausman 2001b; Way et al. 2004). However, coyote use or avoidance of developed areas may also vary seasonally (Grinder and Krausman 2001b) or by time of day, with an increase in urban use during the night (Tigas et al. 2002; Grubbs and Krausman 2009).

In the CCSA, many (31%, *n* = 119) coyote territories were located within natural area fragments or contained substantial (>95%) amounts of natural area (figures 7.3, 7.4). However, there was some variability among coyotes, with some territories composed of relatively little natural area and varying amounts of residential or urban land use (figure 7.5). The smallest natural area fragment in which a coyote pack was able to maintain an annual territory for multiple years was 250 ha. The smallest territory size maintained by a coyote pack for multiple years was within that fragment, and was 98 ha, or just less than 1 km², which is small for a coyote territory. Within territories, coyotes used natural areas most heavily, but they exhibited significant selection in their use of other land use classes. Among all sex and age groups, coyotes consistently avoided areas of residential and urban land use, while preferring patches of undeveloped land, regardless of the landscape composition of their territories. This pattern of use and avoidance was consistent across years, as well as between territorial groups and solitary nomads. Indeed, even those coyotes with the highest composition of urban land use within their home ranges continued to exhibit avoidance of urban land (Gehrt et al., 2009).

In the VCSA, through the first 5 years of the study

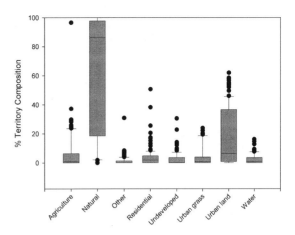

Figure 7.3. Box plots of the composition of coyote annual territories (*n* = 119) by land use class in the Chicago metropolitan area, 2000–2006. Horizontal lines in the boxes are means, each box represents 1 standard deviation about the mean, whiskers represent 95% confidence intervals, and points are outliers outside the confidence interval. Land use classes include "urban land" (commercial and industrial use, impervious surfaces), "urban grass" (usually mowed, or maintained areas), "undeveloped" (fragments of vegetation, usually too small for development, but not preserved from development or managed for natural habitat), and "other" (small fragments of a mixture of maintained and natural vegetation, such as golf courses and cemeteries); other land use classes are self-explanatory.

(1996–2000), radio-collared coyotes were mostly relying on natural habitat. On average, coyote home ranges consisted of about 75% natural area (Riley et al. 2003), although as in CCSA, there was considerable variation between individuals. From 2001 to 2003, the study was extended to the west into even more highly fragmented areas (figure 7.1, *opposite*, study area C) where some habitat fragments used by coyotes were as small as 80 ha. In this more fragmented part of the landscape, a greater proportion of coyote home ranges was made up of altered (see previous "Home Range" section) or developed areas, such that for the entire study, resident coyote home ranges consisted on average of 67% natural area (*n* = 53; National Park Service [NPS], unpublished data). Transient coyotes (*n* = 8) had somewhat higher levels of urban association; on average, their home ranges were made up of 61% natural habitat, 17% altered areas, and 22% development.

However, examination of radiolocations shows that even transient coyotes and residents in the most fragmented parts of the landscape in Southern California used natural, or when those weren't available, altered areas to a great extent. Overall, 77% of locations were in natural areas, for both resident and transient coy-

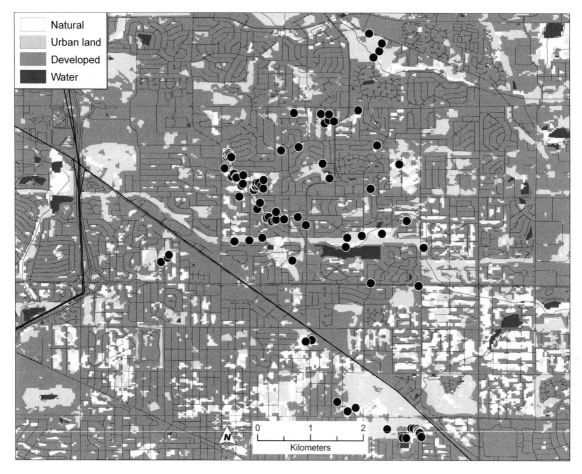

Figure 7.4. Radiolocations for an adult, male coyote with a territory in an urban landscape within the Chicago metropolitan area, 2006.

otes. For example, resident male coyote 109 (C109; fig-ure 7.6, *above*) moved through developed areas between a number of fragments and peninsulas of natural and altered areas. C109's home range was 78% urban and only 14% natural, but only 15% of his radiolocations were in urban areas versus 50% in natural areas. Simi-larly, female coyote C028 dispersed out into the San Fernando Valley, a vast sea of development north of Los Angeles (figure 7.6, *opposite*). However, she was virtually always located on a community college cam-pus or at the Sepulveda basin, an area of parks, golf courses, and natural habitat around a Los Angeles river catchment: 75% of her home range, but only 7.5% of her locations, were urban.

How much habitat is necessary for a coyote to main-tain a territory and successfully rear offspring within an urban landscape is a fundamental question that currently remains unknown, and it may be that the answer is changing through time as an ever-increasing

number of generations of coyotes are born in cities. In the CCSA, 8% of coyote annual territories were com-posed of <1% of natural area (Gehrt et al. 2009). It remains to be seen whether coyotes can successfully maintain territories, survive, and experience reproduc-tive success in areas of complete development, while still avoiding people.

Diet

Diet studies of coyotes in urbanized areas have typi-cally reported diets dominated by small mammals (e.g., rodents, lagomorphs; table 7.3). Comparisons across studies indicate variability in the frequency of human-related food in the diet. For example, human-related foods ranged in frequency from 2% to 35% (table 7.3), and mostly occurred with a frequency below 20%. The studies from the Tucson and San Di-ego areas were focused primarily on single sites of 1 and 14 km², and the most urban site for the initial diet

Figure 7.5. A picture of the radio-collared coyote whose territory is represented in Figure 7.4, illustrating the type of land use dominating the composition of his territory: urban land and urban grass. Max McGraw Wildlife Foundation.

study from the VCSA (Fedriani et al.2001; figure 7.2b, study area A) contained a large trash dump, which likely affected the frequency of human-related food in that study. A later diet study in the VCSA (table 7.3; NPS, unpublished data) occurred in more highly fragmented areas (see figure 7.1, *opposite*, study areas B and C). In these areas, human-related foods also made up 25%–30% of coyote food item occurrence, although in all the Southern California diet studies, the most common anthropogenic item was domestic fruit. Trash, pet food, and domestic cats all represented <6% of food items in both studies. In the recent work, the coyote diet was generally more diverse in the less fragmented, core natural areas, including a higher percentage of food items such as birds and woodrats, while rabbits were consistently important in fragmented areas.

Interestingly, one of the lowest frequencies of human-related food occurred in the diet for coyotes in the CCSA, where scats were collected from an area of relatively high urbanization and human use. In that study area, even natural areas typically received tremendous levels of human use, with some forest parks in the area receiving between 1.5 and 3 million visitors annually. Moreover, Cook County forest preserves had a policy of keeping lids off trashcans and Dumpsters, so picnic areas had a nearly constant flow of anthropogenic foods available, especially during spring and summer months. Despite easy access to refuse, coyotes nevertheless primarily consumed prey or vegetation (Gehrt 2004b; Morey et al. 2007). The lack of refuse in the diet, despite its availability, was supported by radiotelemetry data that determined coyotes were not focusing their movements around refuse areas, in stark contrast to another generalist carnivore, the raccoon (Gehrt 2004b). More dietary studies, including those

that sample across the landscape variation within metropolitan areas, are needed to clarify diets of coyotes in the city.

Within-study comparisons have yielded consistent patterns of increasing frequencies of human-related foods with proximity to residential areas. Studies in Los Angeles, Chicago, and Seattle metropolitan areas collected scats along urban gradients, and each study reported higher frequencies of anthropogenic items in the diet in more urbanized areas (Quinn 1997b; Fedriani et al. 2001; Morey et al. 2007). For example, Fedriani et al. (2001) reported up to 24% occurrence of anthropogenic foods in the study site with the most urban edge, whereas that frequency was only 0%–3% in the more distant sites. However, it is important to note in the CCSA that sites classified as natural areas nevertheless had tremendous levels of human activity and presence of refuse (Gehrt 2004). In some ways, there was more opportunity to exploit human-related foods in the natural areas than in residential areas; yet, human-related diet items remained at low frequencies in the diet. Fedriani et al. (2001) and Morey et al. (2007) also found a positive relationship between local urbanization and diet breadth, reflecting the flexible foraging behavior of coyotes (Bekoff and Gese 2003).

Coyote predation on pets is a major cause of human-coyote conflicts, but the occurrence of domestic cat or dog in coyote diets across studies is consistently very low (table 7.3). The highest published occurrence for domestic cat in the coyote diet was 13% for an urban area in Washington state (Quinn 1997b), although this study recorded only the "dominant" item in each scat, thereby leading to greater representation for those items than if all food items were counted. For every other study, the occurrence of cats has been 1%–2% or less (table 7.3). However, a pack of coyotes in Tucson was observed regularly killing and consuming cats (Grubbs and Krausman 2009). Nevertheless, the consistently low prevalence of pets in diet studies indicates that coyotes may not always consume cats or dogs that they kill, and most important, that coyotes in urban areas are not dependent on pets as food (Gehrt 2007).

ECOLOGICAL ROLE IN URBAN SYSTEMS

As a top predator in most urban systems, coyotes have the potential to have profound impacts on wild and domestic animals (Crooks and Soulé 1999). Although they may influence prey species and fellow predator species in many ways, few studies have rigorously eval-

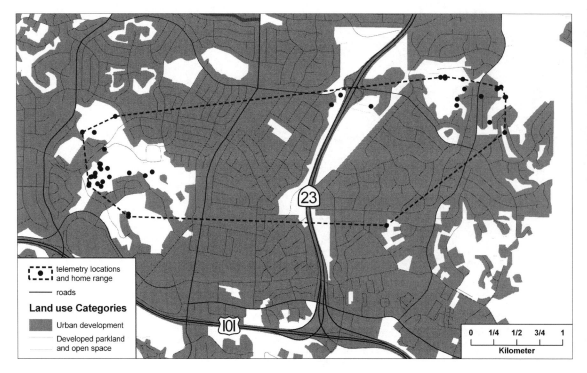

Figure 7.6. Home ranges and radiolocations of coyotes in urban areas in Southern California. *Above,* Locations for an adult male coyote (C109) that used patches of natural habitat or landscaped parkland within a highly fragmented landscape in Thousand Oaks, California. (*Opposite,* Locations for an adult female (C028) that dispersed after initial capture out into the San Fernando Valley, in Los Angeles, California. She was almost always located either at Pierce College (to the west), a community college with vegetated (mostly landscaped) areas and a small demonstration sheep farm, and the Sepulveda basin (to the east), the area around a dam on the Los Angeles River that includes golf courses, playing fields, and a small natural wetland.

uated these relationships or determined the extent to which they exist in urban areas.

Coyotes may have their greatest impact on urban systems through their role in trophic cascades. In a famous study among the natural fragments within a Southern California metropolitan area, Crooks and Soulé (1999) documented evidence of mesopredator release, in which medium-sized predators increase numerically or their functional response becomes stronger in the absence of coyotes. They observed that coyote presence in these fragments was negatively correlated with some mesopredators, notably domestic cats, and positively correlated with diversity and abundance of songbirds. Conversely, in fragments without coyote activity, there were increases in mesopredator activity and lower nesting success by songbirds (Crooks and Soulé 1999).

Competitor

Coyotes may impact other mammalian predators indirectly through competition or directly through intraguild killing or predation. Coyotes have consistently been reported to kill fox species or otherwise influence their distribution (Gosselink et al. 2003; Kamler et al. 2003; Farias et al. 2005; Thompson and Gese 2007). Coyote predation on domestic cats may be a combination of true predation and a response to intraguild competi-

tion. Coyotes killing cats in Tucson were apparently consuming them, but there was no diet analysis reported for that pack (Grubbs and Krausman 2009).

Whereas the relationship between coyotes and cats is relatively well accepted, for other mesopredators, it is less so. Crooks and Soulé (1999) documented mixed results in correlations between coyotes and other mesopredators, and there is little evidence to date that coyotes in urban areas influence the population dynamics or behavior of raccoons (Gehrt and Prange 2007) or striped skunks (Prange and Gehrt 2007; see chapter 14 for more details).

Predator

Considerable research attention has been devoted to coyote impacts on certain prey and competitor species

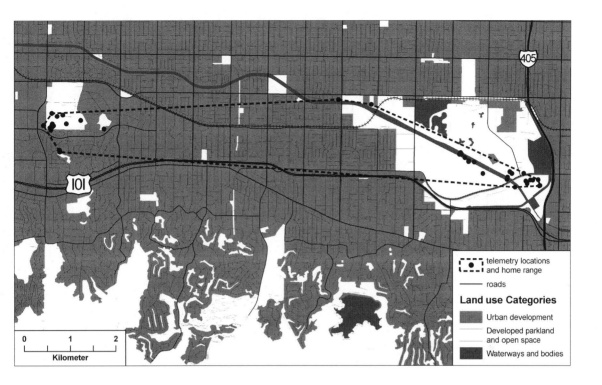

Table 7.3. Diet studies from scat analysis for suburban and urban coyotes, expressed as percentages (occurrence or frequency of occurrence) for selected diet items

Study site	No. scats	Natural items			Human-related items					Source
		Leporid	Rodent	Deer	Cat	Trash	Pet food	Domestic fruit	Other	
San Diego, CA	97	14	8	0	2	16	0	13	11	MacCracken 1982[a]
Los Angeles, CA	250									Fedriani et al. 2001[a,b]
Dry	137	13	39	3	<1	13		11	1	
Wet	113	16	42	2	1	5		3	5	
Los Angeles, CA	238									NPS, unpublished data (collected 2001–2003)[a]
Core areas	70	13	24	8	0	6	0	20		
Fragments	168	17	20	2	1	4	2	22		
Tucson, AZ	667	22	28	3	1	0	0		35[c]	McClure et al. 1995[d]
Chicago, IL	1,429	18	42	22	1	2	0	23		Morey et al. 2007[d]
Cleveland, OH	50	20	28	20	0	0	0	0	2	Cepek 2004[?]
Cleveland, OH	944	8	27	24	<1	0	0	17	0	Bollin-Booth 2007[?]
Albany, NY	274	40	14	34	<1		0	0	<1	D. Bogan and R. Kays, unpublished data[d]

Note: Cat refers to domestic cat.

[a] % occurrence of the diet item (no. occurrences of diet item / total no. occurrences of all diet items).

[b] This sample is from the most urbanized site in their study, which also contained a large trash dump.

[c] Did not discriminate among human-related items.

[d] % scats with diet item (no. scats with item / total no. scats analyzed).

in natural or in rural systems, and those results may provide useful predictions for urban systems. However, ecological relationships may be altered within urban landscapes with extreme fragmentation and high levels of human activity. For example, coyotes are the primary predator of Canada goose nests in the Chicago metropolitan area, typically taking nearly half of the nests in the region and taking the occasional adult attempting to defend a nest. Population modeling suggested that the nest predation by coyotes may be high enough to serve as a biocontrol for limiting the population growth rate of the urban population of Canada geese (Brown 2007). This urban situation stands in contrast to the relationship between coyotes and waterfowl nesting in rural areas, where waterfowl nesting success is positively related to the presence of coyotes (Sovada et al. 1995). In rural systems, waterfowl nests are likely less concentrated, and coyote territories are probably larger than in urban systems. Consequently, coyote control of foxes is thought to reduce the predation pressure on those nests in rural landscapes.

In rural areas, coyotes can be important predators of white-tailed deer fawns (Whittaker and Lindzey 1999), and they may continue this role in urban systems. Fawn survival was monitored in two urban parks in the Chicago metropolitan area, and in each case, coyotes were the primary predator, with predation rates of 20%–80%. Given the challenges of managing white-tailed deer in urban areas, coyote predation may be seen as a welcome aid for resource managers (Gehrt 2006). Likewise, coyotes may assist cemetery and golf course managers in controlling woodchucks and ground squirrels. There has been no rigorous evaluation of the possible limiting role coyotes may have on ground squirrels, particularly in urban areas. However, there is anecdotal support for such a relationship, specifically testimonies by various managers over the years. In addition, Henke and Bryant (1999) demonstrated the limiting role coyotes may have on rodents in a rural area in Texas.

It is important to emphasize the lack of information on the implications of coyotes asserting their role as top predator in urban systems. Evidently, coyotes in urban areas do not simply eat garbage and harass pets but are actively engaged in a predatory role in sometimes unpredictable ways. That coyotes are able to perform their role largely unnoticed by the public, despite the predation acts often occurring literally in their backyards, is amazing.

MANAGEMENT AND CONFLICTS

Management of coyotes in urban landscapes inevitably focuses on conflict, which has been the case for more than a century in rural systems in the western United States. Coyotes differ from most other urban wildlife species in that they can be deemed worthy of removal simply by being seen, rather than after they have caused some damage or inconvenience for human residents. Fear often dictates the public's response to the presence of top carnivores in developed areas. For example, homeowners in the Chicago metropolitan area ranked coyotes as the wildlife species perceived as the greatest threat to human health and safety (Miller et al. 2001), despite relatively few attacks on pets and no record of any attacks on people in that area. Coyotes illustrate quite well that management action dictated by fear is often the price carnivores pay for occupying the top predator role in areas occupied by people.

Threat to Pets

Free-ranging domestic cats seem particularly vulnerable to coyotes, although few studies have actually quantified the extent of coyote predation on cats. In residential areas where cats are presumably more available than in other landscape types, there is a slight trend for cats to appear more frequently in the diet of coyotes (Quinn 1997a; Morey et al. 2007), although diet studies may present a misleading picture of cat predation. Coyotes' frequency of killing may be substantially higher than the presence of cat remains in the diet if coyotes do not always eat the cats they kill, which is often the case with, for example, kit foxes killed by coyotes (Cypher et al. 2001). In the CCSA, we have frequently observed the carcasses of domestic cats apparently killed by coyotes but not always consumed. Thirty-six direct observations of coyote-cat interactions in Tucson, Arizona, revealed that half of the interactions resulted in successful attacks on cats, and coyotes consumed cats at least to some extent after all observed fatal encounters (Grubbs and Krausman 2009).

Although most pet loss involves cats, urban coyotes also occasionally attack and kill dogs. Coyotes appear to attack small dogs most often, but rarely a pair or group will attack larger dog breeds. It is likely these attacks result from competition as well as predation and are an expression of the competitive response that coyotes exhibit toward smaller foxes. Dogs may be attacked at any time of year, but incidents may peak during the mating season and during the rearing of pups. These patterns are consistent with resident, territorial coyotes' dog attacks rather than solitary individuals. For example, a survey of press accounts of coyote attacks on dogs within the Chicago metropolitan area yielded 60 incidents during 1990–2007. A peak of attacks occurred during December–February when mating typically occurs, and another peak fell during April, which corresponds to parturition (S. D. Gehrt and L. A. White, unpublished data). Small breeds were most common in the sample, including Jack Russell terriers (5) and Shih Tzus (6), but larger breeds included a Dalmatian (1), Golden retrievers (3), and a Shetland sheepdog (1). Large breeds are usually attacked by two or more coyotes, whereas the smaller breeds can be taken by individual coyotes (figure 7.7).

Threat to People

Unfortunately, coyotes have been documented attacking people, resulting, in some cases, in injuries and in the death of a young child in 1981 (Howell 1982). The actual number of attack incidents is unknown owing to the lack of a national standardized reporting system and the difficulties of ascertaining which incidents constitute an actual attack. Most reported attacks have occurred in Southern California, and these reports have been investigated by Howell (1982) and Timm et al. (2004). They suggested the frequency of attacks was increasing as coyotes became established in metropolitan areas; however, they did not consider different causes for attacks (e.g., pet attacks, disease, or physical condition). Indeed, while Timm et al. (2004) provides some useful information on the details of

some attacks, their work is of limited value because of inconsistent interpretations of attacks and inaccurate information regarding the ecology and behavior of urban coyotes.

White and Gehrt (2009) classified 142 U.S. and Canadian reports of coyote attacks on 159 victims into the following behavioral categories: predatory (37%), investigative (22%), pet related (6%), defensive (4%), and rabid (7%). They determined that victims of predatory attacks were primarily children, whereas the victims of nonpredatory attacks (defensive, investigative, rabid cases) were typically adults. More attacks occurred during pup-rearing season (May, June, July), but this was apparent for predatory attacks; investigative attacks (typically a minor bite or nip of an unsuspecting victim) were likely to occur at any time. Attacks were equally likely to occur during day and night, and most attacks occurred on residential property. More information is needed on the characteristics of coyotes that attack pets or people to determine whether attacks are likely to occur with coyotes of specific sex, age, or social status.

Coyotes seem more likely to attack people and pets after some level of habituation, although this has yet to be definitively established. As we have seen, coyotes in urbanized areas avoid human activities by shifting to nocturnal activity and avoiding residential areas within their territories. Coyotes do not seem to have an inherent attraction to human activity, but this may change with habituation. Unfortunately, it is difficult to determine the factors that lead to habituation or attacks, but the intentional or unintentional feeding of wildlife likely plays an important role in altering

Fig. 7.7. A coyote captures (*left*) and kills (*right*) a domestic dog (Chihuahua). Photos by Sean Gonzalez.

coyote behavior. During the Chicago study, five coyotes (of 150 radio-collared) became nuisances, and in each case, they were taking advantage of food associated with people and their yards. In one case, bird feeders attracted a coyote as well as squirrels and other rodents. Another possible source of habituation may be unintentional positive reinforcement of increasing aggressiveness in coyotes; that is, when coyotes begin to approach areas with human activity, people often avoid them or leave the area, lending dominance to the coyote.

Successful management of coyotes in metropolitan areas will involve public education to reduce the opportunity for habituation as well as to interpret coyote behavior correctly and to discriminate between threatening and nonthreatening behaviors. Indeed, some educational programs, such as the Stanley Park Ecology Program (Vancouver, British Columbia), have successfully reduced coyote conflicts through a combination of harassment of coyotes and public education (www.stanleyparkecology.ca). The Stanley Park Ecology Program uses various noisemakers, such as coffee cans filled with coins, and has volunteers to chase coyotes that occur too frequently in yards. However, much more research is needed to determine the effectiveness of harassment techniques. Preliminary data from harassment treatments in the Chicago study have showed promising results, in which coyotes quit using yards immediately after motion-sensitive lights or noisemakers were put in place. But questions remain as to how long the harassment effect persists, how much variation in response occurs among individual coyotes, whether relative effectiveness differs among different devices, and most interestingly whether there is a level of habituation beyond which no harassment will change coyote behavior.

When coyotes are habituated and exhibit increasingly aggressive behavior, without better information regarding harassment, it is usually necessary to remove them before people, especially children, are injured. Timm et al. (2004) presented a sequence of increasing aggressive behavior by coyotes. However, their sequence was based on a survey of attack incidents and not on evaluations of individual coyote behavior. Individual coyotes may, or may not, follow their sequence of increasing aggression. Moreover, a number of the phenomena listed in their sequence, such as coyotes taking advantage of pets, trash, or backyard fruit as food sources or being active during the day, often reflect typical behavior for omnivorous, opportunistic,

and flexible predators such as coyotes. The actual likelihood of such behaviors leading to advanced stages of aggression is unknown, and in fact, it is likely that most coyotes that exhibit these behaviors never behave aggressively toward people.

In all cases, coyotes that lose their fear of humans and become threatening should be lethally removed through trapping or sharpshooting by professionals. The appropriate method of lethal control will depend on a combination of local terrain, human density, and legal restrictions, such as municipal codes and state wildlife regulations. Aggressive coyotes would include those that fail to run from people and that growl or bark when approached. Obviously, coyotes that attack pets in the presence of people are unusually aggressive.

Relocation of nuisance coyotes is often a preferred alternative by the public because it can allay conflict between those demanding control and those opposed to it. However, relocation of habituated coyotes will probably not change the behavior of the coyote, so the problem is moved to another area. The movement of individuals also may facilitate disease transmission and intraspecific conflict, and translocation of coyotes frequently results in mortality within days of release. Nevertheless, translocation remains a preferred alternative for some municipalities in response to local sociopolitical conflicts regarding lethal control.

As with livestock protection programs (Blejwas et al. 2002), removal programs are likely most cost-effective if they target offending individuals, rather than attempting a general removal of the local coyote population. For example, during the Chicago study, a resident adult male from an urban park became a nuisance by moving into adjacent neighborhoods at night and killing domestic animals. We never observed a member of his pack leaving the park and accompanying him into the neighborhood. After he was removed, no member of that pack (including the subsequent alpha male that replaced him) was recorded foraging outside the park for at least 5 years following the removal.

CONCLUSION

As evidenced by generally small home ranges and perhaps relatively high densities and survival, coyotes are one of the most successful urban carnivores in North America, especially among the larger carnivores. However, coyotes generally use and prefer natural or remnant habitat fragments, even in highly developed areas. Although coyotes will consume anthropogenic foods, they typically represent a small portion of the diet, indicating that urban coyotes rely on nonhuman food items. Coyotes come into conflict with people by attacking pets, and even people, although based on the available studies, most coyotes in urban areas show no tendency to behave aggressively toward humans. Of 260 coyotes radio-tracked in the Chicago (CCSA) and Los Angeles (VCSA) area studies, to our knowledge, none exhibited aggression toward humans, and five (1.9% overall, 3.3% in that study) from the Chicago study became nuisance animals. Summaries of coyote-human interactions do not necessarily inform us of urban coyote ecology and behavior—they provide limited information about coyote-human interactions. Conflicts may become more common overall as coyotes colonize more urban areas. Therefore, education about not feeding urban coyotes and maintaining coyotes' fear of humans, combined with safe and efficient lethal removal of habituated coyotes, will be important to minimize conflicts.

ACKNOWLEDGMENTS

We appreciate the thorough comments from two reviewers that improved the manuscript. The Cook County, Illinois, research was supported by Cook County Animal and Rabies Control and the Max McGraw Wildlife Foundation. Most of that work also involved close collaborations with multiple agencies and people but especially Chris Anchor of the Forest Preserve District of Cook County and Dr. Dan Parmer (deceased) and Dr. Donna Alexander of Cook County Animal and Rabies Control. The research would not have been possible without the dedication and hard work of numerous students and technicians. We thank Santa Monica Mountains National Recreation Area, the Los Angeles County Sanitation District, and Canon U.S.A. Company for funding the Ventura County study. Countless technicians, interns, volunteers, and students provided invaluable help with fieldwork and analyses for these studies; we would especially like to thank K. Asmus, C. Bromley, S. Kim, S. Ng, P. Roby, C. Schoonmaker, L. Tigas, and J. Sikich for Ventura County. The late E. York provided superior coyote capture work, good humor and guidance, and boundless energy to the Ventura County study.

8

Striped Skunks and Allies (*Mephitis* spp.)

RICHARD ROSATTE

KIRK SOBEY

JERRY W. DRAGOO

STANLEY D. GEHRT

WITH THEIR striking black-and-white coloration, striped skunks (*Mephitis mephitis*) are perhaps one of the most recognizable carnivores in urban areas (figure 8.1). If they are not seen, their presence is occasionally detected by an odor that most people easily recognize. That skunks are frequently cited as nuisance animals in most North American cities indicates a certain level of success in urbanized areas, but there have been few urban studies on striped skunks and even less research directed toward their close relatives.

Comprehensive urban studies on striped skunks have only occurred in metropolitan Toronto, Ontario, and Chicago, Illinois. Metropolitan Toronto, which includes the city of Scarborough, has a human population of about 3 million. During a study in that urban complex (1984–1997), more than 2,000 skunks were captured of which 57 were fitted with radio-collars. Chicago has a human population of 9.1 million. During the 1999–2005 study, 146 skunks were captured and 90 were radio-collared. In this chapter, we summarize what is known about urban skunks based on these studies. However, given the paucity of research, it is evident that there is still much to learn about how skunks respond to urbanization.

SPECIES DESCRIPTION AND LIFE HISTORY

Striped skunks are medium-sized carnivores with a characteristic black-and-white-striped pelage and an odorous musk. Striping patterns vary, including a narrow stripe, a broad stripe, two stripes, or a short stripe.

Figure 8.1. Characteristic black-and-white pattern of a striped skunk. Photo by J. Dragoo.

Coloration is highly variable within the species and can range from almost completely black to almost completely white (nonalbino). This unique and noticeable coloration serves as a warning to most mammalian predators. Because of their short legs, skunks appear to waddle as they walk and are generally poor climbers. However, they have strong forefeet and long, curved claws, which make them excellent diggers. Skunks have a keen sense of smell and excellent hearing, but their eyesight is considered poor (Godin 1982; Rosatte 1987). They also have two scent glands in the anal area and are capable of expelling odorous musk for several meters. The smell is derived from seven major and several minor sulfur compounds called thiols and thiol acetates (Wood 1999).

There is considerable individual and geographic variation in size and in weight of striped skunks, although adult males tend to be larger than adult females in total body length (520–770 mm for both) in both urban and rural habitats of North America (Verts 1967; Wade-Smith and Verts 1982; Rosatte et al. 1991). Adult skunks generally weigh 1.8–4.5 kg (Verts 1967; Rosatte et al. 1991). In Toronto, Ontario, mean fall (October/November) weights of adult male and female striped skunks were 4.5 kg and 3.3 kg, respectively, with juveniles averaging 3.0 kg (Rosatte et al. 1991). In general, Toronto skunks were 1.0–1.5 kg heavier than their rural counterparts (Rosatte et al. 1991). A similar pattern occurred in a Chicago area park, where there was a trend for heavier skunks in the urban area relative to a rural population (Gehrt 2005). Whether heavier weight in urban areas indicates access to anthropogenic resources remains to be determined.

Skunks are not capable of true hibernation but rather become inactive in ground dens during ex-

tended periods of extreme winter weather at northern latitudes. Consequently, skunks typically experience severe weight loss over the winter (14%–65% of their summer or fall body weight; Hamilton 1937). In Toronto, striped skunks sampled in the spring had lost 37%–41% of their fall weights (Rosatte et al. 1991), whereas in Chicago, over-winter weight loss of male and female striped skunks was 19% and 28%, respectively (Gehrt 2005). In the rural portion of that study in Illinois, male and female skunk weights declined 41% and 32%, respectively, over winter (Gehrt 2005). Substantial weight loss, parasite loads, and resultant poor body condition may compromise skunk survival during winter and early spring (Gehrt 2005).

Because of a relatively short life span (Verts 1967), turnover in skunk populations is substantial, resulting in a large percentage (65%–79%) of populations represented by juveniles (Rosatte et al. 1991). Sex and age ratios for skunks in urban areas are highly variable, but in Toronto, 67%–74% of the skunks captured from July to November 1985 were juveniles. Similarly, during 1989, 79% of the skunks captured in Scarborough, Ontario, a suburb of metropolitan Toronto, were juveniles. By comparison, about 65% of the skunks captured in rural Ontario habitats were juveniles (Rosatte et al. 1991).

The life cycle of striped skunks in cities such as metropolitan Toronto includes a breeding period during February and March, with parturition occurring in April, May, and June (Rosatte et al. 1991). Striped skunks are monestrus, with breeding in other urban and rural areas generally occurring from mid-February to mid-April (Wade-Smith and Richmond 1978a, 1978b; Greenwood and Sargeant 1998). Juvenile skunks can be sexually active during their first year of life (Schowalter and Gunson 1982) and will even "practice" breeding behavior when they are about 3 months old. Male striped skunks are capable of breeding with several females during the reproductive period. In Toronto, 87% of sampled skunks examined during June and July were lactating or had suckled young. By comparison, about 75% of skunks in rural Ontario had successfully bred, suggesting reproductive success may be higher in urban skunks (Rosatte et al. 1991). In rural Ontario, mean litter size for striped skunks was 4.8, and 5 skunks were found in one maternal den in Toronto (Rosatte et al. 1991).

Male skunks remain solitary after the breeding period and attempt to rebuild fat reserves, while females are restricted to the vicinity of their maternity den to care for the young (Verts 1967; Larivière and Messier

1998; Rosatte and Larivière 2003). Young skunks are weaned at about 6–8 weeks of age and may disperse during July through October. Depending on winter severity in northern latitudes, skunks usually use winter dens from late October/November until February/March (Rosatte et al. 1991; Gehrt 2004a).

Survival/Mortality

The life span of striped skunks is normally 2.0–3.5 years (Verts 1967). However, captive (pet) striped skunks have been reported to live considerably longer, averaging 8–12 years or longer. In urban areas, skunk mortality is typically high from a variety of factors, including collisions with vehicles, disease, euthanasia by wildlife control companies, and predation by domestic dogs, coyotes, and red foxes (Rosatte et al. 1991; Rosatte and Larivière 2003; Gehrt 2004a). Urban and rural striped skunks also may die of starvation due to substantial weight loss during severe winters (Sunquist 1974; Rosatte et al. 1991; Gehrt 2005). Juvenile skunks have become severely anemic and died because of flea infestations (J. W. Dragoo, unpublished data). Skunks in both cities (Rosatte et al. 1992a; Schubert et al. 1998) and in rural areas (Davidson and Nettles 1997; Rosatte et al. 2007a, 2007b, 2007c) often die from diseases such as rabies and canine distemper, but the impact on those populations is largely unknown. Gehrt (2005) suggested that in the absence of rabies, a variety of pathogens may be the primary cause of mortality in urban and rural skunk populations in and around the Chicago area.

Mortality rates for striped skunks in urban areas are highly variable, both among areas and among years. In Scarborough, Ontario, annual mortality during a 2-year period was 54% and 34% (Rosatte et al. 1991). In another study, 39% of radio-collared skunks in metropolitan Toronto succumbed during the first year (Rosatte et al. 1991). In the Chicago study, survival of striped skunks declined dramatically during the winter and early spring, but there was no difference detected in survival between skunks in rural and urban habitats and few differences in cause-specific mortality. Survival of males was less (0.36) than for females (0.41) when area data were pooled (Gehrt 2004a, 2005). In that study, disease or poor condition was the most common cause of mortality in both urban and rural populations, although this mortality factor was relatively higher in the urban population. Interestingly, both populations had similar rates of mortality due to vehicles despite dramatic differences in road traffic. The urban study

area was bounded and bisected by roads with 4 to 10 lanes of traffic (Gehrt 2004b), yet vehicle mortality rates were <10% and not different from that of a rural population exposed to a few roads and little traffic. Predation mortality rates did not differ between urban and rural populations in the Chicago study. Potential predators of skunks in the urban park included occasional owls, red foxes, and a high-density coyote population (Gehrt 2004b, Prange and Gehrt 2007); nevertheless, predation rates were <5% during 3 years of monitoring (Gehrt 2005), and only 4% of 106 radio-collared urban skunks died from predation during the full 6 years of study (Gehrt 2004a). During radio-tracking, skunks were observed repelling coyotes (Prange and Gehrt 2007), and predation did not appear to be an important cause of mortality.

In the Chicago study, juvenile skunks were captured by hand from dens and monitored with glue-on transmitters until they reached a size appropriate for radio-collars. By tracking juveniles during their first few weeks of independence, a different pattern of cause-specific mortality emerged for juveniles compared with adults or even older juveniles. Of 38 urban juveniles captured from dens, 5 (11%) were killed by vehicles and 7 (18%) by disease or unknown mortality. Most vehicle-related mortality occurred during summer as juveniles became independent, whereas disease/condition mortality occurred during late winter or early spring, which suggests that young skunks may be vulnerable to roads in urban areas, although there are no comparable juvenile data from rural areas.

Annual survival rates for skunks in Toronto and Chicago were similar to estimates for rural skunks in Minnesota (0.67 for adult skunks and 0.17 for juveniles; Fuller, Berg, and Kuehn 1985) and in Texas (0.40–0.48 during two consecutive years; Hansen et al. 2004).

Habitat Use

In urban areas, striped skunks are found in residential areas, fields adjacent to industrial sites, forested parks, vacant buildings, golf courses, ground burrows, and dumps (Rosatte et al. 1991; Rosatte and Larivière 2003). In Toronto, the highest skunk densities were in field areas during some years, whereas in other years, industrial and commercial sites supported high skunk densities. However, it is believed that fields in proximity to those sites were the draw for skunks.

Large backyards with brushy vegetation are prime skunk habitat in New Mexico. Many of these homes will have a diversity of resources for skunks, including birdbaths, bird feeders, garden hoses, and pet food. Along the Rio Grande Valley in Albuquerque, many residents maintain small flocks of chickens in their backyards. Many of these homes are surrounded by alleyways, fields, and culverts that can be used by skunks to move from one location to another.

Den Types

Striped skunks use many types of dens, depending on the time of year, and den solitarily or communally. Den types include maternal dens during the spring, summer/fall resting sites, and winter solitary and communal dens (Storm 1972; Larivière and Messier 1998; Doty and Dowler 2006). Skunks in urban areas will use any sheltered area, including under buildings, commercial sites, residential porches, and culverts (Rosatte and Larivière 2003). Den sites are usually near areas where readily available foods sources abound. During the summer, skunks may use aboveground resting sites along fence rows, under shrubs, in thickets, along waterways, in shaded areas in tall grass, and in standing crops (Storm 1972; Rosatte 1987; Larivière and Messier 1998; Doty and Dowler 2006). In the late fall and winter, striped skunks in urban as well as rural habitats usually use underground or under-building communal dens (Allen and Shapton 1942; Rosatte 1984, 1987; Rosatte and Larivière 2003). Suitable winter dens are vital to over-winter survival in temperate climates. Consequently, the availability of winter dens may explain why some urban communities seem to have a consistent presence of skunks, whereas it is more variable for other communities, particularly for urban areas at northern latitudes.

Population Density

Densities for striped skunks are highly variable and dependent on factors such as geographic location, habitat quality, season, disease outbreaks, and mortality rates. In most areas, population density is at its highest in the early summer as juvenile skunks become mobile and is lowest during early spring because of winter mortality. Striped skunk densities for urban habitats of Toronto during the summer averaged 2–7 skunks/km² (Rosatte and Larivière 2003). However, field habitats in proximity to industrial buildings in that urban complex supported skunk densities as high as 36/km². Conversely, skunk density in forested-park areas of Toronto was only 1/km² (Rosatte et al. 1991). During a rabies control program in Niagara Falls, Ontario, skunk density in urban habitats was 2.2/km² compared with 0.6/km²

Table 8.1. Skunk density and capture success in urban and rural trapping cells during a rabies control program in Niagara Falls, Ontario, during 2002

Habitat	Trapping cells (*n*)	Skunks / 1,000 trap nights	Skunks / km² mean (95% CI)
Urban[a]	30	9	2.2 (1.7–3.2)
Rural[a]	46	3	0.6 (0.4–0.9)

[a]Size of urban area = 122 km²; size of rural area = 556 km².

in rural habitats during 2002 (R. Rosatte and K. Sobey, unpublished data; table 8.1).

Capture data were used to develop a model that predicted metapopulation structure for skunks in Scarborough, Ontario, during 1987–1996. Ten subpopulations were identified over a 72-km² urban area with average density being 1.7 skunks/km². However, skunk density in productive habitats was 7.2/km² (Broadfoot et al. 2001). Comparatively, densities ranged 1.4–2.1/km² in rural areas of Ontario (R. Rosatte, unpublished data). In northeastern Illinois, there was no significant difference in skunk density between an urban (2.1 to 5.9/km²) and a rural (2.5 to 5.5/km²) population (Gehrt 2005). Reported striped skunk densities for other rural areas have been highly variable, with estimates varying from 0.1/km² to 13–26/km² within and across studies (Rosatte and Larivière 2003; Hansen et al. 2004). It appears that skunk density is highly variable across both urban and rural landscapes.

Home Range and Movements

Home range and movements for striped skunks in some urban areas are fairly restricted in area. In Toronto, Ontario, home range estimates for striped skunks averaged 0.51–0.64 km² (Rosatte et al. 1991). The small home range size was attributed to an abundance of food and denning sites in this urban environment. In rural southern Ontario, home range estimates for striped skunks were larger than those of urban skunks and varied between 1 and 3 km² (Rosatte and Larivière 2003). Skunks also had extremely small ranges of 0.08 to 0.5 km² in the Presidio of San Francisco (Boydston 2005). However, Gehrt (2004a, 2004b) found no difference in home range size between urban and rural skunks in Illinois, with urban ranges of about 1–1.7 km². No sex and age class differences in summer or in winter home ranges were noted in the urban study area, but extensive spatial overlap of ranges was observed (Gehrt 2004a). In that study, the movement patterns and spatial organization of skunks appeared relatively unaffected by extremely abundant anthropogenic foods, especially when compared with the dramatic home range reduction and spatial congregation of co-occurring raccoons in that park (Gehrt 2004a, 2004b; Prange and Gehrt 2004). The home ranges and movement patterns of skunks in the Chicago study appeared to be affected by roads more than human food. Most skunks (>50%) did not cross roads with extensive traffic (Gehrt 2004a, 2005). This may explain why vehicle-related mortality did not increase with urbanization in that study. In other rural habitats of the United States and Canada, home range estimates have ranged between 1.0 and 12.0 km² (Rosatte and Larivière 2003). Bixler and Gittleman (2000) found that home range size of striped skunks in rural Tennessee was influenced by season and body weight, and ranges were larger in the spring than in winter.

Generally, striped skunks are fairly sedentary animals, especially in urban environments. For example, the majority (98%) of skunk movements in one metro Toronto study area (the city of Scarborough) were <1 km annually; 2% moved 2–5 km (Rosatte et al. 1991, 1992b; figure 8.2). In another study, movement distances of skunks recaptured in urban and rural trapping cells in Niagara Falls, Ontario, were different from 1997 to 2003. The majority (>93%) of skunks recaptured in urban Niagara trapping cells had moved less than 1 km, similar to movement distances observed for urban areas in Toronto (R. Rosatte and K. Sobey, unpublished data; figure 8.2). In comparison, only 82% of skunks recaptured in rural Niagara Falls trap cells had moved less than 1 km and nearly 15% had moved distances between 1 and 5 km. The remaining 3% of the recaptured skunks in rural cells had moved distances >5 km (figure 8.2). Comparatively, most skunk movements in rural habitats are 1–3 km annually (Bjorge et al. 1981; Rosatte and Gunson 1984a; Larivière and Messier 1997a), with some exceptional movements of 70 to 119 km being documented (Sargeant et al. 1982; Rosatte and Gunson 1984a; Rosatte and Larivière 2003).

Diet

Both urban and rural striped skunks are opportunistic omnivores. In general, skunks feed primarily on insects, such as beetles, grasshoppers, and crickets but will also consume earthworms, snakes, snails, clams, crayfish, frogs, mice, voles, moles, rats, squirrels, wild fruits, grains, corn, nuts, bird's eggs, carrion, and garbage (Hamilton 1936; Verts 1967; Rosatte 1987; Greenwood et al. 1999). However, to our knowledge, there are no published studies of striped skunk food habits in urban areas. Nevertheless, it is evident that skunks continue to forage for invertebrates in developed areas by the damage they cause. They often dig up lawns looking for grubs and they will invade gardens for ripe fruit and vegetables but mostly for the insects associated with garden plants. Skunks also are known to forage on fallen fruit around trees, such as apple, pear, and cherry trees. Outdoor feeding of household pets (dogs and cats primarily) can also attract skunks. Skunks located at a pet food bowl will screech and fuss at one another, trying to bump other skunks out of position; they will even stand in the food to prevent access by others.

An important issue for understanding the urban ecology of skunks is to what extent they, as typical omnivores, adjust their foraging in response to human foods. Rigorous comparisons of diets between urban and rural skunks have not yet been published, although skunks are known to consume garbage, and skunks have been observed consuming pet food placed by residents for wildlife feeding. However, skunks were not observed altering their foraging in response to abundant, highly clumped anthropogenic food resources at the urban park in Chicago, in stark contrast to co-occurring raccoons (Gehrt 2004b). Skunks foraged near picnic groves with trashcans, but visual observations while radio-tracking indicated that they were primarily foraging in mowed areas near the groves rather than seeking garbage cans. Alternatively, Hamilton (1936) speculated that skunks near villages weighed more than rural skunks because of the amount of garbage they consumed.

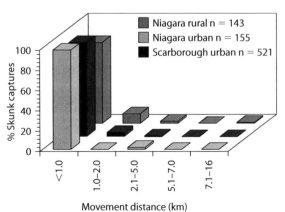

Figure 8.2. Observed distances moved by striped skunks in urban and rural trapping cells in the Niagara Falls, Ontario, area (1997–2003) and in Scarborough, Ontario (1984–1989).

Behavior

Striped skunks are generally considered to be crepuscular or nocturnal (Rosatte and Larivière 2003), although the influence of urbanization on skunk activity patterns has yet to be determined. Skunks may occasionally be seen during the day (Larivière and Messier 1997), especially late summer when the juveniles begin to explore their surroundings. Skunks are generally lethargic but are capable of traveling at speeds of 10–15 km/h (Verts 1967; Godin 1982; Rosatte and Larivière 2003). Rosatte discovered just how fast skunks can travel during a radiotelemetry study in metro Toronto during 1983. While chasing a radio-collared skunk, he fell from a 3 m fence and broke his femur in four places. He spent 2 months in traction in the hospital because of this experience and vowed never to try to outrun a skunk again.

Skunks are normally nonaggressive and docile and often flee if a human approaches (Verts 1967; Rosatte 1987; Rosatte and Larivière 2003). However, if provoked, a skunk will stand its ground and make a series of noises. They will take a defensive posture, often arching the back, erecting the tail hairs, chattering their teeth, and stomping the front feet while shuffling backward. Juveniles will practice many of these behaviors during play when they are outside the den. They may, as a defense mechanism, discharge their musk (3–5 m) at their adversary if a threat persists (Verts 1967; Godin 1982; Larivière and Messier 1996b). During his career, Rosatte has often witnessed this behavior. While trapping skunks in southern Alberta during the early 1980s, he was ejected from a Lethbridge motel for "stinking up the place" and had to sleep in his truck for the night.

Behavior of skunks during the summer and fall includes foraging to accumulate fat reserves for sustenance during the winter denning period in northern latitudes. A peculiar behavior of skunks while foraging is that they will roll caterpillars on the ground to remove the hairs before consuming them. Toads are rolled on the ground to remove poisonous skin secretions, and skunks will also roll beetles that emit a defensive scent. Rolling causes the beetle to deplete its scent before being consumed. Skunks will scratch on a beehive and snag bees as they leave the hive. They will defecate and return to their feces to capture ground beetles. They will open bird eggs by throwing them with their front feet between their back feet until the egg breaks on a hard surface. In rural habitats, skunks have been observed stalking and flushing insects from crops, and they commonly walk slowly to flush grasshoppers from piled alfalfa. Adult skunks are adept at this foraging strategy, but it is a learned skill. Grasshoppers have been observed crawling out from behind juvenile skunks that had pounced unsuccessfully on the insect.

Although skunks are not particularly social, striped skunks will den together communally during the winter (Rosatte 1984a; Gehrt 2004a; Hwang et al. 2007). Rosatte found a domestic cat huddled among a group of skunks under a granary in southern Alberta, trying to keep warm in –20°C weather; although the skunks were not too social, the cat was attempting to be. In the Chicago study, some skunks began denning communally in October, with an increase during November, peaking during winter months and declining through March. During the peak of communal denning, nearly all radio-collared females denned with at least one other skunk, and most males shared dens. Although the overall frequency of communal denning for adult males was lower than for other skunks, some males denned extensively with females and with juvenile males on occasion. However, no adult males shared dens with other adult males. Skunks denned with as many as 11 different individuals during a winter, and dens contained one to seven skunks simultaneously (Gehrt 2004a). Although few members of family units denned together in winter, many skunks sharing dens were nevertheless related (e.g., half-sibs or third-order relatives). Communal denning has not been observed in more southern regions of the range of striped skunks (Hass 2003; Doty and Dowler 2006).

During winter denning, skunks do not hibernate but merely go into a deep sleep and are sometimes active, especially during a mild spell (Godin 1982). They usually become active in January and February to prepare for breeding, when male skunks seek out females at den sites (Rosatte and Larivière 2003). Male striped skunks may defend their harem of females by hitting another male skunk with their shoulders or biting their legs (Rosatte 1987).

In the Chicago study, skunks shifted to a solitary resting system during spring, and parturient females denned alone except with their young. Notably, familial bonds dissipated quickly after the young became mobile and moved from natal dens. Data obtained for 24 mother-offspring dyads (from 8 litters) and 25 sibling dyads (from 6 litters) revealed that nearly all skunks were completely solitary by the end of July (Gehrt 2004a).

Diseases and Parasites

Determining how urbanization affects the host-parasite relationship for skunks is poorly understood. Striped skunks are known to harbor a multitude of ectoparasites and endoparasites in both urban and rural habitats. They are usually infested with fleas, lice, ticks, and mites (Webster 1967; Davidson and Nettles 1997; Gehrt 2005). Striped skunks have also been reported with various helminth infestations (Neiswenter et al. 2006). Larvae and eggs of the intestinal roundworm (*Baylisascaris* spp.) of skunks may be a potential health risk to humans (Davidson and Nettles 1997). Lung flukes (*Paragonimus kellicotti*) have also been recovered from skunks in rural Ontario (Ramsden and Presidente 1975). In Albuquerque, skunks are known to harbor intestinal worms such as *Physaloptera maxillaris* and *Macracanthorhnchus ingens* (J. W. Dragoo, unpublished data). Lungworms (*Filaroides* spp.) also have been found in striped skunks, are also host to coccidia (*Eimeria mephitidis*), a parasitic protozoan.

In the Chicago skunk population, prevalence and intensity of macroparasite infections were notable and may have affected the population dynamics of the skunks. Infection rates were not different between the urban and rural study populations (S. D. Gehrt, unpublished data). Individual skunks were infected with an average of 4.1 species of endoparasites (range 1–11, n = 37); the most common parasites are *Sarcocystis neurona* (78%), *P. maxillaris* (68%), and *Baylisascaris columnaris* (57%). However, 35% of skunks had moderate to severe infections from one or both of *P. maxillaris* and *Skrjabingylus chitwoodorum*, to the point that the parasite loads directly or indirectly contributed to mortality.

Skunks are a primary vector of rabies in some urban and rural areas of North America (Rosatte and Gunson 1984b; Pybus 1988; Gremillion-Smith and Woolf 1988; Greenwood et al. 1997; Krebs et al. 2003; Leslie et al. 2006; Hass and Dragoo 2006; Rosatte et al. 2007a, 2007b). Skunks account for about 20% of the annual rabies cases in the United States and Canada (Rosatte 1988; Rosatte and Larivière 2003). Striped skunks have also been found to be infected with adiaspiromycosis, histoplasmosis, leptospirosis, listerosis, pulmonary aspergillosis, sarcocystis, streptococcus, toxoplasma, and tularemia infections (Albassam et al. 1986; Hwang et al. 2002; Dubey et al. 2003, 2004). Serological evidence also suggests that striped skunks have been exposed to West Nile virus in several states (Bentler et al. 2007). Antibodies to St. Louis encephalitis and Powas-san virus were also detected in a sample of skunks in Ontario (Artsob et al. 1986).

Serologic results for urban and rural skunks in the Chicago area study resulted in 55% positive antibody response for canine distemper, 82% for canine parvovirus, 60% for toxoplasmosis, and 17% for *Leptospira interrogans* serovars, with no difference between populations. Indeed, most skunks in the Chicago study were observed to be suffering from multiple maladies by the time they emerged from their first winter dens or before they reached 1 year old.

SOCIOECOLOGICAL AND ECONOMIC ISSUES

As skunks are a primary vector of rabies, in both urban and rural habitats of North America, they are important economically because of costs associated with human rabies treatments, pet vaccinations, and disease control (Rosatte et al. 1986; Rosatte 1988; Rosatte and Larivière 2003; Rosatte et al. 2007a, 2007b). Striped skunks are also considered economically valuable by the garment industry, and thousands are trapped each year in both urban and rural areas of North America (Novak et al. 1987; Rosatte and Larivière 2003). On another positive note, striped skunks consume insects, including June beetles (*Cotinis nitida*), budworms (*Heliothis virescens*), scarab beetles (Scarabaeidae), potato beetles (*Leptinotarsa decemlineata*), armyworms (*Cirphis unipuncta*), cutworms (*Chorizagrotis* spp.), and moths (Sphingidae), making them an important asset to residential property owners in urban and rural areas (Godin 1982; Rosatte and Larivière 2003).

Unfortunately, skunks occasionally damage lawns because they dig for insects and earthworms with their claws (Rosatte 1987; Rosatte and Larivière 2003). They also may dig where fertilizers such as fishmeal, bonemeal, or blood meal have been applied to lawns and gardens. Striped skunks are also capable of having significant impact on bird populations, especially waterfowl, through the destruction of nests (Larivière and Messier 1998). Their musk can also represent a conflict with people, especially when pets or people are sprayed.

In some states, it is legal to own skunks as pets, but these are usually domesticated breeds raised on commercial farms and have been bred for various color patterns. Some states will only allow pet skunks to be non-black and white to distinguish them from wild skunks. In states where it is illegal to keep skunks as pets or where it is illegal to rehabilitate skunks, many citizens will take matters into their own hands when they find orphaned skunks. These people will try either to raise the animal to release or to keep as a pet (regardless of legal issues).

CONSERVATION AND MANAGEMENT

Historically, striped skunks were managed as a nuisance in both urban and rural areas of North America, primarily through trapping, shooting, and poisoning (Rosatte 1987; Pybus 1988; Rosatte et al. 1986, 1992a, 2001, 2007a, 2007b). Trap-vaccinate-release has been used in urban areas such as Metropolitan Toronto and Flagstaff, Arizona, to control rabies in urban skunks (Rosatte et al. 1990, 1992a, 1993; Engeman et al. 2003). Population reduction has been used also to control rabies in skunks in rural habitats and is part of contingency plans for some urban areas (Rosatte et al. 1986, 1997, 2001). In addition, research is progressing toward using oral rabies vaccine baits to control the disease in skunks (Vos et al. 2002; Hanlon et al. 2002; Rosatte et al. 2007a). Models with habitat variables can be used to predict striped skunk occurrence, as well as for disease management (Carter and Finn 1999; Baldwin et al. 2004; Broadfoot et al. 2001). Roadway underpasses have been used in California to increase connectivity for species such as skunks in urban landscapes (Haas 2001).

Nuisance skunks are a symptom of a problem; they are not the problem. Urban areas are attractive to skunks (figure 8.3). Outside of the breeding season, skunks look for three things: food, water, and shelter. City recreational (parks, golf courses, etc.) and residential areas provide ample water as well as numerous den sites. Plenty of food resources are available from improperly contained garbage and pet food. Recognizing and solving the problem will help eliminate the symptom. As striped skunks are economically valuable because of the insects they consume, public education to inform people of their benefits should be started in urban areas. When skunks are a nuisance in residential areas, proactive efforts can be initiated to prevent skunk entry to buildings with predator-proof fences; sealing off entry holes under porches, buildings, and foundations; and removing trash and rock piles that may serve as den sites. Wire-mesh fencing 1 m in height, with 15–30 cm of the bottom buried underground, is the preferred method for excluding skunks from garden areas. Motion-activated sprinklers can be effective at

Figure 8.3. Jerry Dragoo attempting to remove a skunk from an urban area. Photo by J. Dragoo.

deterring skunks from residential buildings or gardens. Biological insect control in lawns may prevent skunks from digging for insects and chemical insecticides may be used to control grubs in lawns if it is legal to do so but are not recommended because of the toxic effect on the environment. Domestic pets should be confined during the evenings, so they do not come in contact with skunks or their musk and to prevent disease and parasite transmission. Skunk presence around buildings can be reduced through minimizing their access to garbage and pet food (see chapter 15). Other management practices to discourage skunks include removing fruit from trees that has fallen on the ground and controlling the density of rodents and insects around residences.

If a skunk is found denning under a building, it can be encouraged to leave by playing a radio or shining lights under the building. Flour, baking soda, or cornstarch can be sprinkled at the entrance to the den to determine whether the skunk has exited the building (look for track direction). If it is unclear whether all skunks have left the building, a one-way hardware cloth door can be placed over the exit hole. This will allow skunks to push their way out but prevents them from reentering. If that does not work, the first order of business is to seal off all entry and exit holes except one. Once the skunk has left for the evening, the hole can be sealed with wire mesh or other materials. Every effort should be made to ensure all skunks, including young, have been removed before the hole is sealed. Young skunks may be left in the den unattended during May and June; therefore, it is prudent to avoid sealing buildings or foundations at this time to ensure they do not starve. If a skunk is trapped in a Dumpster or window well, a board placed at 45-degree angle should be inserted to allow the animal to climb out. Then, a screen should be placed over the window well to prevent the problem from recurring. Animal removal companies can also be hired to remove skunks from residential properties. The wisest and most economical solution is to fix the problem at the source and coexist with skunks.

SKUNK ALLIES

Striped skunks are 1 of 12 species in the family Mephitidae (Dragoo and Honeycutt 1997; Wozencraft 2005). Other skunk species include the hooded skunk, four species of spotted skunks (*Spilogale*), four species of hog-nosed skunks (*Conepatus*), and two species of stink badgers (*Mydaus*; Dragoo and Honeycutt 1997; Wozencraft 2005). The stink badger occurs in Borneo, Sumatra, and a few other Philippine islands. The other three genera of skunks are distributed over most of the New World, from southern Canada to the Strait of Magellan in South America (Rosatte and Larivière 2003). The hooded skunk (figure 8.4) occurs from southwestern Texas, southwestern New Mexico, and southeastern Arizona, throughout Mexico, into Guatemala, Honduras, Nicaragua, and into Costa Rica (Hwang and Larivière 2001). Spotted skunks (figure 8.5) range from southern British Columbia south through Mexico to Costa Rica and west to south-central Pennsylvania down the Appalachian chain to Florida, west to the Continental Divide, and south to Tamaulipas, Mexico (Kinlaw 1995; Verts et al. 2001). The striped hog-nosed skunks are found throughout South America from the Straits of Magellan all the way north to

Figure 8.5. Western spotted skunk. Photo by J. Dragoo.

Figure 8.6. White-backed hog-nosed skunk. Photo by J. Dragoo.

Veracruz, Tabasco, and Yucatán, Mexico (Wozencraft 2005). White-backed, hog-nosed skunks (figure 8.6) occur from Arizona to the Gulf Coast of Texas south to Veracruz and southwest to Nicaragua. They also have been recorded in the United States as far north as Colorado (Dragoo et al. 2003).

Hooded skunks use similar habitats as striped skunks but are found more often in scrub and urban habitats (Hass and Dragoo 2007). Hooded skunks have previously been reported to avoid human habitation (Patton 1974). However, hooded skunks may be more common in residential areas than once believed (Hass

Figure 8.4. Hooded skunk. Photo by J. Dragoo.

2002). It is likely that they have been misidentified as striped skunks (Dragoo et al. 2004).

Home ranges of hooded skunks are smaller in urban than in rural areas (Hass 2003). Ranges vary from 2.8 km^2 to 5.0 km^2 (Ceballos and Miranda 1986). When females are not nursing, they tend to stay at a den site longer and move shorter distances from one den site to another than males (Hass 2003). Hass (2002) reported hooded skunks translocated from their urban site were able to find their way back home when they were released up to 10 km from their den.

Rabies is rarely reported in hooded skunks (Aranda and Lopez-de Buen 1999); however, Hass and Dragoo (2006) documented a case of rabies in this species. There is no reason to suspect that hooded skunks do not contract the virus, but rather they go unnoticed or misidentified as striped skunks (Dragoo et al. 2004). Hooded skunks have been trapped in areas where other skunks (striped, hog-nosed, and spotted) occur. In urban areas, they likely interact with dogs, cats, and raccoons, much like striped skunks. There is considerable overlap in den sites between striped and hooded skunks in urban areas where they co-occur (Hass 2003).

Data regarding the other skunk species in urban environments are few, but spotted skunks have been removed from houses in uptown Albuquerque, as well as urban areas in other parts of New Mexico (J. W. Dragoo, unpublished data). A hog-nosed skunk was collected in downtown San Antonio, Texas (Dragoo et al. 1989), and these skunks have been diagnosed with rabies (Hass and Dragoo 2006). The distribution of four of these other skunk populations overlap with that of striped skunks. So, they could have an indirect effect in urban areas by interacting with striped skunk populations. As urban environments expand, it is likely that more direct contact with these species will occur.

CONCLUSION

Despite the ubiquitous distribution of skunks and their common occurrence in many metropolitan areas, our understanding of their urban ecology is essentially limited to ecological studies in two metropolitan areas. Compared with other carnivore species known as regular denizens of urban landscapes, many aspects of the urban ecology of striped skunks remain poorly documented. Although striped skunks are commonly observed in developed areas and are frequently removed as nuisances, results have been mixed regarding how skunk populations are affected by urbanization.

In some cases, densities increase and home ranges decrease with urbanization, whereas there is no change with urbanization for other areas. Other ecological aspects, such as survival and cause-specific mortality, do not appear to change. Overall, striped skunks do not appear to respond as dramatically to urbanization as compared with other Carnivora with a generalist lifestyle, such as raccoons (Gehrt 2004b; Prange and Gehrt 2004).

Striped skunks are successful in urban environments compared with other species of skunks, despite euthanization by animal control agents and rabies control personnel, as well as disease, collisions with vehicles, and fur trapping. Other genera of skunks are in decline in some areas. Therefore, increased urbanization may jeopardize the survival of skunk species other than striped skunks. This is critical, as skunks are an important part of both the urban and rural environment. They are vectors of diseases such as rabies, and they are often in conflict with humans. Skunks are also furbearers, as well as ecologically important mesopredators consuming insect pests. As such, long-term data sets for skunks are needed to determine the importance of intrinsic and extrinsic factors on the dynamics of skunk populations (Gehrt et al. 2006). Genetic studies (e.g., Bixler 2000; Hansen et al. 2003) should be initiated on endangered or threatened skunk populations to determine whether there are genetic bottlenecks or low genetic diversity. If there are, relocations and restorations should be considered to improve genetic diversity. Education programs noting the benefits of skunks should be implemented in urban areas to encourage the public to live in harmony with skunks as opposed to viewing them only as nuisances.

9

Eurasian Badgers
(*Meles meles*)

STEPHEN HARRIS

PHILIP J. BAKER

CARL D. SOULSBURY

GRAZIELLA IOSSA

A LTHOUGH THE mustelids are a widespread and adaptable group of mammals, only Eurasian badgers and stone martens have successfully colonized urban areas. The Eurasian badger is found throughout Europe and Asia but achieves highest densities in Britain and Ireland (Griffiths and Thomas 1993; Neal and Cheeseman 1996). For this review, we have treated the Japanese badger as conspecific, although some authorities consider it a separate species (*Meles anakuma*; Wilson and Reeder 2005).

The badger is a medium-sized predator; its body weight varies with the season—highest in autumn and lowest in spring (Neal and Cheeseman 1996)—and location—smaller in the south of its range. Thus, average weight in Donana, Spain, was 7.3 kg for males and 6.9 kg for females (Revilla et al. 1999), whereas autumn weights of males in Russia are up to 30–34 kg (Novak et al. 1962). Average weights in pastoral landscapes in southwest England were 11.1 and 10.2 kg for adult males and females, respectively (Neal and Cheeseman 1996), which is representative of badgers in temperate regions.

Eurasian badgers are omnivorous, consuming a wide range of animals (earthworms, insects, mammals, birds, amphibians, reptiles) and plant food (cereals, fruits, green material); their diet composition varies with the season, weather conditions, and habitat (Neal and Cheeseman 1996). This dietary adaptability enables them to exploit a wide range of habitats, including semideserts, steppes, Mediterranean areas, sand dunes, mountains, and tundra. However, they are most abundant in low-lying agricultural habitats in Britain (Neal and Cheeseman 1996), where earthworms (*Lumbricus* terrestris) are the major dietary component. They collect worms from the surface of short pasture fields and in woodlands on warm wet nights (Kruuk 1978).

The badger's pattern of social organization is equally adaptable, ranging from the solitary intrasexual territorial system typical of the Mustelidae (Sandell 1989), through to large mixed-age and mixed-sex groups that can, in exceptional circumstances, contain up to 35 individuals (Kruuk 1989; Neal and Cheeseman 1996; figure 9.1). These large social groups are predominantly recorded in Britain (Griffiths and Thomas 1993), although badgers typically live in multimale-multifemale groups of five or six adults across much of their range. Badgers living in social groups jointly defend a territory. They mark territorial boundaries with a combination of feces (often deposited in pits, several of which can be clumped into "latrines"), urine, and secretions from the subcaudal and anal glands (Neal and Cheeseman 1996). Conspicuous paths connecting boundary latrines are actively patrolled. Territories are also physically defended by fighting: most observations are of intra- rather than intersex fights (Kruuk 1989). Bite wounds can be severe, and often lead to extensive tissue loss, especially around the base of the tail; on occasion, badgers are killed or die from fight injuries (S. Harris, P. J. Baker, C. D. Soulsbury, and G. Iossa, unpublished data). In low-density populations, territories are often not contiguous, and territorial boundaries are not demarcated with boundary latrines and paths.

Badgers construct underground burrow systems called *setts*. These are relatively simple structures where badgers are solitary. Social groups typically have several shared setts within their territory, and the largest (the main or breeding sett) can have more than

Figure 9.1. Group of badgers feeding in a garden in Essex, United Kingdom.

twenty entrances, although several entrances are often redundant at any given time. The underground tunnels of main setts may total hundreds of meters in length (Roper 1992), and these structures are long lasting: the history of many rural main setts can be traced back for decades or even centuries (Neal and Cheeseman 1996). Main setts are built in specific sites influenced by such factors as slope, cover, and soil, and the availability of these sites can be limiting, such that badgers may be absent from areas with good foraging habitat but no suitable locations for the construction of main setts (Wilson et al. 1997). Badgers also have a number of smaller setts within their territory (Cresswell et al. 1990; Wilson et al. 1997); these collectively referred to as "outliers" for this review. Site selection for these is more variable, and in urban areas, they can be under garden sheds and occupied buildings (Harris 1984).

Setts are an important resource for badgers, which spend most of their life belowground. In addition to using setts for sleeping and breeding in winter, badgers spend extended periods belowground (Kowalczyk et al. 2003), which, in northern areas at least, can be associated with lowered body temperatures and semi-dormancy (Fowler and Racey 1988). In temperate regions, badgers enter a period of reduced aboveground activity, associated with inappetance, which lasts a few weeks around January (Kruuk 1989). During this period, they generally only emerge for a short time each night, often only to use latrines near the sett (Harris 1982). However, they remain active belowground; in addition to digging (which is carried out in most months of the year) and changing bedding, a range of social behaviors occur underground, although we know remarkably little about the underground behavior of badgers.

Badger reproduction has been studied most intensively in high-density populations in Britain. Peak births are from mid-January to mid-March, and litter size consists of typically two or three cubs (range from 1 to 5). Cubs do not emerge aboveground until 9 or 10 weeks old, and lactation lasts a minimum of 12 weeks (Neal and Cheeseman 1996). The two periods of mating are immediately postpartum (February to May) and in the summer or autumn. However, badgers show delayed implantation; the fertilized eggs develop to the blastocyst stage but do not implant until the turn of the year. Because of the extended mating season and delay in implantation, superfetation and polygynandry occur, with mixed-paternity litters in high-density populations (Evans et al. 1989; Carpenter et al. 2005, Yamaguchi et al.

2006; Dugdale et al. 2007). Body condition and social status have a significant influence on breeding success, and generally only dominant males and females breed. Females that fail to mate have a lower body mass, less body fat, larger adrenal glands, poorer health, and a larger number of bite wounds from intraspecific fights than sows that mate (Cresswell et al. 1992). Dominant sows also probably kill the cubs of subordinate sows (Cresswell et al. 1992); boars may also be responsible for infanticide (Neal and Cheeseman 1996). We know a lot less about the reproductive behavior of badgers living in low-density populations.

URBAN BADGERS

Urban badgers are generally rare outside Britain, and although they have been studied in Japan and Scandinavia, most data come from British cities. Drawing general patterns is complicated by the lack of data and because the two main studies in Britain (in Bristol and Brighton) were undertaken 25 years apart. It is unclear whether differences in range sizes, habitat use, and other behaviors between these two populations simply reflect changing human attitudes to urban wildlife, particularly food provisioning, during the intervening period.

Outside Britain, most setts are found in less urbanized areas, such as wooded landscapes and other open areas on the edge of cities. For example, in Copenhagen, Denmark, badgers only persist in the northern areas where many parks, woodlands, and large gardens are available (Aaris-Sørensen 1987). In Hinode, Tokyo, Japan, 91% of the study area was wooded or agricultural (Kaneko et al. 2006). Even in Britain, relatively few cities have extensive badger populations within the built-up area; where there are urban badger populations, most setts are in patches of woodland or disused land covered in scrub enclosed by urban development (Harris 1984; Davison et al. 2008a; Huck et al. 2008a; table 9.1). While setts are built in gardens, they are less favored sites (Teagle 1969; Harris 1984; Huck et al. 2008a).

An early analysis suggested that most British badger populations, such as those in Bristol and southeast Essex, were relicts that had survived urban encroachment (Harris 1984; Skinner et al. 1991a). Many badgers whose setts are in rural fringes also forage in nearby residential areas, which does not mean that badgers will not colonize urban areas. The number of badgers increased in rural Britain during the 1990s (Wilson et al. 1997), and there have been increasing numbers of reports of badgers in urban areas. These may represent recent colonization events. In Britain, new setts can be dug in small gardens, in both older (Harris 1984) and modern housing developments (Smith 1982; Harris and Skinner 2002). In Trondheim, Norway, an appreciable increase in badger numbers occurred in the most urban parts of the city from the 1970s to late 1980s; in the central area, optimal sett sites were vacant houses with low cellars and habitat corridors leading to food patches (Bevanger et al. 1996). In Britain, badgers have also been recorded living under the floorboards of occupied houses.

As in rural areas, urban badgers select specific geological features for sett construction. In London, most natural setts were found in sand; London Clay was avoided (Teagle 1969). Badgers also avoided cold clay soils in Bristol, where their distribution probably reflects patterns of habitat selection preurbanization (Harris 1984; figure 9.2). An analysis of sett distributions in several U.K. cities showed that the most important factor predicting sett location was habitat type,

Table 9.1. Frequency of occurrence of badger setts within different habitats in Bristol and London, UK

Habitat	Bristol (from Harris 1984)	London (from Teagle 1969)
Wooded areas	143 (41)	151 (68)
Waste land	65 (19)	—
Gardens, allotments, school grounds	50 (14)	20 (9)
Fields, golf courses, sports grounds, parks	32 (9)	16 (7)
Factories, railway banks, gravel pits, quarries	18 (5)	12 (5)
Under sheds, buildings	11 (3)	—
Others	27 (8)	22 (10)
Total	346	221

Note: Figures in parentheses are percentages.

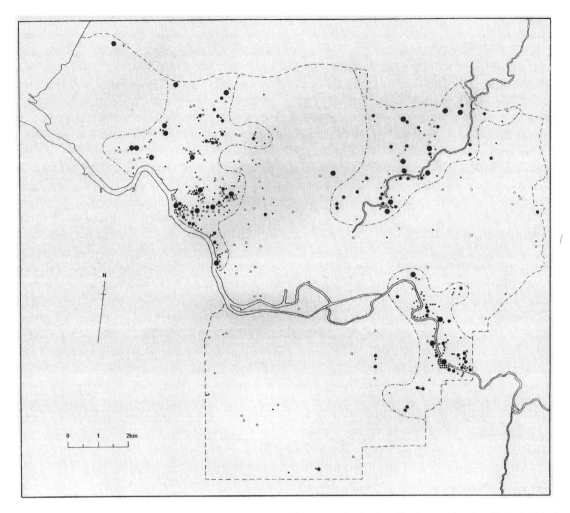

Figure 9.2. Distribution of badgers in the city of Bristol, United Kingdom; this is the most thoroughly surveyed city for badgers and highlights the patchiness of habitats suitable for badgers to build their setts. With increasing size, the symbols denote setts with less than 5 holes, 5 to 10 holes, and 11 or more holes. The dot-dash lines mark the limits of the three populations referred to in the text. Triangles mark the localities from which corpses were collected; squares show sightings and other casual records from outside the three main populations. (From Harris 1984)

followed by slope, although soil was not included in these analyses; badgers also preferred areas of intermediate human densities (Huck et al. 2008a). The selection for intermediate human densities has also been highlighted for badgers in Bristol and parallels the situation for urban foxes (Harris and Rayner 1986a); it probably reflects the amount (and probably quality) of food provisioning they receive from households in more affluent urban areas. In Trondheim, sett sites were selected to provide concealment and early melting of snows in the spring (Bevanger et al. 1996), and setts in Hinode were dug where tree bases or large rocks would stabilize the sett structure (Kaneko et al. 2006).

Most urban setts are small (table 9.2). For example, in urbanized areas of South Yorkshire (Middleton and Paget 1974) and Copenhagen, setts typically have one to three entrances (Asferg et al. 1977). In Yamaguchi City, Western Honshu, Japan, only 6.9% of setts had

more than five entrances; and 25.0% had three to five entrances (Tanaka et al. 2002). Main setts in urban areas of Britain have significantly fewer entrance holes (mean 12.8) than rural main setts (mean 19.8), whereas mean sizes of urban and rural outlier setts were not significantly different (3.0 vs. 3.8, respectively; Davison et al. 2008b). However, very large urban setts are comparable to those seen in rural areas. One London sett had 79 entrances spread over two gardens, with another 24 entrance holes in eight widely separated setts in the same two gardens: a maximum of nine badgers

Table 9.2. Size of setts recorded in Bristol and London, UK

No. of holes	Bristol (from Harris 1984)	London (Teagle 1969)
1–5	266 (77)	88 (60)
6–10	50 (14)	24 (16)
11–15	13 (4)	15 (10)
16–20	9 (3)	8 (5)
21–25	3 (1)	4 (3)
26–30	2 (1)	3 (2)
31–35	1 (<1)	1 (1)
36–40	1 (<1)	—
41–45	1 (<1)	2 (1)
Over 45	—	1 (1)
Total	346	146

Note: Figures in parentheses are percentages.

were seen using this sett (Teagle 1969). The largest active sett of the 346 recorded in Bristol, in a wooded area enclosed by housing, had 37 holes (Harris and Cresswell 1988), although there were larger disused setts (table 9.2).

POPULATION DENSITIES

Although urban badgers are patchily distributed, they can locally achieve very high densities where they do occur. Badger setts provide a good measure of badger density since the number of setts (of all types) per social group remains largely constant at different densities; in rural Britain, for instance, there was an average of four to five setts per social group across a range of habitats (Cresswell et al. 1990; Wilson et al. 1997). Mean density of badger social groups in rural Britain during the 1990s ranged from 0.017/km² in upland areas to 0.493/km² in lowland pastoral areas (Wilson et al. 1997). In Brighton, Hastings, Swindon, and Yeovil, main sett (i.e., badger social group) densities were 2.0, 2.4, 0.3, and 2.1/km² (Huck et al. 2008a), although these densities are unlikely to represent British urban areas in general. In Greater Manchester, for instance, main sett density was just 0.1 setts/km², and not all of these were active (Wright et al. 2000). The most intensive sett survey was in Bristol, where 346 setts were located in 129.4 km² (i.e., 2.7 setts/km²). However, 308 setts were confined to three areas of the city covering just 32.5 km², where sett density was 9.5/km² (Harris 1984).

Relatively few studies have estimated actual densities of urban badgers. Bristol had 46 setts, of which 37 were active, in 129.4 km². These contained 111 resident adults and 74 cubs, with an additional 33 itinerant adults. Mean density across the city was, therefore, 1.1 adult badgers/km²; mean group size was three adults and two cubs. However, density in the three areas with resident populations was 3.4 adults and 2.3 cubs/km² (Harris and Cresswell 1988). Population density can also differ significantly between years: density in one part of Bristol ranged between 4.4 and 7.5 adult badgers/km² during the period 1981–1984 (Harris and Cresswell 1987). In Brighton, genetic data showed densities of around 32 badgers/km² in a small area; mean group size was 4.1 adults and 3.7 cubs per social group (Huck et al. 2008b). Although British urban badger groups appear to be smaller on average than rural social groups, large groups do occur: the maximum size of one group in Benfleet, Essex, in midwinter before the birth of the cubs, was 30 individuals (D. Hunford, personal communication). They were heavily provisioned, but so are many other badger social groups in British urban areas.

Urban badger population densities outside Britain appear much lower. In northern Copenhagen, in the 1970s, density was 0.3 to 0.5 setts/km², with a total population of around 100 badgers (Asferg et al. 1977). A decade later Aaris-Sørensen (1987) estimated around 60 badgers in the same 500 km² (i.e., 0.1 badgers/km²). In Trondheim, density was 0.6 badgers/km² (Johansen 1993).

SOCIAL ORGANIZATION

The social organization of urban badgers appears to be similar to the pattern seen in nearby rural populations. In Britain, animals communally defend an area but forage alone, with an individual's home range typically smaller than that of the whole group (Kruuk 1989). However, urban areas are more heterogeneous than rural areas, with small, often high-quality, habitat patches divided by garden and other fences, making access to some areas difficult or impossible. The two radio-tracking studies in British urban environments both suggest a shift toward a more loosely organized, less territorial, and less stable social system in the highly fragmented and heterogeneous urban environment (Cresswell and Harris 1988a; Davison et al. 2008a). This was reflected in three key differences from rural populations. In Bristol, but not Brighton, there were a number of solitary animals: these were usually young animals that had dispersed from their natal group and had taken up a solitary existence, often in atypical habitats such as the city center or heavily built-up areas. Their periods of activity, distances traveled, and home ranges were significantly smaller than those of group-living animals (Cresswell and Harris 1988a). In contrast, young badgers in Trondheim had larger ranges in the spring, approaching 18 km², probably because they were looking for vacant territories (Bevanger et al. 1996).

Second, there was considerable movement between social groups, with 80% of males and 44% of females recaptured in Bristol more than a year later having left their original social group. Most movements were simply to an adjacent social group, but some animals moved distances of up to 7.8 km (Harris and Cresswell 1988). A study in Brighton using genetic markers showed a higher frequency of intergroup movements than in British rural populations, although in this population, there was a trend for more females to make intergroup movements (Huck et al. 2008b).

Third, territorial boundaries in both Bristol and Brighton were not well defined, and, unlike rural badgers, there were no regular visits to group boundaries or boundary latrines and paths (Harris 1982; Davison et al. 2008a). The change in scent-marking behavior may be because it is difficult for badgers to patrol a boundary in a fragmented urban habitat. Latrines (dung pits) were clustered around setts: in Bristol, 20.8% were within 10 m of a sett, 46.5% within 50 m, and 64.6% within 100 m. In contrast, in rural areas, latrines were more widely dispersed, mainly occurring within 50 m of a territorial boundary (Cresswell and Harris 1988a).

In other urban areas, badgers appear to show intrasexual territoriality typical of mustelids (Sandell 1989). In Yamaguchi City, each adult female maintained an individual home range that did not overlap with other adult females but was shared with her offspring for up to 2 years. Mean home range size for adult males (158 ha) was 3.6 times larger than those of females (44 ha), and male ranges overlapped those of two to three adult females (Tanaka et al. 2002). A similar pattern was recorded in Hinode, where mean home range size of adult males (40.8 ha) was 3.9 times that of adult females (10.5 ha; Kaneko et al. 2006). Data from other urban areas are sparse, but the solitary lifestyles and large ranges recorded in Scandinavia (Johansen 1993; Bevanger et al. 1996) suggest that the badgers were also showing intrasexual territoriality.

RANGING AND DISPERSAL BEHAVIOR

In rural areas of Britain, the size of group territories depends on habitat type, ranging from 15 to 309 ha (Neal and Cheeseman 1996); substantially larger territories have been recorded elsewhere in Europe. In Brighton, group ranges were the smallest so far recorded for badgers, averaging 9.3 ha (range 5.2 to 15.4 ha), with individual ranges averaging 4.9 ha (Davison et al. 2008a). In Bristol, group ranges were largest in spring (mean 50.8 ha), declining thereafter (summer 30.1 ha, autumn 33.9 ha, winter 10.3 ha) (Cresswell and Harris 1988a). There was no correlation between group range size and the number of badgers in the group, with individual range size (excluding winter) averaging 25.0 ha. There was no difference between the sexes, but individual range size was significantly greater for animals over 2 years old (27.1 ha) than younger animals (14.2 ha), and older animals tended to travel farther (average total distance traveled per night in spring, summer, autumn, and winter for animals less than 2 years old vs. those over 2 years old were 0.9 vs. 1.8, 1.2 vs. 2.1, 0.9 vs. 1.4, and 0.3 vs. 0.8 km) and faster (average speed of travel for spring, summer, autumn, and winter 10.2 vs. 12.2, 11.0 vs. 11.4, 8.5 vs. 9.6, and 5.5 vs. 7.9 m/min; Cresswell and Harris 1988a). In Brighton, badgers traveled faster in summer and autumn than in winter and spring (Davison et al. 2008a).

In Trondheim, home ranges varied from 187 to 1,778 ha in spring and 47 to 1,329 ha in autumn, with the larger ranges in the central part of the city. There was no difference between the sexes, although female ranges were 5.5 times larger in the spring than in autumn (Johansen 1993).

Dispersal movements in Bristol were most common in sexually mature animals, that is, those more than 2 years old. Although movements into and out of Bristol were rare, the division between rural and urban badger populations was unclear: one of the longer recorded badger movements in Bristol was a male badger that moved 5.3 km to the southwest of the city into the surrounding rural area, which necessitated swimming the tidal stretches of the river Avon, a distance of more than 100 m, depending on the state of the tide. Another male badger drowned while attempting to swim the same river (Cheeseman et al. 1988).

Little is known about gene flow either within urban badger populations or between urban and rural populations. In Brighton, mean allele richness was 3.4, at the lower end of the scale for British badger populations (mean 4.0, range 3.2–4.5), although higher than in some European rural populations. The overall relatedness value for badgers living in the same group (0.17; 0.12 for adults only) was similar to that from a long-term study in rural Gloucestershire, where the overall mean was 0.15 (Huck et al. 2008b). Finally, there was a high rate of extra-group paternity, comparable to that seen in rural populations (Huck et al. 2008b).

HABITAT USE

In Britain, the proportion of gardens used regularly by badgers declined significantly when they were 50 to 100 m from woodland and larger gardens were more likely to be used (Baker and Harris 2007), suggesting an inverse relationship between the degree of urbanization and the suitability of the habitat for badgers. In Brighton, gardens were used mainly for foraging, and availability of gardens determined range sizes; however, habitats such as scrub and allotment were used to travel from one part of the range to another (Davison et al. 2008a). In Trondheim, spring and autumn range sizes increased with increasing proportion of residential area within the home range and were also larger where woodland patches, the preferred habitat in both spring and autumn, were more fragmented (Johansen 1993). In Hinode, badgers selected rural habitats in preference to a new residential area (Kaneko et al. 2006).

In spring, badgers in Bristol selected gardens where food was provided; for the other three quarters of the year, the amount and diversity of fruit production and the size of the lawn influenced selectivity. The size of vegetable patches, diversity of vegetable crops grown, and to a lesser extent, the presence of compost heaps were positively correlated with badger activity in the summer; in autumn, badgers foraged in more predictable directions because fruits were dominant in the diet (Cresswell and Harris 1988a). In Trondheim, the use of different habitats was identical for both sexes and varied little throughout the year; home ranges increased in size with both increasing area of, and number of patches of, woodland and green areas (Bevanger et al. 1996).

ACTIVITY AND SETT USE

In rural areas, badgers are largely nocturnal or crepuscular, whereas in urban areas, they seem to modify their behavior to avoid disturbance by people. In Bris-

tol, the badgers emerged 60 to 90 minutes later than rural badgers and tended to focus their activity in the immediate vicinity of the sett until the traffic on nearby roads declined. However, since British urban areas are quieter early in the morning, they returned to their sett at around the same time as rural badgers (Harris 1982). Even in the highly fragmented urban landscape, where cover from buildings, fences, and bushes was readily available, and light pollution from street lights was extensive, they delayed emergence on nights with an absence of cloud cover and bright moonlight and traveled less far and more slowly (Cresswell and Harris 1988b). In Brighton, badgers traveled more slowly through scrub and gardens than habitats such as roads, allotments, and playing fields, which have public access; they also traveled significantly less than expected on open habitats such as roads and fields (Davison et al. 2008a). Sex and age differences in movement patterns are similar to those seen in rural badgers: in Bristol, badgers 6 to 18 months old spent significantly more time close to the sett than older animals during the night, and adult females made significantly more return journeys to the sett than adult males (Cresswell and Harris 1988a).

Nocturnal foraging bouts are typically interspersed with periods of rest, either in a nearby sett, under a garden shed, or on a couch of dry vegetation under a bush or in a flowerbed. In Bristol, on 35.5% of occasions, the badger returned to a main sett, on 37.6% to an outlying sett, on 22.7% they laid up in a couch, and on 4.2% under a shed or in an old building (Harris 1982). In Hinode, couches were mainly in deciduous forests and on forest edges, predominantly on the periphery of home ranges (Kaneko et al. 2006).

In Britain, both rural and urban badgers living in social groups generally return to the main sett at the end of each night, although this pattern varies slightly with social status (Kruuk 1989). Thus, in Brighton, social groups had 1 to 8 (mean 4.7) setts per social group; 11 of 18 radio-tracked badgers never used outlying setts, whereas the other 7 badgers spent between 2% and 42% of days lying up in between one and four outliers. Mean stays in outlying setts varied from 1.4 to 8.0 days; animals shared main setts but never shared outlying setts. However, overall use of outlying setts in this urban badger population was lower than in rural areas, suggesting that main setts are a particularly important resource in urban areas (Davison et al. 2008b).

Where badgers are solitary or less social, the pattern of sett use is more flexible. In Trondheim, badgers generally used just one sett in midwinter, but this increased to an average of more than five per badger in August; this was mirrored in a decline in the number of days spent in each sett before moving on (Bevanger et al. 1996). In Yamaguchi City, adult males and females had 32–71 and 20–41 setts, respectively, in their range, with a mean of 13.5 setts used by each individual each year. There were no differences in the number of setts used by males and females or seasonally, although breeding females used the same setts for more than 9 weeks when rearing their cubs (Tanaka et al. 2002).

In Bristol, nocturnal activity periods were longest during July–September, and nocturnal rest periods were longest during December–February and shortest during May–July, when the nights were shortest and food sources such as earthworms and fruit not available. During November–February, it was rare for more than half the animals to be active aboveground at any one time, and during January, 20% of the badgers failed to emerge aboveground on any one night. Badgers were only active for 8.4% of the night in January, which rose to 65.3% in July (Harris 1982). Nutritionally stressed adult females increased activity during the shorter summer nights. In winter, increased wind speed led to increased speed of movement, presumably because of the chill factor associated with a combination of low temperatures and wind. Under these conditions, the badgers tended to be more restless and did not forage in a small area for long periods (Cresswell and Harris 1988b).

Similar patterns of behavior have been recorded in other urban badger populations. In Brighton, badgers spent more time aboveground in summer and autumn than in winter and spring (Davison et al. 2008a). In Yamaguchi City, the pattern was similar, although badgers were often active during the day in spring, especially breeding females, whereas males were largely nocturnal; aboveground activity virtually ceased during January–February (Tanaka 2005).

POPULATION DYNAMICS

In Bristol, comparing estimated productivity with the number of emergent cubs showed a considerable mortality of young cubs (38.4%) while they were still underground; including this preemergence mortality, 64% of badgers died in their first year (Harris and Cresswell 1987). This was little different from a nearby rural badger population (Cheeseman et al. 1987). The sex ratio at birth was 1:1, with little difference in male and

female mortality rates for the first 2 years. However, from the third year onward, male mortality was higher (40.3%) than for females (28.1%), and the adult population was female biased (Harris and Cresswell 1987). Trapping data from Bristol showed that only 2.5% of males and 10.2% of females lived to 6 years (Harris and Cresswell 1987); it was rare for any animal to live 10 years. In Trondheim, annual rates of mortality for badgers aged 1 to 6 years were approximately 35%, and although this rate of mortality was higher than in rural areas, the population was stable following increases in the 1970s and 1980s. The mortality of adult males >2 years old was higher in spring and early summer, adult females in late summer and autumn (Bevanger et al. 1996).

There are few data on the factors leading to interannual variations in social group or population size in urban badgers. Group size in Bristol fluctuated in the absence of significant changes in range size or habitat availability, and there were significant interannual variations in both the number of cubs produced and cub survival, as occurs in rural populations (Cheeseman et al. 1987; Harris and Cresswell 1987). Starvation can be an important cause of mortality of cubs postweaning in rural areas, especially in hot, dry summers (Macdonald and Newman 2002); whether this is also the cause of interannual variations in cub production in urban areas is unknown, although provisioning by local residents may ameliorate the effect of interannual climate variation.

Although road deaths are probably the main cause of mortality in both urban and rural badger populations (Harris et al. 1992), there are no data on the relative importance of different causes of mortality in urban badgers, nor their diseases or parasites. Bovine tuberculosis was the cause of death for one badger living in Bristol in the 1980s (S. Harris, P. J. Baker, C. D. Soulsbury, and G. Iossa, unpublished data). There is no evidence that the disease persists in the urban badger population or poses a risk to local residents.

DIET

The cosmopolitan diet of rural badgers is also seen in urban areas. In Bristol, earthworms constituted 18.1% by volume, other invertebrates 20.0%, vertebrates 2.4%, vegetables 5.1%, fruit 30.2%, and "scavenged" food (predominantly food deliberately supplied by householders) 24.4% (Harris 1984). There was, however, marked temporal variation in the relative importance of different food groups: earthworms were most common from December to May, other invertebrates from April to July, vertebrates in May and June when passerine birds (the main vertebrate food item) were breeding, and vegetables in May and June, when damage to garden crops was most severe; fruit was important from August to November, when it formed 48% to 61% of the diet; and "scavenged" food was most common from January to May, when it formed 32% to 42% of the diet.

Similar results were obtained from a Tokyo suburb (Kaneko et al. 2006), with badgers specializing on earthworms in spring and summer and tree fruits in the autumn, although small sample sizes make direct comparisons difficult. Dietary diversity in Bristol was greatest from May to July and lowest from August to November, when the badgers specialized on fruit. The rate of travel while foraging also increased with dietary diversity because of the widely scattered distribution of food sources (Harris 1982).

CONSERVATION AND MANAGEMENT

While some urban badger populations are thriving, and even expanding, in other cities, there has been a slow decline, particularly from more heavily built-up areas. Urban badgers face three key problems: (1) population fragmentation, (2) destruction or disturbance of sett sites, and (3) road mortality. As urban areas develop, population fragmentation is believed to be a fundamental problem, particularly as rates of immigration into urban areas can be low (Cheeseman et al. 1988). Such fragmentation led to the disappearance of some well-known setts in central parts of London during the early part of the twentieth century (Fitter 1945; Teagle 1969), and badgers only survive in the inner parts of the city where they are associated with large areas of seminatural open space, such as Richmond Park and Wimbledon Common, which can support a number of contiguous social groups (Teagle 1969; Harris 1984). A similar gradual disappearance was recorded in the built-up areas of West Yorkshire, with the final extinction probably occurring in the 1960s (Middleton and Paget 1974), and Copenhagen, with badgers disappearing almost entirely from the inner city during the twentieth century (Asferg et al. 1977).

The destruction or disturbance of sett sites can potentially have significant effects on badger populations, particularly since badger population densities appear to be related to the availability of suitable localities to dig their setts, especially main setts, rather than food availability. The distribution of badgers within urban areas is a major factor affecting their susceptibility to sett disturbance. For example, in northwest Bristol, setts in the central area were predominantly in or close to gardens and thus protected by local residents; in this area, only 12% of setts were inactive, and 35% of holes were not in use. In the southern area, where setts were accessible to the public, 38% were inactive, and 50% of holes were not in use. In the northern area, the setts were in a large public park, with high levels of disturbance; 52% of these setts were inactive, and 64% of holes were not in use (Harris and Cresswell 1988).

In Essex, proximity to houses also afforded setts some protection from illegal disturbance (Skinner et al. 1991a), and in the Bradford area of Yorkshire, setts were more likely to persist if they were in open habitats and visible from roads and nearby buildings (Jenkinson and Wheater 1998), where they received some protection from illegal activities. Public access alone did not reduce sett persistence. In contrast, Aaris-Sørensen (1977, 1987) reported a one-third decline in badger numbers in Copenhagen in just 10 years; the major factor contributing to this decline appeared to be human disturbance through increased access and use of woods and parks for recreation and sport, rather than hunting or habitat change.

In Britain, high traffic loads appear to discourage badgers from crossing roads (Clarke et al. 1998), potentially exacerbating population fragmentation effects in developing urban areas. In the Netherlands, increased traffic mortality and increased fragmentation due to urban development and spread, an expanding road network, and higher traffic volumes were considered key factors contributing to the decline in badger numbers between 1960 and 1980 (van Wijngaarden and van de Peppel 1964; Wiertz and Vink 1986). Further analysis showed that the decline in badger numbers was most closely related to the number of roads (van der Zee et al. 1992). Similarly, in rural Britain, road deaths may account for around two-thirds of all adult and postemergence cub mortalities (Clarke et al. 1998). In Essex, badgers avoid proximity to roads (especially busy roads) for sett sites (Skinner et al. 1991b).

Urban spread has had a negative effect on badger populations in Britain and in the Netherlands through both loss of habitat and population fragmentation (van Wijngaarden and van de Peppel 1964; Wiertz and Vink 1986; Cresswell et al. 1989; Wilson et al. 1997), and it is likely that it will lead to similar population declines in other countries. An additional problem for urban badger populations is that, generally, they manage to survive in habitat patches, often disused land, that are relatively free of disturbance and development. Even though these habitat patches are often small, they provide a secure sett site from which the badgers can forage at night. Current housing policies in Britain and elsewhere favor compact cities through infill development and building new housing on brownfield (previously developed) sites, leading to higher housing densities (Breheny 1997). This leads to the progressive loss of greenspace (Pauleit et al. 2005) as well as to the loss of many established urban badger setts (Harris 1984); reductions in garden size may exacerbate this problem (Baker and Harris 2007).

RELATIONSHIPS WITH URBAN RESIDENTS

Urban badgers appear to be generally popular with their human neighbors. Even though they can potentially cause extensive and costly damage, problems caused by badgers appear to be relatively infrequent (Harris et al. 1994). In Britain, the most common problems faced by private homeowners are nuisance caused by the foraging and territorial activities of badgers rather than by structural or other issues caused by sett digging (Neal and Roper 1991). The most extensive survey was in a 2.5-km² suburb of Bristol, with a high density of urban badgers; 1,596 questionnaires were collected from 96.4% of households with private gardens. Homeowners reported that damage was regular but tolerable other than in drought years, when damage to garden crops was extensive (Harris 1984). The most frequent problems were rifling through garbage cans (16.4% of households) and breaking fences (10.1%); other problems, for example, digging in lawns or flowerbeds and digging dung pits in gardens, each affected <5% of households (Harris 1984). Of various garden crops, 283 households (17.7%) reported losses of carrots. Losses of any other crops were each reported by <5% of households. Damage to garden crops was most severe in May and June, especially in hot, dry summers when invertebrate prey was not available (Harris 1984).

Even in instances where damage is significant, this is often only experienced by a very few people. For example, in Saltdean, Sussex, U.K., the majority of householders living in the vicinity of a sett that had generated a lot of complaints to local and national authorities reported that they had actually experienced little or no damage (75%) or only (14%) moderate damage; only 11% of the 190 people questioned reported serious damage, although this was substantial for the individuals affected. This was reflected in their attitudes to the local badgers, with 66% actively liking the badgers, 24% having no strong views, and 9% actively disliking the badgers (Harris and Skinner 2002).

There has been considerable debate in Britain about the perceived effect of urban badgers on other local wildlife, notably slowworms *Anguis fragilis* (Stanton 1997), a nocturnal legless lizard common in gardens in western Britain, and hedgehogs *Erinaceus europaeus* (Young et al. 2006). However, there is no evidence that badgers have any effect on slowworms; a questionnaire survey of 3,779 households across mainland Britain showed that slowworms were actually 2.4 times more likely to be seen in gardens also visited by badgers (Ansell et al. 2001). However, hedgehogs were 2.5 times less likely to be seen in gardens visited by badgers, and significantly more people with both badgers and hedgehogs in their gardens believed that hedgehogs had declined over the preceding 5 years compared with people with just hedgehogs in their gardens (Ansell et al. 2001). The probability of seeing hedgehogs and their density on suburban amenity grassland were negatively related to badger sett density (Young et al. 2006). Whether these differences reflect actual predation on hedgehogs is unclear; badgers and adult hedgehogs can be seen feeding together in urban areas without any avoidance behavior or other interactions and predation appears to be a rare event when food is readily available (M. Clark, personal communication; S. Harris, P. J. Baker, C. D. Soulsbury, and G. Iossa, unpublished data). It may be that hedgehogs in urban areas tend to avoid habitats frequented by badgers, at least when breeding, since the young rather than the adults are vulnerable to predation.

Most costs incurred by urban badgers are due to damage to buildings or digging in places such as cemeteries (figure 9.3); such problems can be extremely expensive to resolve. Since 1991, badger setts have been protected in Britain, and a license is required to interfere with, or damage, a badger sett; this includes urban development (Clements 1992; Harris et al. 1994; Anon. 1995). Between 1994 and 2004, there was an increase in license applications from English urban areas, although it is unclear whether this was due to a greater awareness of the need to have a license or an increase in perceived problems (Delahay et al. 2009). A number of

Figure 9.3. Cemetery in Essex dug by badgers; preventing such damage is difficult and can be expensive and is distressing when human bones are exhumed.

practical means to reduce problems proved successful, although it is particularly difficult to exclude badgers from main setts in urban areas (Delahay et al. 2009).

In urban areas, badgers benefit from anthropogenic food sources. In Hinodecho, Tokyo, the normal pattern is for body weight to decline from April to June and then increase steadily to autumn. Females were more heavily dependent on supplementary food provided by local human residents than males, and in spring and summer, those badgers that used these additional food sources on a weekly or monthly basis were heavier than those that did not, although there were no differences in the autumn; female badgers that depended on anthropogenic food sources daily were heavier throughout the year than those that only ate natural food items (Kaneko et al. 1996; Kaneko and Maruyama 2005). In Brighton, urban badgers were significantly heavier than badgers caught in a nearby rural population (Davison et al. 2008a), possibly as a consequence of patterns of provisioning.

CONCLUSION

Urban badger populations are thinly scattered around the world. Other than in some British cities, badgers generally live on the fringes of rural areas, and most of the habitats within their ranges are typically rural, for example, woodland and agricultural areas. Where there are established urban badger populations, they often reflect some aspect of past history, such as having survived urban encroachment by having their setts in habitat patches that were unsuitable for development. Even where badgers may be recent colonists, they rely

heavily on disused land to establish their setts. Despite their flexible food habits and social organization, their vulnerability to disturbance and specific habitat requirements for digging their setts mean that badgers are susceptible to urban development. Gradual declines have been reported for many urban badger populations, although a number of urban badger populations are thriving in Britain. The continued persistence of urban badgers is likely to be dependent on sympathetic patterns of urban development.

ACKNOWLEDGMENTS
We are grateful to the Dulverton Trust, the Ministry of Agriculture Fisheries and Food, the Natural Environment Research Council, the Nature Conservancy Council, and the Royal Society for the Prevention of Cruelty to Animals for funding our studies on urban badgers, and to the editors of this volume and referees for their helpful comments on an early draft of this chapter.

10

Bobcats (*Lynx rufus*)

SETH P. D. RILEY

ERIN E. BOYDSTON

KEVIN R. CROOKS

LISA M. LYREN

WIDESPREAD ACROSS several ecosystems, bobcats are perhaps the most adaptable cat species in the Western Hemisphere. They occur throughout much of North America from Canada into Mexico, from sea level to the mountains, and in deserts, forests, swamplands, plains, and shrublands. Although they are strict carnivores, bobcats consume a relatively broad array of animal prey. Bobcats often focus on lagomorphs (Anderson and Lovallo 2003) or larger rodents such as cotton rats (Beasom and Moore 1977), but they can also rely on small mammals (Knick 1990; Riley 2001) and can make significant use of ungulates in certain systems or seasons (Dibello et al. 1990). Finally, because of their smaller size, relatively solitary nature, and reliance on natural prey, bobcats are less likely to be involved in direct conflicts with humans than larger felids, such as mountain lions, or more omnivorous carnivores, such as coyotes or raccoons.

Significant urban impacts on bobcats can be expected, but these wild felids also have the potential to adapt to and even thrive in urban areas. While the habits of this widespread species have been studied in a number of regions and ecosystems, urban studies remain relatively few. Here we review and report findings on bobcat behavior and ecology primarily from our research in California near the large and densely populated cities of Los Angeles, San Diego, and San Francisco, and from research conducted on the heavily developed resort island of Kiawah in South Carolina. While we have studied bobcats near major cities, they were not present in the highly developed centers of these major metropolitan areas, and most of

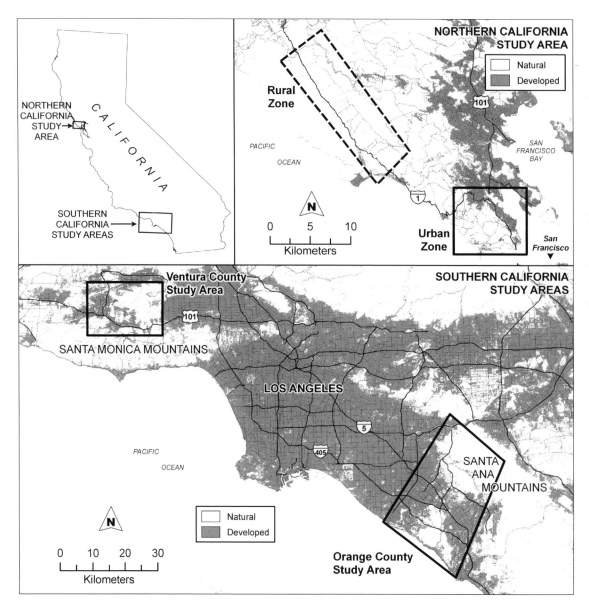

Figure 10.1. Study areas for radio-tracking studies of urban bobcats in California. In Southern California, the northern polygon near the Santa Monica Mountains is the Ventura County Study Area, and the southern polygon near the Santa Ana Mountains is the Orange County Study Area. Developed areas are shaded in gray.

our research was in less heavily developed areas (figure 10.1). Nonetheless, we found and tracked bobcats in surprisingly urban landscapes (table 10.1).

Our study areas in California differed in their landscape configurations, an important consideration in evaluating results across sites. In Northern California, the natural habitat is protected as a national park and bordered on two sides by the ocean, on one side by more parkland, and only on one side by urbanization. Southern California, the region of our ongoing proj-

ects, has a highly fragmented landscape where many areas of remaining habitat are completely surrounded by development. The intensity and types of development in these study areas also differed (table 10.1). In Northern California, development outside the park began immediately at the park's boundary and was relatively continuous. In Southern California, there are areas of high-density housing (1 house per 0.5 acre or less) and areas of very low housing density (1 house per 2+ acres) such as gated communities with extensive

Table 10.1. Human population and landscape characteristics in study areas for urban bobcats

	Human population density (people / km²)	Housing density (houses / km²)	Percent developed and altered	Percent developed	Percent intensively[a] developed
Ventura County study area	534	193	40	24	7
Orange County study area	896	354	51	38	22
North California study area (urban zone)	307	159	24	15	6
North California study area (rural zone)	4	3	6	1	0
Kiawah Island (SC) study area	40[b]	122	29	21	

Note: Housing and population statistics are based on the 2000 census, and in California, land use is based on GIS classification using the National Land Cover Data layer from 2001 computed within polygons around the study areas (see figure 10.1).

[a] Intensively developed refers to areas with >50% impervious surfaces.

[b] Population density for Kiawah Island is based on year-round residents (1,100), but many thousands of people visit the island each year (as is obvious from the higher housing density). Roberts (personal communication) estimates that there were more than 10,000 people on the island at some times of year (pop. density 357 people / km²) and generally at least 2,500 (pop. density 89 people / km²).

landscaped and even natural vegetation. In fact, much of the landscape is made up of relatively open but altered areas such as low-density housing, golf courses, office parks, school grounds, landscaped parks, and landfills. These areas may have substantial value to native wildlife, including bobcats (Riley et al. 2003). Kiawah Island appears to be more similar to Southern than to Northern California, with highly fragmented natural areas and extensive open altered areas such as golf courses.

Developed areas offer both risks and opportunities for bobcats. We have seen that bobcats can move through, hunt in, and even reproduce in these suburban or low-density urban areas. Whether bobcat populations can persist in the long term in urban areas, and at what level of urbanization, remains an open question. The studies described here took place in mild, coastal climates. Fortunately, urban studies are planned or under way in Arizona and Texas, and information from urban areas in other regions, especially at higher (colder) latitudes, would also be helpful. As perhaps the most urban associated felid species, bobcats have great potential for the study of urban carnivores and ecosystems and as a vehicle for the exposure to, and appreciation for, wildlife among urban residents.

PHYSICAL DESCRIPTION

An adult bobcat is roughly twice as large as a typical house cat and about 10 times smaller than a mountain lion. Weights of adult bobcats average 9.6 kg (21.1 pounds) for males and 6.8 kg (15 pounds) for females,

although we find average weights in our study areas to be slightly less. Bobcat tails, white with a black tip and black bars, are longer than might be implied by the "bob" in their name, averaging 14.8 cm (5.8 in) in length or about one-fifth a bobcat's body length. Still the tails are much shorter than those of mountain lions and most house cats, which are as long as the bodies of the cats. Bobcat coat color is reddish, golden, or grayish brown with fairly well-defined spots to small barely visible spots. There are black spots on the white underbelly and distinct black bars on the white fur of the inside upper legs. Bobcats have narrow black tufts on the tips of their ears and a distinctive facial ruff of elongated fur extending from their cheeks.

URBAN BOBCAT ECOLOGY

Landscape Use

Habitat / Land Use

To our knowledge bobcats have been radio-tracked in four urban areas: on Kiawah Island (Kiawah Island study) and in three areas in California (figure 10.1), in the urban zone of Golden Gate National Recreation Area in the San Francisco Bay Area (Northern California study), north of Los Angeles in Ventura and Los Angeles counties (Ventura County study), and south of Los Angeles in Orange County (Orange County study). Further south in the San Diego area, another study involved extensive camera, track, and scat surveys.

In the Northern California study, radio-collared bobcats were never found in developed areas outside the national park, and there were no reports of bobcats in backyards or on sidewalks. Adult male bobcats were radio-tracked across a major freeway (likely using a vegetated ridgeline over a freeway tunnel) and out to the borders of developed areas, as well as around office buildings and residences within the park. Home ranges of adult females, however, were along riparian valley bottoms in the interior of the park and away from developed areas (Riley 1999, 2006).

In the Ventura County study, bobcat populations have been radio-tracked for more than 10 years in a highly fragmented landscape bisected by a large freeway. In contrast to the Northern California study, bobcats used the entire landscape, including developed areas, although they still predominantly used natural habitat. Between 1996 and 2006, on average 66.2% of the home ranges (measured using 100% minimum convex polygons [MCPs]) of bobcats and 79.3% of their radiolocations were within natural areas ($n = 90$ bobcats with >20 radiolocations). Similar to Northern California, female bobcats, adult females in particular, were less urban associated than males (see also Riley et al. 2003). On average, 29.4% of the home ranges and 16.5% of the radiolocations for adult females were in developed or in altered areas, versus 37.7% of home ranges and 26.9% of the points for adult males. More adult females (37.5%) than males (23.5%) were never found in developed areas. The use of altered open areas, especially low-density residential areas and golf courses, was extensive in the Ventura County study for both male and female bobcats. As much as 69.2% of the radiolocations and 62.5% of the home range for some bobcats was in this habitat type. Although females on average were less urban associated than males, notable exceptions included adult females denning and giving birth near houses (see the "Reproduction and Dispersal" section).

Bobcats were tracked using global positioning system (GPS)-telemetry collars starting in 2002 in Orange County. For 16 males and 13 females tracked for 3–6 months each, we acquired an average (±SE) of 717 ± 235 spatially accurate GPS locations per individual. On average, 86% ± 3% of each bobcat's GPS locations were within natural areas, and 77% of MCP home ranges were composed of open-space preserves and undeveloped land (E. E. Boydston et al., unpublished data). Although results from the Orange County study did not reveal a significant difference between the sexes in use of urban, altered, or natural areas, most females had very low levels of urban association. However, two females showed much higher use of developed areas.

In fact, the bobcat with the highest level of urban association in the Orange County study was a female with 35% of her locations in urban areas (figure 10.2, *left*), which was twice as many as the next highest individual. Overall, collared males in Orange County had an average of 14% (range: 0% to 44%) and females an average of 13% (range: 0% to 67%) of their GPS locations within urban or altered areas (E. E. Boydston et al., unpublished data). Excluding the 2 most urban-associated ones, the other 11 females had an average of 5% GPS locations within urban or altered areas (range: 0% to 11%). That females might exhibit such widely different strategies with respect to space use is interesting, although the reasons are not yet clear.

Bobcats have been studied since 1996 on Kiawah Island, South Carolina, a 350-km² coastal barrier island with extensive resort development, particularly on the west end. Bobcats were radio-tracked on Kiawah for 1 year in 2000 (data for 2 males and 6 females; Griffin 2001) and for another year in 2004–5 (8 males, 8 females; Roberts 2007). Both male and female bobcats used a variety of developed areas, including residential areas and golf courses, but natural habitats were still the most important. Griffin found that east end animals (3 females and 1 male) focused on undeveloped dunes and maritime forest and that even at the more developed west end, female core areas "centered around large undisturbed wooded or dune areas" (Griffin 2001, 32). Roberts (2007) determined habitat selection for 13 bobcats and found no evidence of second-order selection (i.e., in home ranges relative to the study area) but found that shrub and dune habitats were selected within home ranges. Interestingly, open-altered areas (golf courses, parks) were avoided during the day, and developed areas were selected at night.

Perhaps because Kiawah is a developed island with limited bobcat movement to and from it, Roberts (2007) notes that all the bobcats in his study were using developed or residential areas within their home ranges. This contrasts with Northern California (Riley 2006) and parts of the Ventura County (Tigas et al. 2002; Riley et al. 2003) study area where bobcats were mostly in natural areas adjacent to development. For the 13 Kiawah Island bobcats with sufficient data, on average 25% of their home ranges were developed (vs. 12% for females and 17% for males in Ventura County and 16% for females and 14% for males in Orange County), there was no difference in urban association between sexes, and no animals had 0% urban association (S. Roberts, unpublished data; although for 2 bob-

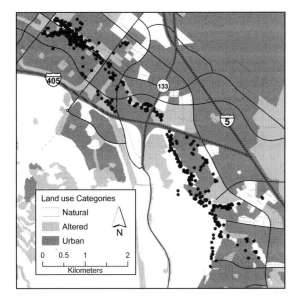 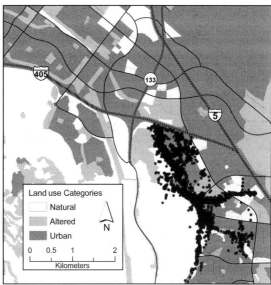

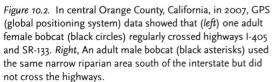

Figure 10.2. In central Orange County, California, in 2007, GPS (global positioning system) data showed that (*left*) one adult female bobcat (black circles) regularly crossed highways I-405 and SR-133. *Right*, An adult male bobcat (black asterisks) used the same narrow riparian area south of the interstate but did not cross the highways.

cats, less than 1% of their locations and about 1.5% of their home ranges were in developed areas). However, some areas in Southern California may approach an island-like situation, resulting in extensive use of development because of a lack of better habitat nearby. Recent animals in the Ventura County study (see next paragraph; figure 10.3, *top*) have been found in increasingly fragmented parts of the landscape, resulting in higher levels of urban association. It may be that when bobcats are constrained by the landscape, they will use developed or altered areas more, but otherwise, they will avoid them. Nielsen and Woolf (2001) measured the proximity of radio-collared bobcats to houses and other buildings in rural Illinois and found that human structures were generally avoided, particularly within bobcat core areas. In a recent study in Iowa, bobcats also generally avoid urban areas, although they have been documented using the edge of urban areas, and one dispersing animal was even seen by the public in the city of Ames (Clark and Gosselink, unpublished data).

Detection methods such as camera, track, and scat surveys have also provided information about how bobcats respond to urbanization. In the Ventura County study, bobcats were detected across a broad range of fragment sizes, including in fragments smaller even than a single female's home range (Tigas et al. 2003).

Subsequent capture and radio-collaring of bobcats in some of these areas has shown that up to three different bobcats may use small fragments simultaneously and that they will move between them, even moving across a major eight-lane freeway (figure 10.3, *top*). Using camera and track surveys in landscape elements, ranging from small urban habitat fragments to large core natural areas in San Diego, Crooks (2002) found that bobcats were absent from most (27 out of 29) isolated fragments smaller than 1 km² but present in all nine sites surveyed larger than 2.5 km². Both isolated urban fragments that bobcats visited had close connections with larger natural areas, with a 6-ha fragment connected by a golf course and a 74-ha fragment separated by a narrow strip of vegetation and several small surface streets.

Home Range Size and Overlap

Although many factors can affect home range size, including the model used, sample size, duration of study, region, and sex, it is still an important descriptor of how animals interact with their landscape. In general, urban radio-tracking studies have found small home ranges relative to those of bobcats studied elsewhere in less-developed landscapes: In the Ventura County study, on average, female home ranges (measured as 95% MCPs) were 2.3 km² and male home ranges were 5.2 km². Although sample sizes were low, in the urban zone in Northern California, female home ranges averaged 1.3 km² and male home ranges averaged 6.4 km² (table 10.2). Home ranges on Kiawah Island were

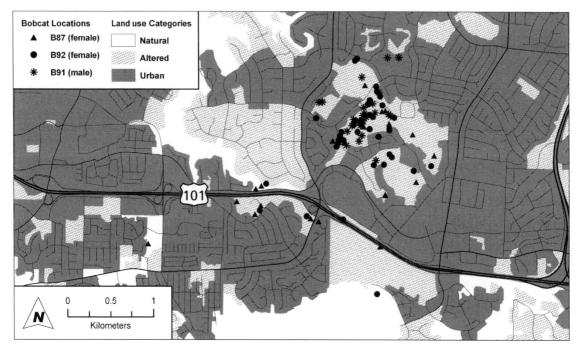

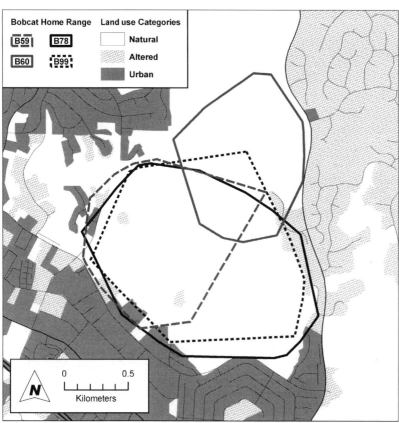

Figure 10.3. *Top*, Radio-locations for three bobcats in the Ventura County, California, study using small habitat patches and crossing the 101 Freeway. *Bottom*, Home ranges (95% minimum convex polygons) for female bobcats exhibiting extensive home range overlap next to developed areas.

also small, particularly in the later, larger study (adult males 4.4 km², adult females 3.1 km², juvenile females 4.0 km²; Roberts 2007). Orange County home ranges were similar in size to these other studies for both sexes (females, average = 3.2 km²; males average = 5.6 km²; table 10.2).

The smaller urban home ranges are comparable only to a few other studies, one also in Southern California (San Diego County; Lembeck 1978; Lembeck and Gould 1979), where two adult female home ranges averaged 1.0 km² and male home ranges averaged 3.1 km², and three in the southeastern United States (Marshall and Jenkins 1966; Hall and Newsom 1976; Miller and Speake 1979), where female home ranges varied from 1.0 km² to 1.1 km² and male home ranges varied from 2.6 km² to 5.2 km². Overall, the home range sizes from the urban studies are small relative to those from most nonurban studies (table 10.2). However, since the urban studies were in the relatively mild climates of California and South Carolina and the other stud-

ies with small home ranges were conducted in areas with similarly mild climates, it is impossible to separate climatic influences from urbanization. Bobcat home range size, for females in particular, is often thought to be related to the density of prey resources (Buie et al. 1979; Knick 1990; Anderson and Lovallo 2003), and the diversity and year-round abundance of bobcat prey may be greater in areas with milder climates. However, in a long-term study in Mississippi, Benson et al. (2006) found that variation in population density actually had a greater influence on home range size than food availability or body weight, so high population densities in urban areas may also be contributing to smaller home ranges.

Urbanization could be affecting home range size in two different ways that may initially seem contradictory. Development, especially the severe habitat fragmentation seen in Southern California, could lead to smaller home ranges and greater home range overlap if animals are restricted in their regular movement across

Table 10.2. Mean bobcat home range sizes in km² (SD) in four urban study areas and selected nonurban areas

Location	Sex	*n*	100% MCP	95% MCP	95% FK
Urban studies					
Orange County, Southern California	M	16	6.8 (3.8)	5.6 (3.0)	4.9 (2.7)
	F	13	4.4 (4.4)	3.2 (3.1)	3.5 (3.1)
Ventura County, Southern California	M	43	7.1 (6.0)	5.2 (4.2)	6.4 (5.7)
	F	54	3.5 (2.3)	2.3 (1.7)	2.1 (1.7)
Northern California (urban zone)	M	3	8.4 (3.7)	6.4 (4.8)	6.4 (0.4)[a]
(Riley 2006)	F	5	2.1 (0.9)	1.3 (0.3)	1.5 (0.3)[a]
Kiawah Island, South Carolina	M	5	6.4 (4.2)	4.4 (2.9)[b]	5.8 (2.9)[b]
(2004–5 study; Roberts 2007)	F	5	3.8 (1.0)	3.1 (0.9)[b]	3.7 (1.0)[b]
Nonurban studies					
Minnesota (Fuller et al. 1985)	M	18	61		
	F	8	32		
Maine (Litvaitis et al. 1986)	M	4	71.1		
	F	6	32.5		
Pennsylvania (Lovallo et al. 2001)	M	17	42.2[c]		
	F	17	17.2[c]		
Idaho (Bailey 1974)	M	4	42.1		
	F	8	19.3		
Mild climate nonurban studies					
Alabama (Miller and Speake 1979)	M	6	2.6		
	F	6	1.1		
San Diego County (Lembeck 1978)	M	11	3.1		
	F	5	2.1		

Note: FK = fixed kernel; MCP = minimum convex polygon.

[a] These home ranges are adaptive kernels, and they have been edited to remove areas of the Pacific Ocean.

[b] The Kiawah Island home ranges were modified to remove open water (rivers and ocean) and salt marsh areas >150 m from upland.

[c] Median values.

the landscape (assuming that resources are sufficiently dense to support all individuals). However, if urban areas represent lower-quality habitat than natural areas, we might expect bobcat home ranges that include developed areas to be larger. Both effects could also be occurring simultaneously: animals whose ranges abut, but do not include, developed areas may have smaller ranges, while animals whose ranges include development may use more area as the amount of development increases.

There is some evidence from within specific studies, and therefore from within a single climatic region, of these potential urbanization effects. In support of the smaller adjacent-home range idea, in Northern California, the home ranges of female bobcats in the urban study area were significantly smaller and more overlapping than those in a rural study area just 10 km to the north (Riley 2006). In the Ventura County study, bobcat home ranges that bordered on hard boundaries such as roads and development were smaller for both sexes and more overlapping for females (figure 10.3, *bottom*; Riley et al. 2006).

In support of the larger urban ranges idea, in the Ventura County study, bobcat home ranges were significantly larger when they included more developed and altered areas (Riley et al. 2003). This relationship between size and urban association was positive for all groups of bobcats evaluated, although not significantly so for adult females, but because adult female bobcats had very low levels of urban association, any effect of developed areas on overall home range size would be hard to detect. Among Orange County females, larger home ranges were associated with higher levels of urban association, but no relationship was detected for males (E. E. Boydston et al., unpublished data). Similarly, on Kiawah Island, both studies found larger home ranges in the more developed west end. In 2000, female ranges averaged 7.9 km² on the west end versus 3.4 km² on the east end (Griffin 2001), and in 2004, home ranges and core areas were larger on the west end for both males and females, although again for males the difference was not significant (Roberts 2007).

Urban development could be directly influencing the density of prey species, thereby potentially influencing home range size. In the Ventura County study area, western cottontail rabbits are an important component of bobcat diets (Fedriani et al. 2000; National Park Service [NPS], unpublished data), and rabbit abundance may be higher along urban edges and in landscaped parklands (NPS, unpublished data). More studies that evaluate bobcat home range size in (1) one region with varying amounts of development and (2) urban landscapes in other geographic regions would be valuable in helping to determine the relationship between urbanization and home range size and overlap.

Movement Patterns, Road Crossings

Although larger roads often acted as home range boundaries, in the Ventura County study area, bobcats and coyotes moved between habitat patches and across roads, using habitat corridors when available and using small culverts to cross busier roads during times of greater human activity (Tigas et al. 2002). However, bobcats tended to make these movements at night, they made them less often than coyotes, and female bobcats made them less often than males. Ten bobcats were also documented to have moved across an 8–10 lane freeway, likely also using culverts or underpasses (Riley et al. 2006). Some of these animals crossed the freeway multiple times, and in two cases, females eventually settled on the opposite side from which they were captured. In the Northern California study, radio-collared bobcats never moved outside the park into developed areas (Riley 1999) and seemed to avoid crossing roads, which often acted as home range boundaries (Riley 2006). Other than two adult males likely using a vegetated ridge over a tunnel, there was no evidence that bobcats crossed over the large, multilane freeway bordering the urban zone in Northern California.

In Orange County, one of the most urban-associated females had a remarkably linear home range because of her movements along a partially channelized creek bed with narrow riparian scrub habitat (figure 10.2, *left*). She moved back and forth along the creek, including crossing major highway I-405 on multiple occasions, apparently always through the same culvert. This female eventually extended her ranging pattern to the northwest, bred, and gave birth to kittens on the north side of the highway. However, a male who was collared during the same time period did not cross to the north side of this freeway, although his 5,594 GPS locations indicated his movements traced the same creek bed and led to the mouth of the same culvert (figure 10.2, *right*). There was no evidence of any other bobcats in the area, so it is unlikely that the male avoided the other side of I-405 because of agonistic encounters with other males. This anecdote about one male and one female indicates dramatic variation between individuals in response to urbanization for both sexes.

Demography

Population Density

Population density is difficult to determine for wildlife populations, particularly for carnivores such as bobcats (e.g., Ruell et al. 2009), but it is an important parameter with implications for ecology and management. Perhaps not surprisingly, given the small home range sizes, estimates of bobcat density for the few urban studies are generally high relative to nonurban areas. Anderson and Lovallo (2003, table 38.4) list eight density estimates, and six of these are between 4 and 28 bobcats/100 km². For the other two, Witmer and De-Calesta (1986) reported 77 bobcats/100 km² in Oregon, and Lembeck and Gould (1979) reported 115–153 bobcats/100 km² in the rural San Diego study. In the Ventura County study, crude minimum density estimates (number of bobcats radio-tracked divided by the area of their home ranges) were 60 and 68 bobcats/100 km² in two different areas. These estimates are from just before a disease epizootic (see "Survival Rates and Mortality Causes," below), and the estimate from the same areas in 2006 was considerably lower, at only 21 bobcats/100 km². On Kiawah Island, an estimate of resident bobcat density based on radio-collared animals and home range overlap was 84 bobcats/100 km² (Roberts 2007).

Crude minimum density estimates (also based only on collared animals divided by the area of their home ranges) from the Orange County study were lower, averaging 23 bobcats/100 km². However, documented sightings and photographs from remotely triggered cameras showed that GPS-collared animals were only a subset of bobcats in the study area. Thus, these differences in minimum density estimates are likely related to study parameters (e.g., shorter time span, smaller sample of collared animals). Overall, bobcat populations can reach high densities in urban areas. That these urban estimates are all from mild climates confounds potential conclusions. The highest nonurban density estimate is also from coastal California (Lembeck and Gould 1979).

Survival Rates and Mortality Causes

In the Northern California study, the Ventura County study, and on Kiawah Island, survival rates were relatively high (mean = 0.822 in Northern California; mean = 0.779 in Ventura County, excluding the mange epizootic; annual rate = 0.875 on Kiawah Island in 2004–5) compared with nonurban studies. This may be because activities such as hunting, trapping, and poaching, which can cause 14%–100% of mortality in bobcats (mean = 59%; n = 10 study areas; Anderson and Lovallo 2003), are almost entirely absent from urban environments. Survival rates may also be higher because the urban studies were in relatively mild climates that lack severe winter weather and may offer more diverse and stable prey populations. One non-urban study with a survival rate of more than 75% was in Mississippi (survival = 0.80; Chamberlain et al. 1999). Moreover, direct comparisons between the urban and rural study areas in Northern California (0.827 vs. 0.815) and between animals with different levels of urban association in the Ventura County study (Riley et al. 2003) did not reveal differences in survival rates or rates of anthropogenic mortality.

The exception to high survival rates in the urban studies was during a few years in the Ventura County study, when survival rates dropped precipitously because of an epizootic of notoedric mange (Riley et al. 2007). Notoedric mange has been reported in wild felids before (Pence et al. 1995), including in bobcats (Penner and Parke 1954; Pence et al. 1982) but never before as an epizootic. Among radio-collared bobcats in Ventura, 1 animal died from mange in 2001, 9 in 2002, and 11 in 2003, and the mange mortality rate increased from 0.04 to 0.27 to 0.51. Since 2003, bobcats have continued to die from mange, and although the survival rate of the reduced population is currently higher than at its nadir (mean = 0.58, 2004–6 vs. 0.23 in 2004), it has not returned to pre-epizootic levels (average of 0.77). Uncollared animals in the region have also contracted and died from the disease.

The prevalence of mange mites on bobcats or on other potential carriers, such as rabbits, ground squirrels, or domestic cats, is unknown, so it is impossible to know (at present) the source of the disease. Notoedric mange is the most common ear mite in pet cats and may be more common in Southern California than in other areas (Brooks 2000). The mange epizootic may also be related to humans in another way. Every bobcat that died of mange had also been exposed to anticoagulant rodenticides, extremely widespread chemicals used to control rodent populations worldwide. Overall, 90% of tested bobcats were positive for anticoagulant residues regardless of mortality cause (NPS, unpublished data). However, there was a strong interaction between mange disease and anticoagulant exposure of ≥0.05 ppm, as 18 of 19 bobcats that died with severe mange but only 8 of 20 animals that died of other causes had at least this level of exposure

(Riley et al. 2007). Although we do not know the mechanism of a potential interaction between anticoagulant exposure and severe mange, sublethal effects of anticoagulants have been seen in other mammals (Oliver and Wheeler 1978; Littin et al. 2002) and in raptors (Mendenhall and Pank 1980). Bobcats have also died with notoedric mange in urban Orange County, and recently, we learned of an apparent epizootic of mange in bobcats in Santa Clara County, in the San Francisco Bay Area. Between 2002 and 2005, 10 bobcats were documented with severe mange in urban and suburban areas of Santa Clara County, although samples were not available to determine the mite species involved or whether the animals were also exposed to anticoagulants (M. Phillips, County of Santa Clara, personal communication).

Bobcats are also susceptible to direct poisoning by anticoagulant rodenticides, a mortality source that may be more common in urban areas. Animals in both the Northern California study (Riley 1999) and the Ventura County study (Riley et al. 2003; Riley et al. 2007) died from anticoagulant toxicosis, although it was not a major source of mortality for bobcats in either study. This may be in part because certain rodenticides, in particular brodifacoum, the most common residue found in bobcat livers, are up to 100 times less toxic in cats than, for example, in dogs (Roder 2001). Predation, particularly by coyotes, can also be a source of mortality for bobcats in both urban and nonurban populations (Litvatis and Harrison 1989; Fedriani et al. 2000). In fragmented urban landscapes, populations of competing carnivores such as bobcats and coyotes may have increased levels of interaction relative to nonurban areas where there are fewer restrictions on movements or home ranges (see also chapter 14). This may result in increased levels of mortality from interspecific interactions, particularly for female bobcats and for bobcat kittens (see the "Reproduction and Dispersal" section).

Bobcat mortality causes were studied through examination of 49 bobcats found dead from 2005 to 2007 in the Orange County study. To locate carcasses of marked and unmarked animals, researchers sought help from local animal control officers, resource managers, and residents who reported 48 of the mortalities. Six of the reported mortalities were previously captured bobcats that had been GPS-collared (3 males, 1 female) or marked only with an eartag and microchip (1 male, 1 female). The six marked bobcats survived 14 months on average (range: 3 to 48 months) after capture.

Vehicle strikes were the main cause of death in Orange County, accounting for 69% (n = 34) of the known mortalities (L. M. Lyren et al., unpublished data). Only one bobcat was found dead on a major highway, and all other known vehicle mortalities occurred on secondary (i.e., surface) streets and local neighborhood roads. This relationship between road type and bobcat mortality was similar in the Ventura County study, where no radio-collared bobcats were killed on freeways between 1996 and 2007, but 15 died on secondary or smaller roads. An analysis of the locations of vehicle mortalities in Orange County showed that 48% occurred where a road cut across riparian areas, a habitat type frequently used by bobcats (L. M. Lyren et al., unpublished data).

Notoedric mange appeared to be the second highest mortality factor for bobcats in Orange County but to date only in particular areas and seasons. Eleven bobcats (4 males, 7 females) had severe mange at the time of death, and all were on the inland side of freeway I-5 (see figure 10.1), where road density is lower and where larger areas of intact habitat are located. On the coastal side of I-5, mange was not observed on any bobcats captured or found dead. Interestingly, all of the deaths from mange occurred in the dry season between May and early November (L. M. Lyren et al., unpublished data). The spatial and temporal patterns of mange and links to urbanization in Southern California and elsewhere warrant further investigation.

We expected that the different space utilization patterns of male and female bobcats would lead to skewed sex ratios among mortalities, but sex ratios of bobcats found dead in Orange Country were nearly 1:1. This pattern suggests that, despite the tendency for male bobcats to have larger home ranges and to disperse over greater distances, which might lead them to encounter more roads and developed areas, females are not at a reduced risk of mortality compared with males.

Reproduction and Dispersal

Information on pregnancy rates in bobcat populations has sometimes been obtained by examining carcasses from hunters and trappers (Fritts and Sealander 1978). However, this kind of information is generally not available in urban populations and does not provide a good measure of offspring survival rates. Intensive radio-tracking of collared female bobcats during the birth season allows direct measurement of litter size and the number of successful births. From the Ventura County study, 8 years of direct measurement revealed that the proportion of adult females producing litters varied from 0% to 100% and was generally >50% but was not significantly different between females in fragmented versus core natural areas (NPS, unpublished data). No litters were directly observed in the Northern California study, although young animals (<6 months) were captured, indicating that reproduction was occurring.

Despite evidence of female bobcats being more sensitive to urbanization than males (see "Landscape Use";

Tigas et al. 2002; Riley et al. 2003; Riley 2006), some females used open-altered and developed areas extensively, even giving birth behind houses. In the Ventura County study, one adult female bobcat (B075) denned in the 1-m-wide space between the yards behind condominiums in three consecutive years (figure 10.4). In two of these years, a closely related (S. P. D. Riley et al., unpublished data) female gave birth to kittens in the same thin strip of vegetation, within 20 m of B075 and her kittens. It is possible that denning females find backyard areas to be relatively secure den sites, particularly from potential predation by coyotes, the most common cause of bobcat kitten mortality in the Ventura County study (Moriarty 2007). Observations by the public are another potential source of information

Figure 10.4. Den site for three consecutive years for an adult female bobcat, Ventura County, California. She denned in the 1-meter-wide space (*top*) between the yards behind condominiums (*bottom left*) and, in one year, later moved the kittens into a hollow wooden wall (*bottom right*) between backyards. National Park Service.

about reproduction, particularly in urban areas where direct encounters may be more likely. In Orange County, members of the public on several occasions reported female bobcats with young kittens denning in dense shrubbery within backyards. Females and kittens were also documented outside of protected areas through radio-tracking and camera surveys (figure 10.5, *left*), but most evidence of females with kittens near development came from local residents (figure 10.5, *right*).

The Kiawah Island studies also documented extensive bobcat reproduction. Eight dens with kittens were located in 2000 and 2001, including from five of six females in 2000 (Griffin 2001), and six dens were located in 2004 and 2005 (Roberts 2007). A few of the dens were close to development (e.g., in a vacant lot between houses), although all were in natural features. In 2005, two females gave birth around the same time within 85 m of each other.

From a population persistence point of view, once successful reproduction is occurring, the critical question is whether young animals are successfully dispersing and establishing home ranges of their own. Tracking the fate of young carnivores is challenging because of the difficulty in safely radio-tagging young, growing animals. A study in Pennsylvania pioneered the use of implanted radio-transmitters for tracking bobcat kittens from the age of 4–6 weeks (M. J. Lovallo and J. Tischendorf, personal communication), and we have begun to evaluate kitten movements, survival, and mortality causes in the Ventura County study. Kitten survival has generally been low, 40% to 12 weeks of age, largely as a result of coyote predation on all but one of the kittens in a litter (Moriarty 2007). This contrasts with results from rural Pennsylvania, where kitten survival to 6 months was high (M. J. Lovallo and

J. Tischendorf, personal communication). Greater predation mortality may be related to urbanization if female bobcats are less able to avoid predators in isolated habitat fragments, although it may also be related to different carnivore community composition (i.e., different coyote densities) or biome types (scrubland vs. deciduous forest) in the two regions.

In urban areas, isolated subpopulations may depend on dispersal from larger natural areas to maintain genetic diversity and demographic viability. Although dispersal is difficult to study, we have documented some individual cases. The Ventura County study documented seven cases of dispersal and subsequent home range establishment, including five kittens that were radio-tracked starting at 4 weeks old. In one case, a young male dispersed from core natural habitat to a highly fragmented area where he established a home range and lived for 5 years (before dying of mange). After her mother died, a female established a home range at 1.5 years old in the home range previously occupied by her mother (and visited by her as a kitten). Another young female bobcat dispersed from an established home range in core natural habitat south of the 101 Freeway to a new territory centered on a small habitat fragment (17 ha) north of the road (she died of mange without reproducing in the new area).

In the Northern California study, one young male dispersed from the urban zone north to the edge of a rural study zone (figure 10.1) before being killed by a vehicle. Although perhaps not surprising, based on the presence of continuous habitat to the north, this case was important in establishing effective connectivity for the urban zone. Another young female exhibited wide, likely exploratory movements that included areas next to the 101 Freeway and near residences and

Figure 10.5. In coastal Orange County, California (*left*), a camera trap photo of a female bobcat and kittens crossing through a golf cart tunnel and (*right*) a tagged adult female with kittens along a roadway construction zone by a marsh. Photos: *Left*, U.S. Geological Survey and Colorado State University. *Right*, Michael Daugherty.

offices within the park before settling in the former home range of two adult female bobcats that had recently died (Riley 2006). Overall, these individual cases are not sufficient to establish conclusive trends, but they show similarities with nonurban studies in that males tend to disperse farther and that females may stay in or near their natal areas. These dispersing animals also indicate that both male and female bobcats can successfully disperse across fragmented urban areas and establish home ranges in many different parts of the landscape. However, the risk of mortality during dispersal in fragmented urban environments may be high, because dispersers may encounter and attempt to cross multiple unfamiliar roads as they seek suitable habitat.

Food Habits

Bobcats are generally strict carnivores (as are felids generally) that focus on large rodents or lagomorphs when available but can also use and even rely on a variety of prey. In the Northern California study, scat analysis revealed that bobcats in the urban zone were one of few studied populations to rely heavily on a small mammal species (see also Knick 1990), in this case California voles (Riley 1999, 2001). The reliance on voles in this region was interesting but was likely related to high vole densities across the landscape, particularly in introduced grasslands, rather than to any effect of urbanization. Rabbits and larger rodents such as woodrats and pocket gophers were also present, both on the landscape and in the bobcat diet.

There is also extensive information on bobcat diets from the Ventura County study, from two different parts of the landscape and two different periods. In large natural areas bordering urbanization, Fedriani et al. (2000) found that rabbits were the most common prey item for bobcats in both the wet and dry seasons but that other items such as small mammals and birds were also important in the diet. Further scat collection between 2001 and 2004 compared diets across a range of habitat fragmentation, including core natural areas, large fragments, and small fragments. Rabbits were still the most important prey item as the landscape became more fragmented, and in fact, the diet in fragmented areas was particularly dominated by *Sylvilagus* spp. In small and large fragments, 60% or more of bobcat scats contained rabbits during both seasons, and in the dry season, the percent occurrence of rabbits was about three times that of any other item (NPS, unpublished data). In core natural areas, the breadth of

the diet expanded to include more birds, reptiles, and other mammals, such as woodrats and pocket gophers. The reliance on lagomorphs in fragmented areas may be related to high rabbit densities in residential and landscaped areas (based on pellet counts and spotlight surveys; NPS, unpublished data), which may in turn be the result of lush, well-watered vegetation in developed areas, a rarity in the dry, Mediterranean climate of Southern California.

Activity Patterns

Bobcats can be active at any time of the day or night, but they are often thought to be nocturnal or crepuscular, with the most activity occurring in the evening and early morning (Anderson and Lovallo 2003). Through focal-animal radio-tracking in the Ventura County study, Tigas et al. (2002) found that bobcats were most often moving into developed areas or between habitat patches during the night, as opposed to at dusk or during the day. Using motion-sensitive collars as an indicator of activity and based on a larger sample of animals in the same study area, Riley et al. (2003) also found that bobcats were more likely to be in developed or

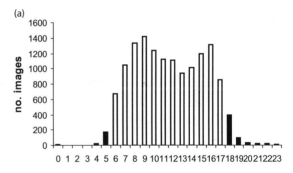

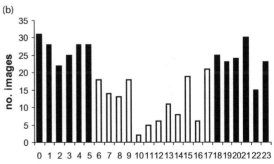

Figure 10.6. Circadian distribution patterns for (a) humans and (b) bobcats as recorded by remotely triggered camera stations across 49 sites from 1999 to 2001 in natural areas in urban Orange County, California (S. L. George and K. R. Crooks, unpublished data). Open bars represent the daytime hours (0600–1759) selected for this study; dark bars represent nighttime hours. Hour bins (in military time) are Pacific Standard Time.

altered open areas at night, specifically between 2200 and 0500 hours when humans are least active.

In Orange County, bobcats photographed by remotely triggered cameras displayed relatively even levels of activity from dusk until dawn with a reduction of activity during midday (figure 10.6; S. L. George and K. R. Crooks, unpublished data). Bobcats tended to have even lower daytime activity in sites with the highest levels of human recreation. In these sites, 21% of all images were recorded during the day (0600–1800 hours Pacific Standard Time), whereas in areas with lower levels of human recreation, daytime activity rose to 33% (George and Crooks 2006). Lyren (2001) examined activity patterns of bobcats and coyotes and their use of culverts under a highway relative to the level of highway traffic. Camera trap photos revealed that bobcats continued to show some daytime activity when traffic levels were highest (figure 10.7a), while coyote activity was the inverse of traffic levels (figure 10.7b).

Movement rates calculated from GPS collar data collected in Orange County revealed considerable variability among individuals in activity patterns, an insight not easily obtainable from camera-trap results. For females, hourly averages of movement rates (figure 10.8, *top*) yielded an activity pattern resembling that from the camera-trap data (figure 10.6b). However, among males, there was no clear depression in daytime activity (figure 10.8, *bottom*), and patterns among males were highly variable. Some males showed rates of movement during daytime hours that matched nighttime activity. Because collared bobcats were tracked for less than 6 months, the data do not allow evaluation of whether individuals might adjust their circadian rhythms seasonally, during different life stages, or in response to anthropogenic disturbance. Avoidance of coyotes may further influence bobcat behavior, for example, with respect to the use of features that may be rare on the landscape such as culverts under highways (Lyren 2001).

Since radio-collared bobcats were never found outside the park in the Northern California study, there was no potential to document a change in activity pattern for animals using developed areas. However, on the basis of direct observations, bobcats were often active during the day, even in areas and at times when large numbers of visitors were using trails, indicating that animals were potentially acclimating to human presence within the park. On Kiawah Island, bobcat movement rates were lowest during the day, perhaps as a response to human activity (Roberts 2007). Overall, bobcats have shown a variety of activity patterns in urban areas. They can exhibit significant amounts of daytime activity, although the amount may be altered in more developed areas because of human recreation, because of traffic volume, or perhaps because of interactions with other carnivores. Severe enough restrictions on activity in urban areas could in turn affect survival, reproduction, and ultimately the presence of bobcats.

Interactions with Other Animals

Interactions with Other Carnivores

Many studies in nonurban areas have documented interactions between bobcats and both larger and smaller carnivores. Bobcats can be chased, killed, and sometimes eaten by larger carnivores such as mountain lions (Koehler and Hornocker 1991) and coyotes (Fedriani et al. 2000). In turn, bobcats have been known to kill and sometimes eat smaller carnivores (e.g., kit foxes; Disney and Spiegel 1992). Interactions between carnivores were directly assessed in two of the urban study areas in California. In the Northern California study, interference competition between bobcats and gray foxes was seen in the form of direct interactions, including bobcats killing (but not eating) foxes and an observation of a bobcat chasing a fox (Riley 2001). There was also evidence of spatial separation between the two species and the potential for resource competition (Riley 2001), although bobcats and foxes were broadly overlapping across both the urban and rural zones. In the Ventura County study area, bobcats overlapped with coyotes in fragmented areas and with coyotes and gray foxes in core natural areas. Bobcats, specifically females, were killed by coyotes in both areas (Fedriani et al. 2000; Riley et al. 2003; NPS, unpublished data), and bobcats in turn killed foxes in the core natural area (Fedriani et al. 2000; Farias et al. 2005).

Interactions with Prey Species

Evaluating the effects of predation on particular prey species is difficult in any landscape, and few studies have effectively addressed this question for bobcats, particularly in urban areas. As with interactions with other carnivores, it is possible that effects on prey populations may be greater or more likely in habitat fragments—many of the best-documented examples of top-down effects from terrestrial mammalian carnivores have been on true islands (e.g., McLaren and Peterson 1994). Certainly, interactions with prey spe-

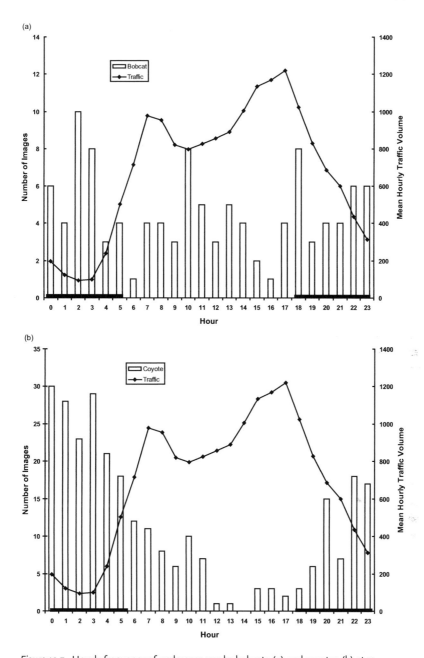

Figure 10.7. Hourly frequency of underpass use by bobcats (a) and coyotes (b) at 13 underpasses relative to mean hourly traffic volume, on highway CA-71 from September 1998 through January 2000 in the Chino Hills, San Bernardino, and Riverside counties, California (Lyren 2001).

cies are important for bobcat populations and individuals, and these relationships are often thought to dictate habitat use and home range size and overlap (Anderson and Lovallo 2003; but see Benson et al. 2006). In urban areas, the movements and habitat associations of primary prey species may affect how often bobcats use developed or altered areas. In the Northern California study area, bobcats preyed heavily on voles, which

were common in the grasslands within park areas. However, the urban areas adjacent to the park generally lacked suitable vole habitat, or even the landscaped lawns that might be attractive to other bobcat prey items, such as rabbits or pocket gophers. However, in the Ventura County study area, there are often yards or other landscaped and irrigated areas (e.g., golf courses, schools, office parks) adjacent to parklands. These

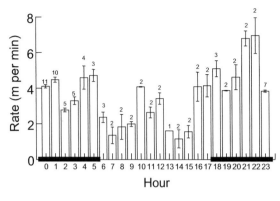

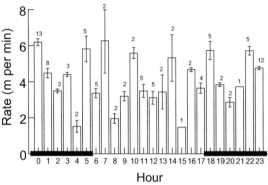

Figure 10.8. Mean hourly rates of movement for GPS-collared bobcats in Orange County, California, during 2002–2006 for (*top*) 13 female and (*bottom*) 11 male bobcats. Sample sizes over standard error bars indicate numbers of individual males or females with data for that hour.

landscaped areas may be attractive to rabbits, the primary bobcat prey.

There were interesting interactions between bobcats and prey, specifically deer, on Kiawah Island. Roberts (2007) argues that bobcats are limiting the population growth of deer on the island through impacts on fawn survival. Bobcats were responsible for 67% (70 of 105) of fawn deaths and 89% of fawn predation. There were apparently differences between individual bobcats in their ability to exploit deer, and one adult male bobcat in particular seemed to be successful at killing fawns and was single-handedly having an impact on fawn survival (Roberts 2007). This is one of few studies to show such a strong effect of bobcats on a deer population, and this relationship may be related to the intense residential and resort development on the island. Bobcats are selecting the scrub habitat, interspersed with development, that remains on the island, and these are also the areas where deer are giving birth and tending young fawns (Roberts 2007).

Interactions with Domestic Animals

Bobcats are sometimes seen as a threat to pets, such as domestic cats or smaller dogs. While there have been anecdotal cases where bobcats have chased or attacked pets in Southern California, bobcats are generally much less of a threat to pets in urban areas than coyotes. Although the importance to the overall coyote diet is still low, domestic cats remains have often been found in coyote scats in urban areas (see chapter 7), but to date not in bobcat scats (Fedriani et al. 2000; Riley 2001; NPS, unpublished data). This may be because coyotes spend more time in developed areas because bobcats are smaller than coyotes or because bobcats view domestic cats as more similar to themselves. In fact, there is sometimes speculation about mating between bobcats and domestic cats, and one author (S. P. D. Riley) observed an interaction in the Northern California study area where a bobcat and a barn cat sniffed each other and then sat 3 m apart for a few minutes before the bobcat walked away.

Interactions between bobcats and domestic animals, specifically cats, could also result in disease transmission. Bobcat populations in nonurban areas have shown evidence of exposure, in the form of positive titers, for a number of different diseases associated with domestic cats, some of which may have the potential for population-level effects. For example, high levels of exposure to feline panleukopenia virus were found in bobcats in New York (Fox 1983), and the disease killed 8 of 17 bobcats in a study in Florida (Wassmer et al. 1988). The potential frequency of interactions between bobcats and domestic cats is high in urban areas where pet cat "populations" can be dense (see chapter 12). In the Northern California study area, bobcats exhibited little to no exposure to a number of viral diseases (Riley et al. 2004), perhaps not surprising since bobcats were not moving outside the park in the urban zone. The one exception was feline calicivirus, a disease that is relatively easily transmitted through respiratory secretions, for which bobcats in a nearby rural area actually had higher levels of exposure (67% vs. 17% in the urban zone). This exposure may have been related to bobcats using the area around ranch houses in the rural zone, and the two seropositive animals in the urban zone lived near residences and a commercial stable within the park (Riley et al. 2004). In the Ventura County study area, where many bobcats visited residential and altered areas, 20% of bobcats tested had been exposed to feline panleukopenia virus, which contrasts with the complete lack of exposure in Northern California, and 8% had been exposed to feline calicivirus. Interestingly, for animals that were both tested and radio-tracked, bobcats with more developed area within their home range were significantly more likely to be seropositive for these two diseases (S. P. D. Riley and J. E. Foley, unpublished data).

Interactions with People

Bobcats are not a threat to human safety—there are very few records of attacks on people, and those few that exist often turn out to have involved rabid animals. In our experience, urban bobcats avoid humans that get too close. For example, in one early morning instance in the Northern California study, one author (S. P. D. Riley) watched from above as a bobcat hunted by walking slowly along a hiking trail, until two loudly conversing people approached. The bobcat made a short detour behind a bush 10 m to the side and then returned to his hunting along the trail after the people had passed. Some bobcats may become habituated enough to human presence to remain closer than they might in nonurban areas. In Northern California, visible daytime activity by bobcats presented an opportunity for people to observe naturally behaving bobcats. On rare occasions, people tried to approach bobcats too closely, although even in these instances we know of no physical encounters. In a broader sense, high visibility can be a positive aspect of urban bobcats because many people that see bobcats behind their houses or in park areas nearby gain a greater appreciation for wildlife and wild areas.

In a study of human recreation and large mammal activity in an urban nature reserve south of Los Angeles (George and Crooks 2006), bobcats appeared responsive to both dogs and human recreation. Bobcats were detected less frequently along trails with higher human activity, including both biking and hiking, and they displayed a relatively wide range of activity levels at sites with the lowest human use, but less activity, and a more restricted period of activity, in sites with high recreation levels. In addition to changes in overall activity, bobcats also appeared to shift their daily activity patterns to become more nocturnal in high human use areas, particularly those frequented by bikers, hikers, and dogs (see "Activity Patterns" section). Various studies have also found that urban coyotes shift their activity to times when humans are less active (Andelt and

Mahan 1980; Quinn 1997; Grinder and Krausman 2001; McClennen et al. 2001).

CONSERVATION AND MANAGEMENT OF URBAN BOBCATS

Perhaps because of their minimal negative interactions with people and pets and their tendency to stay in natural areas more than some carnivore species (such as coyotes or raccoons), bobcats in urban areas have rarely generated much management concern. This has been the case even for some bobcat populations at relatively high densities. If management concerns continue to be negligible where bobcat populations persist in and near cities, it may be easier to address conservation concerns for this species than for those where human-wildlife conflicts are more common (see also chapter 16).

Because bobcats in the city depend largely on "natural" habitat and native prey, for bobcat populations to survive in urban landscapes, connectivity appears to be a key requirement. The California landscapes where we have studied bobcats and urbanization have generally included areas of thousands of acres of contiguous habitat immediately adjacent to heavily developed urban areas interspersed with small parks, golf courses, and narrow stretches of habitat. In these areas, we have found relatively high densities of bobcats compared with those reported for studies in more remote locations, and we have documented bobcats using small habitat fragments and narrow corridors through development. These observations suggest that at least some individual bobcats can persist and adapt to a high degree of fragmentation.

But in these same landscapes, we have found that bobcats inhabiting large, open-space areas tend to stay there rather than cross freeways or disperse into more fragmented areas. Where elevated freeways already cross through potential habitat, a better understanding of how to facilitate movement between the bisected areas and prevent a build up of territories along the man-made boundary appears important to maintaining or restoring gene flow (Riley et al. 2006). This situation of a freeway crossing otherwise undeveloped areas is not uncommon in North America, and the patterns of bobcat movements relative to freeways found here could also exist away from cities and towns. Where smaller open-space parcels are spread like steppingstones of habitat in the urban matrix, bobcats may attempt to cross smaller roads between fragments and face a high mortality risk from vehicles. Landscapes that include habitat islands are common in urban environments, but we currently know little about the configuration and characteristics that make disjointed habitat viable for some bobcats. One conservation challenge is how to attract bobcats to these areas without increasing their mortality risk.

In the urban studies, we found high numbers of mortalities that were mostly attributable to human activities, such as vehicles and application of rodenticides. Despite the wider-ranging habits of males and their greater association with urbanization, males and females are both at risk of mortality sources associated with developed landscapes, although their different behaviors may indicate different mechanisms of exposure to these risks. Overall, if we can minimize human-associated mortality, preserve sufficient amounts of suitable habitat, ideally with some relatively large reserves, and maintain or restore connectivity between these areas, bobcat populations may be able to survive and even thrive in many urban landscapes. If they do, their presence may also benefit conservation efforts more generally.

Wild cats, bobcats included, have a special place in many people's imagination. Bobcats also represent essentially no threat to humans, in contrast to other big cats such as mountain lions, which have the documented, if very rare, potential to attack and harm humans. Moreover, we have studied bobcats in landscapes where mountain lions are present but have a tenuous hold or are absent. Through their presence in urban landscapes, bobcats can become a flagship species for urban communities, serving as local ambassadors for wildlife by providing a sense of wilderness to people who might not otherwise have the opportunity to experience it. For example, in the visitor center for the Upper Newport Bay Ecological Reserve in Orange County, posters and stationery for sale feature photographs of bobcats, in particular a GPS-collared female who was seen several times around the estuary hunting coots and ground squirrels. On Kiawah Island, the nonprofit Kiawah Island Natural Habitat Conservancy has made bobcats the center of its efforts to preserve habitat for wildlife on the island (Roberts 2007). This adaptable felid can become an important symbol of wild areas, even in an increasingly urban world.

ACKNOWLEDGMENTS

We thank T. Gosselink for comments on a previous version of this chapter. We thank S. Roberts for kindly and promptly sharing his dissertation, unpublished data, new analyses, and insights from the Kiawah Island studies. We thank Golden Gate National Recreation Area and Earthwatch Institute for funding for the Northern California study; Santa Monica Mountains National Recreation Area, the Los Angeles County Sanitation District, and Canon U.S.A. Company for funding for the Ventura County study; the Nature Conservancy, the Irvine Company, Orange County Great Park Corporation, and the Transportation Corridor Agencies for funding for the Orange County study. Countless technicians, interns, volunteers, and others provided invaluable help with fieldwork and analyses for these studies; we would especially like to thank G. Brooks and B. Mckeon for Northern California; C. Bromley, S. Kim, J. Moriarty, S. Ng, C. Schoonmaker, L. Tigas, and J. Sikich for Ventura County; and S. Weldy, G. Geye, C. Haas, R. Mowry, R. Alonso, G. Turschak, and D. Newell for Orange County. The late E. York provided excellent capture work, extensive experience, and boundless energy to the Ventura County and Orange County studies.

11

Mountain Lions
(*Puma concolor*)

PAUL BEIER

SETH P. D. RILEY

RAYMOND M. SAUVAJOT

W E HAVE all seen the front-page photographs of a mountain lion treed in a backyard or tranquilized in a garden at the edge of town (figure 11.1). Since 1999, no fewer than six popular books have focused on mountain lion attacks on humans. David Baron's *The Beast in the Garden* (2001) chillingly claimed that mountain lions have lost their aversion to humans and now regard suburban and urban humans as part of the menu. As researchers who spent years studying mountain lions near the urban-wildland interface of Southern California, we have seen no evidence that this is the case.

In fact, we believe that there are no truly urban mountain lion populations. There are indeed populations abutting the edges of urban areas, and dispersing juveniles and the occasional adult will enter them. However, there are no urban mountain lions in the same sense that we have urban foxes, raccoons, and skunks. The mountain lions at the wildland-urban interface spend most of their time on the wildland side of the divide and behave like their counterparts in more remote areas. Doubtless the main reason for this similarity is that cougars survive and reproduce best by selecting habitats and behaving in ways that have served them well for millennia.

On the other hand, few if any mountain lion populations are untouched by human influences because few pristine areas are big enough that a mountain lion, with its vast home range size, could avoid some contact with humans and their artifacts. Probably every adult mountain lion in the coterminous United States has seen humans, crossed paved roads, and encountered human settlements. To avoid the inappropriate term *urban mountain lion*

Figure 11.1. Mountain lion resting in a bathroom in an outbuilding near Chatsworth Reservoir, Los Angeles, California, January 2008. National Park Service.

populations, we will use the term *near-urban* for mountain lion populations or study areas in which about a third or more of the population boundary consists of urban edge, and we will use *nonurban* for populations occurring in more remote areas.

Mountain lions are one of the most widespread carnivores in the world, with a historical range from Alaska to Tierra del Fuego, Chile, and from the Atlantic to the Pacific. Currently, mountain lions are still widespread across the western United States, and they contact urban areas on the Pacific Coast, in the Southwest, in the Rocky Mountains, and in south Florida. In recent years, mountain lions have also begun moving eastward into the developed midwestern United States.

Because there are no truly urban mountain lions, and because no mountain lion population is completely free of urban influences, this chapter necessarily includes information about the biology and ecology of mountain lions in general. We focus on any traits known to differ between near-urban and non-urban mountain lions. Where evidence for or against such differences is lacking, we liberally speculate on likely differences. We focus on two near-urban studies in Southern California (Fig. 11.2), namely, P. Beier's "Santa Ana Mountains study," which radio-tracked 32 mountain lions during 1986–1992, and S. P. D. Riley and R. M. Sauvajot's "Santa Monica Mountains study," which has involved 11 radio-collared mountain lions during 2002–2007. Some data from our studies are presented here for the first time. Ongoing and future studies, including those near Tucson and Prescott, Arizona; near Boulder and Colorado Springs, Colorado; near several cities in Washington, in several regions of California; and in south Florida may soon significantly increase our knowledge of near-urban mountain lion ecology.

Mountain lions, also known as pumas, cougars, or panthers, are the largest carnivore (female weight ≈35 kg, male weight ≈55 kg) covered in this book. As large carnivores, two characteristics of mountain lions are especially relevant in urban areas. First, mountain lions need lots of space and exist at low population densities; this area sensitivity makes it a challenge to conserve mountain lions in the increasingly urban and fragmented western United States. Second, mountain lions are a potential threat to humans and their animals; this makes some people less tolerant of mountain lions and can complicate the conservation and management of near-urban populations.

PHYSICAL DESCRIPTION

Mountain lions are large cats, with adults measuring from 1.8 to 2.4 m in length (body plus tail). On average, mature males weigh 53 to 72 kg and mature females weigh 32 to 48 kg, depending somewhat on latitude. Mountain lions have a muscular, slender body, a relatively small head with a blunt snout and short, rounded ears, and a long tail that is about one-third of the animal's total length and reaches to the ground and beyond. The body of adult mountain lions appears uniformly tawny yellow, brown, or light gray, except for black patches at the tip of the tail, back of the ears, and muzzle. Mountain lions, as all felids, have few teeth, but impressive canines. They also have very large paws.

SURVIVAL RATES AND CAUSES OF DEATH

Causes of death for mountain lions include fights among mountain lions, hunting, killing mountain lions to reduce loss of livestock, removing mountain lions perceived as a threat to human life or property, injuries sustained during attempts to kill prey, vehicle collisions, and starvation (Lindzey et al. 1988; Beier and Barrett 1993; Ross et al. 1995; Logan and Sweanor 2001). In near-urban areas vehicle collisions, accidental poisoning, and killing mountain lions that kill pets or hobby animals may be more important causes of death than in more remote areas. This may be partly

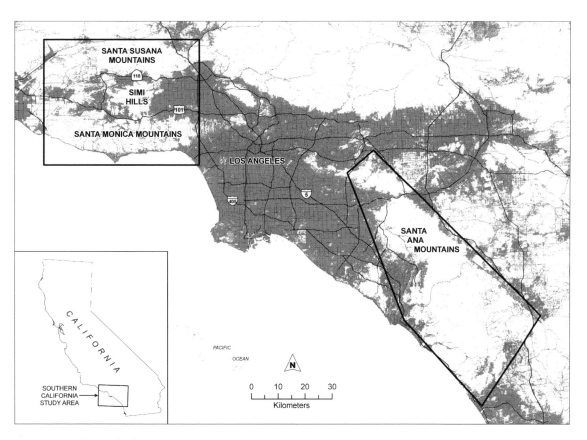

Figure 11.2. Study areas for the Santa Ana Mountains study and Santa Monica Mountains study near Los Angeles, California.

offset by fewer deaths due to sport hunting in near-urban areas.

In unhunted populations, adult male mountain lions regularly kill male and female subadults and occasionally kill other adult males and females (Logan and Sweanor 2001). Females may rarely kill other females (Lindzey 1987). Human-caused mortality is the main cause of death in populations subject to sport hunting or control to limit killing of livestock (Hornocker 1970; Murphy 1983; Logan et al. 1986; Smith et al. 1986; Ross and Jalkotzy 1992; Cunningham et al. 1995; Stoner et al. 2006). Orphaned kittens have a survival rate of about 5%, with most deaths attributable to starvation (Beck et al. 2005).

Average annual survival rates (the proportion of animals surviving 12 months) were 0.52 for juveniles ($n = 19$) and 0.75 for adults ($n = 20$) in the Santa Ana Mountains (Beier 1993, 1996), and 0.74 for juveniles ($n = 4$) and 0.78 for adults ($n = 5$) in the Santa Monica Mountains. These are similar to estimates in nonurban areas of 60%–85% for adults (Lindzey et al. 1988; Anderson et al. 1992; Logan and Sweanor 2001; Beck et al.

2005; Stoner et al. 2006) and 50%–75% for dependent young mountain lions (Hemker et al. 1982; Anderson et al. 1992; Logan and Sweanor 2001).

Mountain lions near urban areas are subject to increases in several type of human-related mortality; these increases are at least partly compensated by decreases in other types of mortality. Vehicle collisions were the leading cause of death in two near-urban populations in the Santa Ana Mountains (Beier 1993, 1996) and south Florida (Shindle et al. 2001). Of 20 deaths of radio-tagged mountain lions in a study area near Seattle, Washington, 18 were directly or indirectly related to human activities, including 7 animals killed by hunters, 5 killed for taking pets or livestock, 4 killed by feline leukemia virus, and 2 road kills (B. Kertson, Washington Cooperative Fish and Wildlife Research Unit, personal communication). In California and Colorado, permits to kill depredating mountain lions are high in near-urban areas (Beier and Barrett 1993; K. Logan, Colorado Division of Wildlife, unpublished data), largely because of attacks on pets and hobby animals (Torres et al. 1996). Together with high vehicle-related mortality, depredation permits can cause high mortality of mountain lions in near-urban areas.

We documented 9 mortalities in the Santa Monica Mountains: 5 from intraspecific strife, 2 from vehicles, and 2 from anticoagulant poisoning. Exposure to anticoagulant rodent poisons may be greater in near-urban areas; 7 of 8 animals in the Santa Monica Mountains tested positive for two or more anticoagulants, and the total toxicant load of bobcats and mountain lions was positively associated with the use of developed areas (Riley et al. 2007; NPS, unpublished data). However, these compounds are also widely used in nonurban areas, and 69% of lion livers across California, including nonurban areas, had anticoagulant residues (R. Poppenga et al., University of California, Davis, unpublished data).

Hunting pressure on mountain lions is low near many urban areas due to restrictions on the use of firearms near dwellings and hunter preference to hunt in remote areas. Urban residents are less sympathetic to hunting, specifically hunting of predators (Teel et al. 2002), and the four most urbanized western states have restrictions on hunting predators on public land (Arizona) or on hunting mountain lions in particular (California, Oregon, and Washington). These measures reduce hunting mortality in both near-urban and nonurban areas of affected states.

REPRODUCTIVE RATES

Outside of captivity, female mountain lions first give birth as early as 17–19 months of age (Maehr et al. 1989a; Logan and Sweanor 2001) but more typically at about 27–29 months (Ashman et al. 1983; Logan and Sweanor 2001). A mountain lion mother that loses her litter may come into estrus and breed again. When litters survive to dispersal age, litters are spaced 17 to 24 months apart on average (Anderson 1983; Maehr et al. 1991; Lindzey et al. 1994; Logan and Sweanor 2001), but occasionally at intervals as short as 12–15 months. Data on 327 litters across many study sites suggest a mean litter size of 2.7, with litters of 2 or 3 most typical. Gestation is about 92 days (Anderson 1983; Beier et al. 1995; Logan and Sweanor 2001).

Near-urban mountain lions probably are similar to nonurban animals in litter size and gestation time, but their rates of pregnancy and production of litters may be more variable. In the Santa Ana Mountains, 4 of the 12 adult females followed for more than a year apparently never produced successful litters, even though some were followed for 2–3 years (Beier and Barrett 1993). In one part of the study area, there was no evident reproductive activity among seven females for more than a year, apparently because of a lack of males. This kind of demographic instability is more likely in isolated near-urban populations where population size is small and one sex may be temporarily rare or nonexistent. Small numbers of either sex coupled with barriers to dispersal causes loss of genetic variation, which can also lead to catastrophic reproductive and health problems, as occurred in the Florida panther population until the population was rescued by infusion of new genetic material (Beier et al. 2003; Pimm et al. 2006). In the Santa Monica Mountains, dispersal barriers and small numbers of adults increase the potential of breeding between first-order relatives. These demographic and genetic effects are the result of small population size and relative isolation, rather than direct impacts of urbanization. However, in all three cases (Santa Ana Mountains, Santa Monica Mountains, and south Florida), urbanization was the ultimate cause of isolation and small habitat area.

DISPERSAL

Dispersal (movement of an animal from where it is born to where it breeds) is important for mountain lion populations because new breeding animals join a population mainly by immigration of juveniles from adjacent populations, while most of the population's own offspring emigrate (or attempt to emigrate) to other areas (Beier 1995; Sweanor et al. 2000; Maehr et al. 2002). Mountain lions become independent of their mother at 10–20 months of age and disperse out of her home range shortly after that (Anderson et al. 1992; Sweanor et al. 2000; Maehr et al. 2002). Regardless of population density, almost all male offspring disperse out of their natal population (Sweanor et al. 2000), typically moving 85–100 km and occasionally more than 200 km (Anderson et al. 1992; Sweanor et al. 2000). Recently, two radio-tagged male mountain lions moved from occupied habitat in the Black Hills of South Dakota eastward across apparently unoccupied habitat. One traveled a straight-line distance of 1,067 km to central Oklahoma (Thompson and Jenks 2005) and the other more than 800 km into northwest Minnesota (Fecske 2006). Typically, 50%–80% of female offspring remain in their natal population, replacing a resident female or establishing a home range that partially overlaps other breeding females (Sweanor et al. 2000).

Dispersing mountain lions in New Mexico crossed up to 100 km of unsuitable habitat in a few days (Sweanor et al. 2000). Emigrants are apparently strongly motivated, crossing rivers and freeways more readily than adults (Beier 1995; Sweanor et al. 2000; Maehr et al. 2002), but they are often killed on roads (Beier 1995). In the Santa Ana Mountains and in heavily urbanized south Florida, dispersing animals entered urban areas, where they were killed by vehicles or firearms or captured and relocated by management agencies (Beier 1995; Maehr et al. 2002). None of the four dispersers killed in urban areas in Beier's study (1995) behaved in a way that threatened human life or property.

Urban areas are largely impenetrable to movement of both adult and dispersing mountain lions; dispersing mountain lions move along the wildland-urban edge with frequent forays into peninsulas of habitat that intrude into urban areas (Beier 1995). This pattern suggests they are searching for a way out of their natal range. Most dispersing mountain lions in the Santa Ana Mountains and south Florida were unable to emigrate from their natal range, and either died in the attempt or returned and attempted to settle in their natal range (Beier 1995; Maehr et al. 2002), a pattern known as frustrated dispersal (Lidicker 1975). Two young males in the Santa Monica Mountains also exhibited frustrated dispersal, moving into the farthest northeast and northwest corners of the Santa Monica Mountains (figure 11.3, *top*) and making forays into inhospitable agricultural and urban areas at night (figure 11.3, *bottom left*). Both animals were eventually killed in fights with adult male mountain lions.

Dispersal barriers, including freeways and development, may affect gene flow and genetic diversity in carnivores. California's agricultural Central Valley and freeways in California, Arizona, and New Mexico coincide with significant gradients in gene flow for mountain lions (Ernest et al. 2003; McRae et al. 2005). In the Santa Monica Mountains, genetic diversity (polymorphism and heterozygosity; $n = 11$ animals) was lower than elsewhere in California (S. P. D. Riley, J. A. Sikich, E. C. York, R. M. Sauvajot, and H. B. Ernest, unpublished data; Ernest et al. 2003). Animals in the Santa Monica Mountains were more similar genetically to mountain lions 500 km to the north than to animals in the Santa Ana Mountains study, less than 100 km to the south but across the highly urbanized Los Angeles Basin.

USE OF HABITAT IN AND NEAR URBAN AREAS

Mountain lions are habitat generalists that seem to require only three things: (1) freedom from excessive human interference, (2) adequate numbers of large ungulate prey, and (3) ambush or stalking cover. Although these three requisites are lacking in most urban areas, suitable conditions—and therefore mountain lion populations—commonly occur in near-urban settings. Sometimes these conditions occur right up to the wildland-urban interface, or even in pockets of habitat within urban areas. Accordingly, some near-urban mountain lions use habitat at the urban edge and within urban areas.

When humans detect mountain lions in urban areas, it often becomes a newspaper or TV story. Quite reasonably, reporters and citizens ask, Does this animal live in our community all the time? Is this behavior alarming? Does every near-urban mountain lion visit town occasionally? Do mountain lions actively seek out urban areas as habitat? On the basis of our work in two areas in Southern California, we offer the following answers: No, the animal does not live in urban areas all the time. No, mere occurrence in or near town is not cause for alarm. Yes, mountain lions regularly visit urban areas, usually without being noticed. No, mountain lions do not seek out urban habitat; to the contrary, they tend to avoid urban and near-urban areas, except in rare cases when the best wild habitat is close to the urban edge.

Our data (tables 11.1, 11.2, 11.3) are particularly relevant to the question of the degree to which mountain lions seek or avoid urban areas. The overarching pattern was that the near-urban animals in our study areas tried to minimize use of urban habitat features. Their home ranges were located in the larger landscape in a way that minimized overlap with urban areas (table 11.1). At a finer spatial scale, mountain lions tended to use those parts of their home range >1 km from urban areas and to make relatively little use of those parts of their home range increasingly close to urban areas (table 11.2). The two most urban animals (table 11.3) made only slightly disproportionate use of near-urban areas.

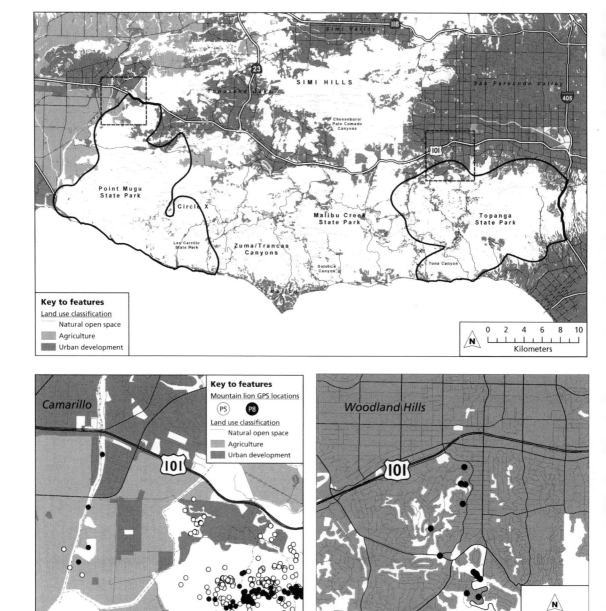

Figure 11.3. Top, Home ranges for subadult male mountain lions P5 (west, *bottom left*) and P8 (east, *bottom right*) in the Santa Monica Mountains Study Area, showing potential "frustrated dispersal" (see text). Dashed boxes are areas shown in the bottom figures. *Bottom left,* Locations for subadult male lions P5 (open circles) and P8 (filled circles), showing movements into developed and agricultural areas around the Santa Monica Mountains. See top figure for area locations. The western (left box) movements of P8 occurred during a 5-week exploration outside the home range depicted in top figure.

The two "city cats" in our studies illustrate two situations that favor relatively intense use of habitats in and close to urban land. The most urban animal in the Santa Ana Mountains study was adult female F15, whose home range was concentrated in the broad riparian zone of the lower Santa Margarita River on U.S. Marine Corps Base Camp Pendleton. Because this preferred riparian area was immediately adjacent to military offices, warehouses, outdoor storage facilities, housing, and other areas classified as urban,

Table 11.1. Radio-tagged mountain lions in Southern California occupied home ranges less urban than the study area at large, except as noted in the last column

Distance to urban area[a]	% of study area (2060 km²)[b]	Mean % of 95% fixed kernel home range[c]	Number of mountain lions for which % of home range exceeded % of study area	Values (% of home range) suggesting selection for proximity to urban areas
Santa Ana Mountains (2,060 km²; n = 15 radio-tagged animals, 1988–1992)				
0 (i.e., in urban area)	5.0	3.8	5[d]	5.2, 6.0, 6.3, 7.9, 8.5
1–100 m	4.4	2.1	1[d]	5.1
101–250 m	6.4	3.6	1[d]	8.5
251–500 m	9.8	7.0	2[d]	11.7, 14.8
501–1,000 m	16.8	15.0	5[d]	17.1, 19.4, 19.6, 23.6, 27.6
>1,000 m	57.5	68.6	13	36, 51
Santa Monica Mountains (792 km²; n = 6 radio-tagged animals, 2002–2007)				
0 (i.e., in urban area)	12.9	3.5	0	
1–100 m	7.3	3.3	1	9.2
101–250 m	6.5	4.3	1[e]	9.0
251–500 m	9.5	7.8	1[e]	13.2
501–1,000 m	17.5	16.7	3[e]	19.4, 21.9, 17.8
>1,000 m	46.0	64.4	5	35.0
Simi Hills and Santa Susana Mountains (768 km²; n = 2 radio-tagged animals, 2002–2007)				
0 (i.e., in urban area)	18.4	11.8	0	
1–100 m	9.0	8.3	0	
101–250 m	8.7	8.7	0	
251–500 m	12.0	12.6	2	12.2, 12.8
501–1,000 m	18.3	20.0	2	19.4, 20.5
> 1,000 m	33.5	38.7	2	

Note: All values contrary to this trend are presented in the last column. No inferential statistics were calculated.

[a] Urban areas included commercial, industrial, and high-density residential areas.

[b] The study areas were delineated as irregular polygons defined by freeways and high-density residential areas. Any other reasonable delineation of the study areas would have resulted in a much higher urban percentage.

[c] Most of the overlap with urban areas was an artifact of the fixed kernel algorithm used to draw a home range polygon that encompasses 95% of the animal's locations; very few radio-locations were in urban areas (table 11.2).

[d] F15 (table 11.3) was one of these animals.

[e] P8 (table 11.3) was one of these animals.

much of her home range was near urban areas (table 11.3). She showed less affinity for urban areas when her radiolocations were compared with her home range (last column of table 11.3). Despite F15's high use of areas 1–250 m from urban and disturbed areas, there were no reports of F15 being seen by humans during the 15 months she was monitored. Indeed, fewer than six sightings of radio-tagged animals were reported to PB in 5 years of monitoring. In the Santa Monica Mountains, P8 showed the most evidence of increased use of areas in or near development (table 11.3). P8 was a subadult male that was apparently attempting to disperse, probably using marginal areas to avoid conflict with the resident adult male. He made a number of forays, always at night, into highly urbanized areas and some agricultural areas (figure 11.3, *bottom left*). As with F15, we had no reports that P8 was seen

by people. Another subadult male in the Simi Hills (P3) made multiple forays into habitat fragments completely surrounded by roads and development, at times even resting during the day and making kills in these patches.

Another animal that made extensive use of urban areas in the Santa Ana Mountains was M5; he is excluded from tables 11.1 and 11.2 because his home range and movement patterns were strongly affected by a broken left hip and right femur that he suffered in a vehicle collision (Beier and Barrett 1993). During the 5 months between the accident and his death, most of M5's locations were <500 m, and typically <200 m, from urban areas, in an area of flat terrain that included three reservoirs with dense riparian shrubbery. His only documented kills during these 5 months were opossums and skunks. Nearby human residents only

Table 11.2. Twelve of 15 radio-tagged mountain lions in the Santa Ana Mountains, five of six mountain lions in the Santa Monica Mountains, and both mountain lions in the Simi Hills and Santa Susana mountains concentrated their activities in portions of their home ranges over 1 km from urban areas

Distance to urban	Mean % of 95% fixed kernel home range	Mean % of locations	Number of animals using habitat in greater proportion than % of home range
Santa Ana Mountains (*n* = 15 radio-tagged animals, 1988–1992)			
0 (i.e., in urban area)	3.8	1.7	0
1–100 m	2.1	1.7	4
101–250 m	3.6	3.4	6
251–500 m	7.0	5.5	2
501–1,000 m	15.0	14.3	4
>1,000 m	68.6	73.4	12
Santa Monica Mountains (n = 6 radio-tagged animals, 2002–2007)			
0 (i.e., in urban area)	3.5	0.1	0
1–100 m	3.3	1.8	1
101–250 m	4.3	3.0	1
251–500 m	7.8	6.7	1
501–1,000 m	16.7	20.1	3
>1,000 m	64.4	68.3	5
Simi Hills and Santa Susana Mountains (n = 2 radio-tagged animals, 2002–2007)			
0 (i.e., in urban area)	11.8	2.7	0
1–100 m	8.3	5.4	0
101–250 m	8.7	7.6	0
251–500 m	12.6	13.7	2
501–1,000 m	20.0	21.8	2
>1,000 m	38.7	48.9	2

Note: Each of the 23 radio-tagged animals made relatively little use of the urban portion of its home range. No inferential statistics were calculated.

Table 11.3. Habitat use by adult female F15, the "most urbanized" radio-tagged mountain lion in the Santa Ana Mountains, and subadult male P8, the "most urbanized" animal in the Santa Monica Mountains study area, California

Distance to urban area	% of study Area	% of 95% fixed kernel home range	% of radiolocations
Adult female F15 (*n* = 131 locations, February 1991–June 1992)			
0 (i.e., in urban or disturbed area)	5.0	**8.5**	8.4
1–100 m	4.4	**5.1**	**7.6**
101–250 m	6.4	**8.5**	**13.7**
251–500 m	9.8	**14.8**	12.2
501–1,000 m	16.8	**27.6**	18.3
>1,000 m	57.5	**35.5**	39.7
Subadult male P8 (*n* = 2,958 locations, October 2005–September 2006)			
0 (i.e., in urban or disturbed area)	12.9	11.8	0.7
1–100 m	7.3	**9.2**	5.8
101–250 m	6.5	**9.0**	**11.0**
251–500 m	9.5	**13.2**	**21.3**
501–1,000 m	17.5	**21.9**	**28.1**
>1,000 m	46.0	**35.0**	**33.1**

Note: Values in boldface suggest selection for habitat in or near urban or developed areas (no inferential statistics were calculated).

twice reported seeing M5; in neither case did M5 appear to threaten humans or pets.

Ongoing studies in other areas confirm our observations that mountain lions are rarely detected when they visit urban habitat. Twelve of 16 adult mountain lions captured and radio-tagged within 10 km of residential or urbanized areas near Payson and Prescott, Arizona, regularly used low-density residential areas. The proportions of locations within residential areas (many of the low-density residential areas do not qualify as urban) ranged from <1 to >96%. During the 24-month study, these "residential" animals generated 0.21 reported sightings per 100 radio days (T. McKinney, Arizona Game and Fish Department, personal communication). In a study area including suburbs of Seattle, Washington, 32 radio-tagged animals generated 0.28 interactions (including sightings and depredation events) per 100 radio days (B. Kertson, Washington Cooperative Fish and Wildlife Research Unit, personal communication). One young male generated three reported sightings during his 3-month presence in and near the small town of Oracle, Arizona (K. Nicholson, University of Arizona, personal communication).

In our studies and ongoing studies in Arizona (K. Nicholson, University of Arizona, personal communication; T. McKinney, Arizona Game and Fish Department, personal communication), mountain lions use urban areas primarily for traveling, exploratory movements, and dispersal. They also regularly hunt and kill wild and domestic prey in urban areas, or areas nearly surrounded by urban areas, but prefer to hunt in more remote areas. In only one study area, namely, an ongoing study near Seattle, Washington, do mountain lions also use urban areas for territory marking and raising and rearing of offspring (B. Kertson, Washington Cooperative Fish and Wildlife Research Unit, personal communication), although the prevalence of such behavior has not yet been quantified.

Many previous studies in both nonurban and near-urban areas similarly show that mountain lions tend to avoid urban and disturbed areas (Logan and Irwin 1985; Van Dyke et al. 1986; Laing 1988; Belden and Hagedorn 1993; Williams et al. 1995; Maehr and Cox 1995; Dickson and Beier 2002; Arundel et al. 2006; Cox, et al. 2006; Kautz et al. 2006).

MOUNTAIN LIONS AND ROADS

Mountain lions tend to avoid crossing roads. Radio-tagged mountain lions in the Santa Ana Mountains crossed paved roads 65% less often than expected based on observations of their turning angles in areas away from roads, and they especially avoided crossing freeways (Beier 1995; Dickson et al. 2005). Similarly, none of the 9 mountain lions monitored over 5 years ever crossed the 8–10 lane freeway (U.S. 101), separating the Santa Monica Mountains from the Simi Hills, although they did regularly cross two-lane roads, often at culverts or bridged streams, and sometimes across the road surface.

Mountain lions are not very good at crossing roads, especially wide roads with heavy traffic. Road kill was the main cause of death in the Santa Ana Mountains. Two male mountain lions died attempting to cross two-lane Malibu Canyon Road in the Santa Monica Mountains. Mountain lions successfully crossed a four-lane New Mexico highway at grade on seven occasions. After the road was widened to six lanes, only two attempted crossings were documented; both animals were killed in collisions with vehicles (Sweanor et al. 2000).

Highway underpasses and overpasses have successfully facilitated mountain lion movement across freeways in Canada (Clevenger et al. 2001; Clevenger and Waltho 2005), California (Beier 1995, 1996), and Florida (Foster and Humphrey 1995). In the Santa Monica Mountains, a male mountain lion used an underpass (3 cases documented, 11 probable) to cross the SR-118 Freeway, and an adult female crossed the same freeway three times, likely using road or trail underpasses. Mountain lions readily use culverts and bridged undercrossings with low openness ratios (Clevenger and Waltho 2005). However, their ungulate prey are highly reluctant to use underpasses less than about 5 m wide and 3 m tall (Brudin 2003; Ng et al. 2004; Clevenger and Waltho 2005). Clevenger and Wierzchowski (2006) recommend that where highways cross suitable habitat, crossing structures for large mammals should be located no more than 1.5 km apart, vegetative cover should be present near entrances to reduce negative effects of lighting and noise, and roadside fencing should funnel animals off the highway and toward crossing structures.

HOME RANGE SIZE AND POPULATION DENSITY

Near-urban populations resemble nonurban ones in home range size, home range overlap, and population density. However, urbanization could alter mountain lion space use in two divergent ways. Initally, mountain lions in areas isolated by urban barriers may, at least in the short term, exhibit smaller and more overlapping home ranges, leading to increased intraspecific interactions. In a later phase, a small isolated near-urban population in a downward spiral toward extirpation would have low density and low overlap among home ranges.

In the Santa Ana Mountains, annual 85% fixed kernel home ranges averaged 93 km² for 12 adult females and 363 km² for 2 adult males (Dickson and Beier 2002). In the Santa Monica Mountains, annual 95% fixed kernel home ranges were 283 to 476 km² for one adult male (followed for 5 years), 430 km² for the other adult male, and 106 to 230 km² for adult females. After 4 littermates became independent subadults when their mother had died and they dispersed, the 2 females had home ranges of 88 and 144 km², and the 2 males had ranges of 189 and 177 km². Similarly, mountain lions in nonurban areas have home ranges averaging about 140 km² (range 30 to 600 km²) for adult females and 275 km² (range 150 to 700 km²) for adult males (Anderson et al. 1992; Pierce and Bleich 2003).

There is extensive overlap among home ranges of adult female mountain lions but usually less overlap among adult male home ranges (summaries in Anderson et al. 1992, 70; Logan and Sweanor 2001, 248). Consistent with their smaller and more overlapping home ranges, adult females typically outnumber males. Resident adults comprise over half of a typical nonhunted population (Seidensticker et al. 1973; Hemker et al. 1984; Logan and Sweanor 2001) and one-third to one-half of moderately hunted populations (Logan 1983). Subadults (about 10–30 months of age) include nonbreeding independent offspring before and after dispersal, and comprise 0% and 50% of populations (Seidensticker et al. 1973; Hemker et al. 1984; Logan 1983). Dependent young range from 0% to 60% of the population (Seidensticker et al. 1973; Hemker et al. 1984; Logan and Sweanor 2001). In the Santa Monica Mountains, 1 adult male has reached approximately 11 years of age, while 5 of the 7 known subadults died before 2 years of age.

Minimum density estimates for the two near-urban populations were 1.05 adults/100 km² in the Santa Ana Mountains area (Beier and Barrett 1993) and 0.8 adults/100 km² in the Santa Monica Mountains. Density of nonurban populations range from 0.45/100 km² (Lindzey et al. 1994) to 3.6/100 km² (Hopkins et al. 1986). Mountain lion populations are ultimately limited more by prey availability and vulnerability (Pierce et al. 2000b; Logan and Sweanor 2001) than by social interactions among mountain lions (Seidensticker et al. 1973).

FOOD HABITS AND INTERACTIONS WITH OTHER WILDLIFE

Ungulates provide most of the prey biomass consumed by mountain lions throughout their range (Beck et al. 2005). In North America, no mountain lion population persists in an area without mule deer, elk, or other large ungulate prey. Smaller prey include rabbits, hares, skunks, porcupines, beavers, bobcats, coyotes, raccoons, opossums, gray foxes, and small rodents (Boyd and O'Gara 1985; Koehler and Hornocker 1991; Beier and Barrett 1993; Logan and Sweanor 2001). Small prey is consumed in an hour or two; a mountain lion may feed on a mule deer carcass for several days (Beier et al. 1995). An adult mountain lion kills about 48 large mammals and 58 small mammals per year (Beier et al. 1995). Rate of killing by an adult female probably increases with the number and age of her dependent cubs (Ackerman et al. 1986; Harrison 1990). Kill rate may also vary by season (Ross et al. 1997; Hayes et al. 2000). Bauer et al. (2005) documented that mountain lions scavenged ungulate carcasses placed as bait in their nonurban study area. In the Santa Ana Mountains, we documented only two instances of apparent scavenging: one involved a rancid mule deer fawn fed on by a 12-year-old female mountain lion a few weeks before she died, and one involved a cow carcass visited by a male disperser that had been recently crippled in a car accident.

In the Santa Ana Mountains, prey carcasses ($n =$ 145) and scats ($n = 178$) indicated that mule deer were the most important prey item based on frequency of occurrence (58.6% of carcasses, present in 53.4% of scats) and by biomass estimates (78% for carcasses, 81% for scats; Beier and Barrett 1993). In the Santa Monica Mountains, investigating clusters of GPS (global positioning system) locations that covered at least 24 hours led us to discover the remains of 278 carcasses eaten by 9 mountain lions. Mule deer composed 95% of the total, followed by coyotes (4%). Our method was biased against finding carcasses of smaller prey, as many smaller prey are likely consumed entirely. Nonetheless, deer were clearly the dominant prey in both near-urban studies.

In Washington, preliminary results from four ongoing mountain lions studies (3 nonurban study areas, 1 near-urban study area near Seattle) suggest that mountain lions in the near-urban study area have a more diverse diet that includes small and mid-sized prey, including beavers, raccoons, mountain beavers, opossums, coyotes, and domestic animals (B. Kertson, Washington Cooperative Fish and Wildlife Research Unit, personal communication). Although black-tailed deer and elk are the primary prey in the near-urban area, for some individuals nonungulate prey constitutes as much as half of the prey biomass consumed.

Although mountain lions prey disproportionately on young, old, or weakened ungulates with low reproductive value (Hornocker 1970; Kunkel et al. 1999; Logan and Sweanor 2001; Husseman et al. 2003), mountain lions kill more than the "doomed surplus." Mountain lions annually killed about 15%–20% of the mule deer on the Kaibab Plateau in Arizona (Shaw 1980), 8%–12% of the mule deer on the Uncompahgre Plateau, Colorado (Anderson et al. 1992), and 2%–3% of an elk herd and 3%–5% of a deer herd in Yellowstone National Park (Murphy 1998). Because some of these animals would have died anyway, and because prey reproduction may increase when predation lowers herbivore density, the actual impact on prey populations is lower than these numbers suggest. Thus, in most cases, the net effect of predation by mountain lions probably reduces ungulate numbers by well under 10%.

In some cases, predation can cause a *predator pit*, keeping an herbivore population far below its food-based carrying capacity. Predator pits driven by mountain lions have been postulated for wild horses in Nevada (Turner et al. 1992), porcupines in Nevada (Sweitzer et al. 1997), and bighorn sheep in California (Hayes et al. 2000; Schaefer et al. 2000) and Arizona (Kamler et al. 2002).

Mountain lions may opportunistically take advantage of increased abundance of some types of prey in near-urban settings. In particular, prey species that eat human refuse may occur in higher abundance in near-urban areas (e.g., coyotes: Fedriani et al. 2001; raccoons: Riley et al. 1998). Beier and Barrett (1993) documented significant consumption of raccoons, opossums, skunks, and coyotes in their near-urban study area; PB's impression is that most of the carcasses or scats containing these species were close to urban areas. Despite any attractiveness of these near-urban prey, data on habitat use did not suggest mountain lions in the Santa Ana Mountains were moving close to urban areas for these feeding opportunities. In the Santa Monica Mountains, two of the mountain lions that had higher percentages of developed or altered areas within their home ranges also had a higher percentage of coyotes among their kills.

Partaking in a food chain in an urban landscape potentially exposes mountain lions to secondary or tertiary poisons. Two mountain lions in the Santa Monica Mountains died of rodenticide poisoning, specifically from anticoagulant toxicity (Uzal et al. 2007; Riley et al. 2007). These were also the individuals with the highest percentage of coyote kills, and coyotes were widely exposed to anticoagulants in this area (Riley et al. 2003; chapter 7). Thus, mountain lions may be exposed to these toxicants by consuming coyotes, rodents or lagomorphs, and perhaps ungulates (Stone et al. 1999; Eason et al. 2002).

Mountain lions also eat domestic sheep, goats, cattle, horses, llamas, pigs, dogs, cats, geese, chickens, and emus (Beck et al. 2005). In a few areas where cows and calves are kept on the range year-round, cattle can be an important prey, but most complaints concern sheep. Mountain lions often kill 5 to 10 sheep at a site but usually eat only one or two of them (Roy and Dorrance 1976).

Scavengers such as coyotes and bears can take mountain lion kills, causing mountain lions to kill more often (Harrison 1990). Any benefit to these other carnivores is likely offset by mountain lions regularly eating coyotes and other small scavengers. The diets of wolves, coyotes, bears, bobcats, and jaguars overlap extensively with mountain lion diets (Beck et al. 2005). As a mountain lion population decreased in the eastern Sierra Nevada, coyote predation on mule deer increased (Pierce et al. 2000a).

Most reports of strong interactions in terrestrial ecosystems between mammalian carnivores and their prey or competitors have been in either true islands (Peterson 1977; Terborgh et al. 2001, 2006) or habitat islands created by urbanization (Crooks and Soulé 1999). Thus, we speculate that interactions between mountain lions and populations of prey and other predators may be intensified in near-urban areas bounded by relatively impenetrable road or development barriers (see chapter 14).

ACTIVITY PATTERNS AND MATING

Mountain lion activity patterns are best known for a particular near-urban population (Beier et al. 1995). On the basis of these analyses, mountain lions travel about 36% of the night, averaging 5.5 km per night, and 9% of daylight hours, averaging about 900 m, when they are not feeding on a kill. Peak activity occurs during the evening crepuscular period. The most common nocturnal activity pattern is associated with hunting, when mountain lions travel about 10 km per night (sum of distances moved per 15-minute interval). When hunting, mountain lions stalk or sit in ambush for about 40 minutes and move about 1.4 km over 1.2 h to another area, repeating this pattern about six times on nights when no prey is killed. When prey is killed, this pattern is suspended until the carcass is consumed (4–6 h for a small mammal, 2–5 days for a large mammal).

Mountain lions minimize loss of ungulate carcasses to scavengers and heat spoilage by dragging the kill 0–80 m to a cache site, burying the carcass under leaves and debris during the daytime, and feeding only at night. During the day, mountain lions usually rest; rest sites are up to 4.2 km (mean 0.4 km) from an ungulate carcass on which the animal is feeding. Mountain lions apparently do not feed on about 40% of days, but when a large mammal is killed, it is fed on for about three nights, creating a "feast or famine" pattern. Mothers with young litters hunt from dusk to midnight and return to the den for the rest of the day; mothers spend increasing amounts of time at greater distances from the den during the first 8 weeks after giving birth (Maehr et al. 1989b; Beier et al. 1995).

We believe that nonurban mountain lions exhibit similar activity patterns. If any differences exist, we suspect that nonurban mountain lions may have greater diurnal activity than near-urban animals.

Mountain lions are generally solitary, except for mothers and cubs in juvenile dependency. Hemker et al. (1984) reported mother and cubs were about 450 m apart when cubs were <6 months old and 800 m apart when 6–12 months old, reflecting changes in nursing requirements, mobility of offspring, and need to defend young.

Mating associations last 2–5 days (Hemker et al. 1984; Padley 1990), during which mountain lions traveled little, vocalized frequently, and apparently did not feed (Beier et al. 1995). On one occasion in the Santa Ana Mountains, a single male bred two females in the same location for several days (Beier et al. 1995). In this same population, pairs of adult females, including eight individuals, remained 0–300 m apart for 1–3 days on 20 occasions at intervals that seemed to coincide with estrus cycles (Padley 1990; W. D. Padley, Santa Clara Valley Water District, unpublished data). During that 12-month period, there was no reproductive activity among these females and no evidence of a breeding male in the area (Padley 1990, 40–43). Although we

cannot explain these unusual behaviors, they may be related to scarcity of breeding males and the near isolation of this population by encircling urbanization.

CONSERVATION AND MANAGEMENT

Conservation of mountain lions is warranted in near-urban areas for the same reasons as in more remote regions. Mountain lions are part of the evolutionary environment, potentially contributing to top-down regulations of ecosystems by structuring herbivore communities and affecting patterns of herbivory. Equally important, living with large predators is good for people because it reminds us that, until recently, we were not masters of the natural world. Of North America's three big predators (wolves, grizzly bears, and mountain lions), in most places only the mountain lion is present to remind us of our roles as plain citizens of the biotic community. The presence of carnivores, especially large ones, greatly enhances the sense of wildness people experience in a landscape. We need carnivores wherever people live—not just in the remotest areas—just as we need monuments, music, and art wherever we live—not just in a few cultural centers.

Fortunately, mountain lions are widespread and resilient. In most areas where they still occur, conserving the species requires only tolerant human attitudes and some restraint in transforming landscapes. Mountain lions tolerate moderate levels of predator control, hunting, and road mortality, and even high human-caused mortality if the population is connected to source populations (Stoner et al. 2006).

The most endangered mountain lion population, the Florida panther, has received new genetic material to reverse serious medical and reproductive problems arising from inbreeding (Beier et al. 2003; Pimm et al. 2006) but is still threatened by urbanization and habitat loss. If there is one conservation lesson to be derived from the Florida panther, it is that preventing isolation and habitat fragmentation is far cheaper and more effective than trying to reverse these trends (Beier 2009).

In other areas, conserving mountain lions will require recolonization. Driven by impressive dispersal abilities, since 1990 mountain lions have begun recolonizing midwestern and eastern North America (Cougar Network website, http://easterncougarnet.org; Thompson and Jenks 2005; Fecske 2006). Fourteen states appear to be regularly receiving cougars dispersing from known breeding populations or can soon expect to receive dispersers (Beier 2009). Policy in each of these states will affect the ability of cougars to recolonize that state or states farther east. In most of these states, mountain lions are not protected, and only Nebraska has a management plan for recolonizing mountain lions (Beier 2009). We encourage other states to develop plans and regularly revise their plans in collaboration with all stakeholders. Perhaps the most important step to facilitate expansion is to educate citizens about cougars; such efforts should be informed by surveys to learn what people know and feel about mountain lions.

Mountain lions can help inform conservation strategies and motivate conservation efforts that benefit many other species. Because they are sensitive to area, connectivity, and other human influences, they can help planners estimate the minimum size of habitat cores and set limits on human activities (roads, fencing, artificial night lighting, use of anticoagulant poisons) that can degrade wildland blocks, buffer zones, and corridors (Beier 2009). Cougars have been documented to use narrow habitat corridors and peninsulas of habitat up to 7 km long in urban areas (Beier et al. 1995); this further justifies their utility as a focal species for corridors in urban areas. We have been involved in collaborative, science-based efforts to conserve wildland linkages that would ensure persistence of mountain lions and other species in Southern California (Beier et al. 2006). These linkage designs have been incorporated into federal and state habitat conservation plans, county and city land-use plans, and the agendas of government agencies and conservation groups.

Most Americans have a positive, protective attitude toward mountain lions, wolves, and grizzly bears and have fewer negative associations with mountain lions than with the other two large carnivores (Kellert et al. 1996). Positive attitudes are also prevalent in urban areas (Belden and Hagedorn 1993; Vickers 2007). Thus, mountain lions can help build public support for efforts to create wildland networks in these areas.

As people move into and recreate within mountain lion habitat, fear of mountain lion attacks poses a major challenge for conservation biologists, wildlife managers, and land managers, particularly in near-urban areas. Nonetheless, mountain lion attacks on humans are rare. Beier (1991) documented 9 fatal attacks on 10 humans and 44 nonfatal attacks on 48 humans in the United States and Canada during 1890–1990. Most victims (66%) were unsupervised children or lone adults. Of the attacks, 30% were in or near (within sight of) de-

veloped areas such as a campground or rural residence, but only two occurred in urban areas (Alberta in 1962 and Colorado in 1970; listed in Beier 1991). Fitzhugh et al. (2003) reported 7 fatal and 38 nonfatal attacks that occurred after Beier (1991) and 10 nonfatal attacks missed by Beier (1991), who had estimated that 12 nonfatal attacks had been missed. Underweight or yearling mountain lions may have a higher probability of encountering and attacking humans (Beier 1991; Ruth et al. 1991). Encroachment by humans into mountain lion habitat and increases in mountain lion numbers are probably the main factors driving a fivefold increase in attacks since about 1970 (Beier 1991; Fitzhugh et al. 2003). During 1990–2003, there were about 3.5 attacks per year in the United States and Canada.

Reports from several national parks (summarized in Beck et al. 2005) suggest that mountain lions may become habituated to human developments and activities. However, there is no clear evidence that habituation increases risk of attack. In one nonurban state park in Southern California, mountain lions avoided areas within 100 m of trails, roads, and camp facilities during the day but not at night (Sweanor et al. 2008). During peak hours of human activity in that state park, about 9% of adult male locations and 19% of adult female mountain lions were within 100 m of a trail. This created ample opportunities for encounters, but no dangerous encounters occurred. When a human approached within 50 m (often <10 m) of mountain lions in a remote area of New Mexico, mountain lions typically responded by moving away (Sweanor et al. 2005). Those that did not flee usually stayed in place without exhibiting threat behavior or showing interest in the human as potential prey. Thus, staying in place should not be interpreted as habituation to humans or lost fear of humans (Sweanor et al. 2005).

Because mountain lions generally avoid urban areas, there is little risk of attack in cities. However, many people living near mountain lion country hike, run, bicycle, ski, and engage in other activities in adjacent wildlands, creating opportunities to encounter a mountain lion. This may increase the risk of encounters per unit of habitat area, but our results suggest that a near-urban mountain lion is not inherently more dangerous than a nonurban mountain lion. Radio-collared mountain lions near human-populated areas in the Santa Ana and Santa Monica mountains studies were rarely observed by people. They behaved similarly to mountain lions studied in nonurban areas. The near-urban lions primarily hunted and consumed mule deer, generally restricted themselves to natural habitats or areas of dense cover, and remained out of sight of humans. The occasional locations immediately adjacent to homes, commercial establishments, golf courses, recreational trails, picnic areas, and campgrounds (e.g., figures 11.3, *bottom left*, 11.4) make interesting anecdotes. But these near-urban animals were not attracted to areas of human activity and were not treating humans as prey. An appreciation of these facts is critical if mountain lions are to survive in near-urban habitats.

As people continue to move into and use mountain lion habitats, mountain lions will interact with humans, livestock, pets, and hobby animals at the urban-wildland interface. Removing depredating mountain lions or individuals perceived as a risk may significantly increase mortality for small near-urban populations already at risk because of limited habitat, fragmentation, and roadway mortality. Simple strategies such as improved pens, bringing pets in at night, and using guard dogs to help protect hobby animals

and livestock can reduce depredation without harming mountain lion populations. In the Santa Monica Mountains, for instance, a male mountain lion attacked and killed several goats, sheep, and a llama over a 2-year period. After husbandry practices were improved, attacks on livestock in this area were eliminated. Beck et al. (2005) advocated modifying the process of issuing depredation permits to emphasize preventive education and requiring property owners to modify husbandry practices to reduce such incidents.

In general, near-urban areas pose more risks for mountain lions than near-urban mountain lions pose for humans and human property. Secondary poisoning, movement constraints from habitat fragmentation and roads, removal of mountain lions for depredation or real or perceived safety risks, and perhaps increased intraspecific strife among mountain lions confined to small isolated habitats pose special challenges for near-urban mountain lions. Mountain lions will persist in near-urban environments if people tolerate their pres-

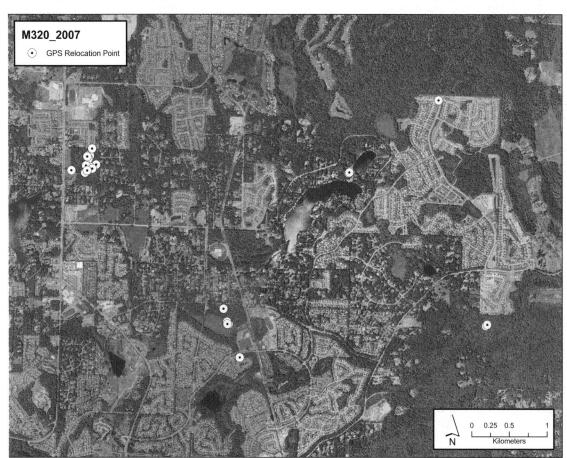

Figure 11.4. GPS (global positioning system) locations of subadult male M320 in a suburban landscape he used for 2 months in the Issaquah-Sammamish Plateau area east of Seattle. Image: B. Kertson, Washington Cooperative Fish and Wildlife Unit.

ence, implement strategies to reduce human-caused mountain lion mortality, and ensure that large blocks of interconnected habitat are protected to support mountain lion populations and dispersal between them.

ACKNOWLEDGMENTS

We thank M. Elbroch and H. Quigley for reviewing an earlier version of this chapter. Brett Dickson calculated many of the statistics in the tables. M. Lotz (Florida Fish and Wildlife Conservation Commission), K. Nicholson (University of Arizona), T. McKinney (Arizona Game and Fish Department), B. Kertson (Washington Cooperative Fish and Wildlife Research Unit), and L. Lyren (U.S. Geological Survey) kindly shared observations from their ongoing studies of mountain lions in near-urban areas.

12

Domestic Cat (*Felis catus*) and Domestic Dog (*Canis familiaris*)

PHILIP J. BAKER

CARL D. SOULSBURY

GRAZIELLA IOSSA

STEPHEN HARRIS

D UE TO their close association with humans, domestic cats and dogs are ubiquitous residents of human settlements. They have been introduced around the world as biological control agents, to aid hunting, for guarding livestock and property, and as a source of food or companionship. Consequently, they are two of the most widely distributed terrestrial mammals, occurring throughout Africa, Asia, Australasia, Europe, North America, and South America. Furthermore, cats are, or have previously been, present on some sub-Antarctic islands (Bester et al. 2000, 2002; Pontier et al. 2002; Nogales et al. 2004) and sled dogs are used within mainland Antarctica. At a global scale, therefore, cats and dogs are probably the most common and widespread urban carnivores, although their densities may occasionally be exceeded locally by other species (see chapter 13). There are approximately 600 million cats and 400 million dogs worldwide (Matter and Daniels 2000; Coppinger and Coppinger 2001; Driscoll et al. 2007).

Despite their global prevalence, however, there have been relatively few studies on the behavior and ecology of cats and dogs in urban areas; in the case of cats, especially, more work has focused on rural populations. In writing this chapter, we have, therefore, drawn on studies across the full spectrum of human habitations where cats and dogs live alongside people, from cities to isolated farms. Although these habitats are all dominated by human activities, ecologically they are very different. Consequently, wherever possible we have highlighted similarities and differences in the ecology and behavior of cats and dogs living in different habitats.

This chapter is divided into four main sections. In the introduction, we outline the history of the domestication of cats and dogs, their physical characteristics, and briefly consider their association with humans. We then give an ecological classification of these species as they occur in urban areas and define a number of key terms. Next we describe key elements of the social behavior of semiferal cats and dogs (i.e., reproduction, ranging behavior, and social interactions). We then review factors affecting survival and mortality, including diseases and parasitic infections, before outlining problems caused by cats and dogs in urban areas. In particular, we consider their impact as predators, their interactions with other species and discuss the practical issues associated with their management. Finally, we emphasize the important role of cats and dogs in the ecology of urban areas and identify a few key areas that we believe require more detailed research.

ORIGINS AND PHYSICAL CHARACTERISTICS

The domestic dog originated from the gray wolf and was the first species to be domesticated (Clutton-Brock 1999). Although archaeological evidence suggests domestication occurred as recently as 14,000 years ago (ya; Olsen 1985), modern molecular techniques imply that the two species started to diverge up to 40,000 ya, with the first true dogs arising in East Asia; these were subsequently transported to Europe and to North America 12,000–14,000 ya (Vilà et al. 1997; Vilà et al. 1999; Leonard et al. 2002; Savolainen et al. 2002; Linblad-Toh et al. 2005). Interbreeding between wolves and early dogs appears to have been commonplace (Vilà et al. 2005), and it was not until nomadic humans started to settle ca. 15,000 ya that the selective breeding of dogs began in earnest (Vilà et al. 1997). During breed development, the selective breeding of males, rather than females, with desirable traits appears to have been particularly common (Sundqvist et al. 2006). Recognizable counterparts to some modern breeds have been unearthed from as long as 4,000 ya, whereas others are much more recent (<200 ya; Sundqvist et al. 2006).

Cats were domesticated approximately 9,000 ya and are most closely related to *Felis silvestris lybica*, a subspecies of wildcat found in Africa (Vigne et al. 2004; Driscoll et al. 2007). The timing of domestication and its location in the Fertile Crescent coincide with the commencement of human agricultural practices in that area. It has been speculated, therefore, that cats first began to associate with humans because of a tendency to feed on rodents living in grain stores (Driscoll et al. 2007).

Cats and dogs are sometimes considered subspecies of *Felis silvestris* and *Canis lupus*, respectively, but are more commonly regarded as true species: *Felis catus* and *Canis familiaris* (Wilson and Reeder 2005). Both reproduce successfully with their wild ancestors, and dogs occasionally breed with other canid species (Vilà and Wayne 1999; Adams et al. 2003). Such interbreeding can be deleterious for these wild populations: introgression and hybridization are recognized threats in the conservation of all subspecies of *Felis silvestris* (Nowell and Jackson 1996; Beaumont et al. 2001) and dingoes (*Canis familiaris dingo*; Corbett 2004), a subspecies of dog introduced into Australia 3,000–4,000 ya, probably by traders (Savolainen et al. 2004).

Selective breeding has led to substantial physical differences, and even physiological changes (Björnerfeldt et al. 2006), between modern cats and dogs and their wild ancestors. Although there are a large number of color forms of cats, there are only around 100 recognized breeds (Morris 1999). Yet even in countries where pedigree breeds are popular, they only make up around 10% of the domestic cat population; in most countries, they constitute <2% (Fogle 2001). In contrast, there are several hundred dog breeds (Morris 1986), although this number is substantially lower than in the 1800s (Fogle 2000). The lower number of cat breeds probably reflects their later domestication but possibly also a need to retain a degree of independence because their dietary requirements have not easily been met by humans (Bradshaw et al. 1996; Bradshaw 2006). Most cat breeds weigh 2.5–7.0 kg, although some exceed 11 kg. The weight range is more pronounced in dogs (1–90 kg). Other pronounced physical changes evident in some breeds include the near or complete absence of fur, elongated fur, the absence of a tail, drooped rather than erect ears, and changes in skull morphology. However, despite our ability to modify the physical appearance of both species through selective breeding, with a more independent existence both cats and dogs tend to revert to a more generalized form because of selection against highly modified breeds.

Historically, humans have placed great value on cats and dogs, but both species have also experienced periods of less favorable association with people. Indeed, it has been suggested that early hunters may have first taken wolf pups they found simply as a source of food (Clutton-Brock 1995), and dogs were eaten in many parts of the world until recently (Serpell 1995), a practice now common only in parts of Asia (Corbett 1995b). Similarly, although revered in ancient Egypt, cats were also killed in large numbers to provide religious followers with bodies for sacrificial offerings; these mummified cats and other animals were so common that they were later removed from tombs and used as fertilizer (Clutton-Brock 1999). Cats were also widely persecuted during the Middle Ages as the proverbial "witch's familiar" and often cruelly put to death (Serpell 1988, 2000).

In recent times, however, cats and dogs have generally been viewed more positively as working animals and companions and have even been used in therapeutic medicine (Beck 2000; Filan and Llewellyn-Jones 2006; Souter and Miller 2007). Unfortunately, this close association with people also means they are often victims of cruelty and neglect and pose a number of management concerns. From an ecological and societal perspective, the most serious problems arise when animals are allowed to stray or are abandoned and have to fend for themselves; these are likely to range more widely and receive less human care, increasing their likelihood of encountering dangers (e.g., catching diseases or being killed by vehicles). The Humane Society of the United States estimates that 6–10 million cats and dogs enter animal shelters in the United States each year (www.hsus.org). It is important to recognize, therefore, that there are different classes of cats and dogs living in modern conurbations and that these vary in the degree to which they are likely to affect urban ecosystems because of differences in their behavior and relative abundance.

AN ECOLOGICAL CLASSIFICATION OF URBAN CATS AND DOGS

Cat and dog populations have tended to be categorized according to the degree to which they depend on or associate with humans, although such attempts to classify animals into discrete categories is fraught with problems and has, therefore, led to the development of a range of descriptive terms. For example, Moodie (1995) recognized three categories of cats (which can also be applied to dogs): (1) pets are individuals living in close connection with a household where all their ecological needs are intentionally provided by humans;

(2) semiferal individuals are animals that rely only partly on humans for provision of their ecological requirements; and (3) feral cats or dogs are free-living individuals that have minimal or no reliance on humans and which survive and reproduce in self-perpetuating populations. However, these divisions are not absolute. For example, pet cats and dogs may range freely within their environment, catching wild prey or scavenging from trashcans (Rubin and Beck 1982; Macdonald and Carr 1995; Hsu et al. 2003; Fielding and Plumridge 2005), implying a degree of independence from humans. With cats, other descriptive terms have been used (e.g., house cat, inside/outside cat) to indicate pets that use their owner's home as a base but which wander beyond the confines of the property. The term *free roaming* has also been used to refer collectively to combinations of pet, semiferal, and feral cats in the same area (Butcher 1999). Confusingly, some terms have been used to describe contrasting behaviors. For example, the term *latchkey dog* has been used to refer to pets that were allowed to wander freely, especially while their owners were out at work, and to describe dogs left indoors during the day (Andersen 2002). Similarly, the term *house cat* now tends to be used to describe pets that are totally confined to their owner's residence.

Thus, cats and dogs exhibit a continuum of dependence on humans between the two extremes of complete reliance (totally housebound pets) and complete independence, which would only occur when humans are absent. Urban cats and dogs fall somewhere in the middle of this spectrum, receiving food, shelter, and medical attention from humans to some degree. Despite these caveats, it is still useful to classify (or to attempt to classify) urban cats and dogs in accordance with their ecology and behavior. This chapter will use the following terminology (adapted from Moodie 1995): (1) *housebound pets* are those that remain totally within their owner's home, although even these may exert some ecological effects through the disposal of their waste (Conrad et al. 2005); (2) *pet cats* live in close association with a household where most of their needs are intentionally provided by humans but which range through their owner's garden and through their local neighborhood and thus are, or are potentially, free ranging; (3) *pet dogs* live in close association with a household where most of their needs are intentionally provided by humans, which range through their owner's garden and through their local neighborhood either (a) only in the presence of their owner (*restricted ranging pet dogs*) or (b) on their own (*free-ranging pet dogs*): note that restricted ranging pet dogs may also influence the ecology of nonlocal neighborhoods, as they are frequently taken for walks by their owners in other areas of the city or surrounding rural habitats (Miller and Howell 2008); (4) *semiferal cats and dogs* are animals that rely only partly on humans for provision of their ecological requirements and which are free to wander unrestricted. Semiferal cats tend to associate in colonies, most obviously at hospitals, museums, universities, and dockyards; such colonies may contain large numbers of individuals, but smaller colonies (typically <10 individuals) are more prevalent within residential areas. Semiferal and free-ranging pet dogs tend to form small packs but may be more solitary. Given the propensity of cats and dogs to seek human assistance, and of human "caretakers" or "patrons" to offer it (Centonze and Levy 2002; Levy et al. 2003b; Ivanter and Sedova 2008; Lord 2008), truly feral individuals are probably absent from urban areas. Consequently, we avoid using this term except where it relates to figures expressed by an original author. The term *free roaming* is used to describe a combination of free-ranging pets and semiferal animals.

In developed countries, across both urban and rural habitats, pet cats and dogs are typically more common than their semiferal and feral counterparts. For example, there are an estimated 60–90 million pet cats versus 10–50 million semiferal/feral cats in the United States (Nassar and Mosier 1991; Mahlow and Slater 1996; Luoma 1997; Clarke and Pacin 2002; Slater 2002; www.hsus.org); comparative figures for Britain are 7.8 million and 0.8 million, respectively (Harris et al. 1995). Conversely, the reverse is true in many developing countries and in Australia, where there are an estimated 18 million feral cats and only 3 million pets (Anonymous 1996; Pimentel et al. 2001; Clarke and Pacin 2002). There are an estimated 40–75 million pet dogs in the United States (Macpherson et al. 2000; www.hsus.org), although the number of owned dogs is probably declining (Patronek and Glickman 1994; Macpherson et al. 2000); the number of semiferal/feral dogs is not known.

SOCIAL BEHAVIOR

The social organization of a species is typically described in terms of the dispersion of individuals in space and time, including patterns of social interaction within and between the sexes and patterns of mating

and parental care. Although the basic reproductive characteristics of cats and dogs do not change between different habitats, many of these other facets of the social behavior of cats and dogs are more plastic and vary in relation to resource availability and population density. In the following sections, we focus on the social behavior of semiferal animals. Pet animals display many of the characteristics described and are generally able to adopt a semiferal existence if abandoned.

Reproduction

Cats are seasonally polyestrus. Females ("queens") become sexually mature at 4–10 months, males at 5–7 months. The male's penis has backward-pointing spines that rake the female's vagina on withdrawal, causing the female's characteristic mating screech and stimulating ovulation. Semiferal females living in colonies typically mate with several males, either from within or outside the home colony. Consequently, multiple paternity litters are common and many kittens are sired by extra-colony males (Natoli and De Vito 1991; Yamane 1998; Say et al. 1999; Natoli et al. 2000; Ishida et al. 2001; Kaeuffer et al. 2004; Say and Pontier 2004; Natoli et al. 2007). Larger males ("toms") are more likely to court females outside their immediate social group (Yamane 1998; but see Ishida et al. 2001). Variance in male mating success is greater when females breed asynchronously because larger males are able to monopolize individual females (Pontier and Natoli 1996; Say et al. 2001). Matings between sexually intact pets and semiferal cats are common (Bradshaw et al. 1999). Gestation lasts approximately 63–65 days. Average litter size is two to five kittens (Mirmovitch 1995; Scott et al. 2002; Nutter et al. 2004a; Schmidt et al. 2007). Kitten mortality is high, and females typically successfully produce <1.5 offspring per year (Warner 1985; Kerby and Macdonald 1988; Nutter et al. 2004a; Schmidt et al. 2007). Males take no part in rearing kittens and may kill them (Macdonald et al. 1987; Natoli 1990; Pontier and Natoli 1999). During weaning, the female will bring live and dead prey to the kittens.

Female dogs ("bitches") are sexually receptive approximately every 7 months, producing one or two litters per year (Macdonald and Carr 1995; Pal 2001); most become sexually mature at 6–12 months, although this varies between breeds. The penis of the male ("dog") is supported by a baculum and has a spherical area of tissue at its base that becomes engorged with blood during penetration, causing it to swell and become trapped inside the vagina; the resultant postcopulatory lock lasts 5–20 minutes. Females appear to mate selectively with males they are familiar with (Daniels 1983a), although nonpreferred males may be able to mate forcibly (Pal et al. 1999). Some females mate with more than one male (Pal et al. 1999). As with cats, matings between pets and free-roaming individuals are common. Gestation lasts 56–72 days, and average litter size is five or six pups (Macdonald and Carr 1995; Pal 2001, 2005). Males may participate in rearing the offspring by guarding pups and provisioning with food, which may be regurgitated (Pal 2005). Pups range independently at 10–11 weeks old (Pal 2008). Patterns of reproduction in pet dogs in U.S. cities are described by Nassar and Mosier (1980); Nassar et al. (1984); and New et al. (2004).

Ranging Behavior

The ranging behavior of semiferal cats and dogs, and their tendency to live in groups, is heavily influenced by the availability of key resources, such as food and secure resting sites (Liberg and Sandell 1988; Macdonald and Carr 1995; Liberg et al. 2000; Macdonald, Yamaguchi, and Kerby 2000; Steen-Ash 2004). For free-ranging pet cats, most of these needs are provided at their owner's home, and this residence is typically an area of core activity within the cat's range (Leyhausen 1979). Free-ranging pet dogs, however, often tend to roam beyond their owner's home specifically because food and water are not provided or because they are deliberately locked out of the house. The availability of sexually receptive conspecifics will also heavily influence the propensity of sexually intact pets to wander beyond their home residence, although in many developed countries many pet animals are likely to have been neutered.

As the distribution of food, resting sites, and mates varies markedly between different towns and cities, and between different areas within the same conurbation, the home range of the different classes of cats and dogs in urban areas is highly variable. Home range size of semiferal cats living in urban areas is typically between 0.2 and 15.2 ha (table 12.1); free-ranging pet cats living in residential suburbs tend to have ranges <5 ha, although few studies have been conducted (table 12.1). In comparison, semiferal cats associated with farms in agricultural landscapes can have ranges of 100–200 ha or more. The home range size of free-roaming urban dogs is typically <11 ha (table 12.2), although again few studies have been conducted. The concept of a contiguous home range area is not readily applied to both free- and restricted-ranging pet dogs where these are transported by their owners to dispersed locations such as parks to be walked.

Unlike many other urban carnivores, cats and dogs are not overtly territorial, although aggressive interactions between individuals are common when they do meet (e.g., Leyhausen 1979). Consequently, the densities of urban cats (>200 individuals/km²; table 12.1) and dogs (>90 individuals/km²; table 12.2) are often substantially higher than those observed in similar-sized wild carnivores (see chapter 13) and rural cat and dog populations (Liberg and Sandell 1998; Liberg et al. 2000). The highest densities of semiferal cats (>700/km²; table 12.1) and dogs tend to occur where food is highly clumped and superrich, such as garbage dumps. Such populations are, however, typically localized. In contrast, pet populations often occur at high densities and over large areas. For example, the mean density of pet cats across 10, 1-km squares in Bristol, United Kingdom, was 348/km² (Baker et al. 2008).

Sightings of free-roaming cats are generally highest at night (Berman and Dunbar 1983), whereas urban dogs tend to be most active in the evening and during the night and early morning (Fox et al. 1975; Berman and Dunbar 1983; Daniels 1983b); the latter may reflect an avoidance of humans and other possible hazards or temporal changes in the availability of key foods. Both species have few specific habitat requirements within urban areas (sensu Crooks 2002), although they will avoid areas frequented by larger predators.

Social Interactions

Although naturally solitary, pet cats readily live together (Leyhausen 1979), and semiferal cats frequently form multimale/multifemale colonies, particularly where food or refugia are clumped, such as garbage dumps (Macdonald et al. 1987; Kerby and Macdonald 1988; Yamane et al. 1997; Say et al. 2003; Crowell-Davis et al. 2004; table 12.1). Within such colonies, smaller groups of related females are typically present (Mirmovitch 1995; Yamane et al. 1997); their ranges overlap extensively with one another and with those of one or more males (Mirmovitch 1995; Natoli et al. 2001). In one population, females tended to disperse between colonies (Devillard, Say, and Pontier 2003), whereas males dispersed between groups within a colony (Devillard, Say, and Pontier 2004). Such patterns may explain, in part, why colonies sometimes contain a preponderance of males

Table 12.1. Summary of studies of home range size (ha), social grouping, and population density in cats associated with human settlements (after Liberg and Sandell 1998; Liberg et al. 2000)

Source[a]	Location	Habitat	Food source	Status[b]	Social grouping[c]	Mean female range area[d]	Mean male range area[d]	Density (cats/km²)
A	Israel	Urban area	Garbage bins	F (S)	G	0.3 (0.1–0.5)	0.8 (0.3–2.0)	>2,300
B	Japan	Fisher village	Fish dumps	F (S)	G	0.5 (<0.1–1.8)	0.7 (0.3–1.7)	2,350
C	Italy	City park	Handouts	F (S)	G	—	—	>1,000
D	Italy	Market square	Handouts, market refuse	F (S)	G	—	—	1,200
E	France	Hospital grounds	Handouts	F (S)	G	0.2 (±0.1)	1.5 (±0.3)	>700
F	Australia	Garbage tip	Garbage	F/S (S)	G	—	—	>700
G	England	Urban area	House feeding	P (P)	S	—	—	417
H	England	Urban area	House feeding	P (P)	S	—	—	>350
I	England	Urban area	House feeding	P (P)	S	—	—	348
J	Australia	Urban area	House feeding	P (P)	S	—	—	330
K	United States	Urban area	Garbage bins	F/P (S/P)	S?	1.8 (<0.1–6.8)	2.6 (<0.1–10.3)	>300
L	United States	Various	—	F/P (S/P)	—	—	—	>240
M	England	Dockyard	Handouts	F (S)	G	0.8 (<0.1–4.2)	8.4 (0.1–24.0)	200
N	Australia	Suburban/rural	House feeding	P (P)	G/S	2.3 (1.4–3.3)	3.0 (1.6–4.5)	200
O	United States	Suburban	House feeding	P (P)	S	—	—	32
P	England	Dairy farm	Milk, some prey	Fm (S)	G	4.0 (0.7–15.0)	—	30
Q	Scotland	Sand dunes	Rabbits, some food scraps	S (S)	G/S	—	—	19
R	Switzerland	Agricultural land	House feeding, rodents	Fm (S)	G	28.0 (10.6–45.3)	6.5 (–)	18
S	Switzerland	Agricultural land	House feeding, rodents	Fm (S)	G	6.0 (1.2–17.8)	7.2 (0.8–16.0)	14
T	United States	Urban	—	F (S) / S (S) / P (P)	—	14.7 / 5.3 / 1.6	—	—
U	England	Dockyard	Handouts, food waste	F (S)	S?	10.3 (2.6–17.6)	15.0 (0.9–56.0)	>10
V	United States	University campus	Handouts	S (S)	G	12.8 (±5.0)	15.2 (±4.2)	—
W	United States	Agricultural land	—	Fm (S)	G	—	—	9
X	United States	Agricultural land	Farm feeding, rodents	H (P)	G	112 (4.8–185)	228 (109–528)	6
Y	England	Dairy farm	Milk, some prey	F (S)	G	—	—	6
Z	Sweden	Farms, pastures, woods	House feeding, rabbits	M (S/P)	G/S	50	370 (84–990)	>3

[a] A: Mirmovitch (1995); B: Izawa et al. (1982), Izawa (1984); C: Natoli (1985); D: Natoli and de Vito (1988, 1991); E: Say and Pontier (2004); F: Denny et al. (2002); G: Sims et al. (2008); H: Gaston et al. (2005); I: Baker et al. (2005, 2008); J: Grayson et al. (2007); K: Calhoon and Haspel (1989); L: Dabritz et al. (2006); M: Dards (1978, 1983); N: Barratt (1997b); O: Kays and DeWan (2004); P: Panaman (1981); Q: Corbett (1979); R: Weber and Dailly (1998); S: Turner and Mertens (1986); T: Schmidt et al. (2007); U: Page et al. (1992); V: Steen-Ash (2004); W: Coleman and Temple (1993); X: Warner (1985); Y: Macdonald and Apps (1978); Z: Liberg (1980, 1981, 1983, 1984).

[b] Letters outside parentheses denote the classification used by the original authors: F = feral, Fm = farm, S = semiferal, P = pet, H = house, M = mixed; letters within parentheses denote the classification used in the current manuscript: S = semiferal, P = pet.

[c] Social grouping: G = group or colony, S = solitary.

[d] Figures in parentheses are range (min.–max.) or ± standard error (SE).

Table 12.2. Summary of studies of home range size and population density in dogs associated with human settlements

Source[a]	Location	Habitat	Home range size (ha)	Population size Density (dogs/km²)	Population size No. of humans per dog
A	Nepal	Urban	—	>2,900	—
B	India	Urban	—	850 (over 3 years)	—
C	Mexico	Urban	—	534	—
		Urban		936	
D	Philippines	Rural communities	—	468	—
E	Australia	Urban	—	401	—
F	Sri Lanka	Rural communities	—	87 (owned)	—
				108 (ownerless)	
G	India	Urban	—	185	—
H	United States	Urban	0.2–11.1	154	—
I	Spain	Urban	7.8–28.0	127–1,304	—
J	Canada	Urban / suburban	—	>90	—
K	Brazil	Suburban / rural	—	77	—
L	Russia	Urban	0.1–2.4	37	236
M	Brazil	Nonmetropolitan	—	>20	—
N	Zimbabwe	Communal lands	—	8–53	
O	Mexico	Urban	—	—	3.4
P	Brazil	Urban	—	—	3.6
Q	United States	Urban	—	—	3.9
R	Nigeria	Urban	—	—	4.1
S	United States	Urban	—	—	4.1
T	Kenya	Urban	—	—	11
U	Tanzania	Urban	—	—	16.8–19.4
V	Zambia	Urban	—	—	45.0
W	United States	Urban	<0.1–8.4		
X	United States	Urban	2.0–9.0	—	

Note: The density of dogs is presented as absolute density or the number of humans per dog.

[a] A: Kato et al. (2003); B: Pal et al. (1999); C: Daniels and Bekoff (1989); D: Childs et al. (1998); E: Grayson et al. (2007); F: Matter et al. (2000); G: Pal (2001); H: Daniels (1983a, 1983b); I: Font (1987); J: Anvik et al. (1974); K: Campos et al. (2007); L: Ivanter and Sedova (2008); M: Alves et al. (2005); N: Butler and Bingham (2000); O: Ortega-Pacheco et al. (2007); P: Nunes et al. (1997); Q: Nassar et al. (1984); R: El-Yuguda et al. (2007); S: Nassar and Mosier (1980, 1982); T: Perry et al. (1995); U: Knobel et al. (2008); V: De Balogh et al. (1993); W: Berman and Dunbar (1983); X: Rubin and Beck (1982).

(Page et al. 1992; Fromont et al. 1996; Denny et al. 2002; but see Scott et al. 2002). Social hierarchies are typically evident for both sexes, leading to sex and age differences in pair-wise patterns of interaction (Kerby and Macdonald 1988; Yamane et al. 1997; Macdonald et al. 2000; Natoli et al. 2001); outcomes of aggressive interactions may also vary in relation to their context, such as the presence or absence of food (Bonanni et al. 2007). Social communication involves visual, auditory, and chemical signals (Bradshaw and Cameron-Beaumont 2000).

Urban semiferal dogs form groups of 2–10 individuals, containing several males, females, or both (Beck 1973; Daniels 1983a, 1983b; Font 1987; Daniels and Bekoff 1989; Macdonald and Carr 1995; Boitani et al. 1995; Matter and Daniels 2000; Ivanter and Sedova 2008); one or more females may breed (Macdonald and

Carr 1995). There is marked variation in the frequency and extent to which group members associate with one another, making it difficult to discern patterns of organization where individuals have limited direct contact (Macdonald and Carr 1995). For example, of 816 observations of free-roaming dogs in Berkeley, California, 91% were of single individuals and only 1% were of groups of three or more dogs (Berman and Dunbar 1983; see also Daniels 1983b; Lehner et al. 1983; Font 1987; Daniels and Bekoff 1989). Conversely, the three feral dogs in Fox et al.'s (1975) study in St. Louis were rarely apart for prolonged periods. However, these differences may be related to the origins of the animals under observation: free-ranging pet dogs are more likely to be observed singly, whereas semiferal dogs are more likely to be observed in groups (e.g., Westbrook and Allen 1979; Lehner et al. 1983).

The composition of dog packs is often highly dynamic because of mortality, removal by humans (as pets or by animal control agents), and through emigration, recruitment, and immigration (including animals abandoned by their owners). Both juveniles and adults disperse (Pal, Ghosh, and Roy 1998); distances moved are typically small (Pal et al. 1998) because of the high density and small home range sizes observed in urban areas. Dispersal is male biased, although males may remain in their natal group (Macdonald and Carr 1995; Pal et al. 1998). As with cats free-roaming dog populations are often male biased (Daniels 1983b; Font 1987; Daniels and Bekoff 1989; Macdonald and Carr 1995; Pal 2001). Groups may defend a territory using a combination of urine and other scent marks and physical confrontations (Bradshaw and Nott 1995; Macdonald and Carr 1995; Pal 2003); aggressive, sometimes fatal, confrontations are most frequent at range boundaries and in the presence of estrous females (Daniels 1983a; Macdonald and Carr 1995).

SURVIVAL AND MORTALITY

Urban cats and dogs are vulnerable to a range of mortality factors, including attacks by other animals (e.g., coyotes; Quinn 1997a), disease (Kalz et al. 2000), and collisions with vehicles (Kolata and Johnston 1975; Childs and Ross 1986; Kalz et al. 2000; Rochlitz 2003a, 2003b, 2004). Age, sex, and traffic volume influence the risk of collision for cats (Childs and Ross 1986; Rochlitz 2003a, 2003b, 2004); presumably, these factors also influence risks for dogs. Although the number of animals killed by vehicles can be substantial (annual mortality rates of owned cats, semiferal cats, and semiferal dogs in Baltimore were estimated as 3.5%, 8.7%, and approximately 10%, respectively; Beck 1973; Childs and Ross 1986), it has been speculated that the single greatest cause of death for semiferal individuals in the United States is euthanasia in animal shelters (Scarlett 1995; Slater 2001).

Juvenile mortality rates in semiferal cats can be as high as 75% (Kerby and Macdonald 1988; Mirmovitch 1995; Denny et al. 2002; Nutter et al. 2004a; Schmidt et al. 2007), although the relative importance of specific causes is generally unknown. Adults sometimes deliberately kill kittens (Macdonald et al. 1987; Natoli 1990; Pontier and Natoli 1999). Adult mortality rates are hard to quantify because of dispersal from study sites, undetected deaths, and removal by humans. For example, 26 of 39 individuals vanished in one 42-month study

(Kalz et al. 2000), and Devillard et al. (2003) recorded 99 disappearances versus 148 deaths in their 8-year study of about 500 cats. Fourteen-month survival rates for semiferal and feral urban cats in Caldwell, Texas, were 90% and 56%, respectively (Schmidt et al. 2007).

Survival rates of semiferal dogs in urban areas are similarly low, with mortality rates in the first year typically >45% (Beck 1973; Daniels and Bekoff 1989; Macdonald and Carr 1995; Pal et al. 1998; Pal 2001, 2008). In Bengal, India, 15%, 16%, 23%, 5%, and 12% of pups were lost to adverse weather conditions, disease, vehicle collisions, predation by adults, and malnutrition, respectively; a further 30% of pups were adopted by local residents (Pal 2001). The life expectancy of semiferal and feral dogs at birth is, therefore, very low in both urban (2.3 years; Beck 1973) and nonurban populations (1.1 years; Butler and Bingham 2000), although annual survival rates for adult animals are greater. Consequently, urban populations can contain low numbers of juveniles and larger proportions of adults. For example, Daniels and Bekoff (1989) reported approximately 16% juveniles (<1 year), 74% adults (1–7 years), and 2% (>7 years), with 8% of unknown age.

In comparison, pet cats and dogs are given easy access to food, are immunized against diseases, and receive vaccinations and emergency medical care. Consequently, the life expectancy of pet cats and dogs is typically 9–15 years (Egenvall et al. 2005); the oldest known pet cat and dog were reputedly 38 and 29 years old, respectively.

Diseases and Parasitic Infections

Cats and dogs are susceptible to many viral, bacterial, and parasitic infections (Macpherson et al. 2000; Samuel et al. 2001; Williams and Barker 2001; Robertson and Thompson 2002; Luria et al. 2004; Martínez-Moreno et al. 2007; Mendes-de-Almeida et al. 2007), some of which are potentially transmissible to humans and other species (Biek et al. 2002, 2006; Kravetz and Federman 2002; Munson et al. 2004; Riley et al. 2004). As pet animals tend to be given a high degree of preventive and prophylactic veterinary care, disease is typically a minor cause of death for prime age animals. Notable diseases that significantly affect semiferal cat populations and which can be transmitted to pet cats and vice versa, include feline immunodeficiency virus (FIV, or "feline AIDS"), feline leukemia virus (FeLV), and feline panleukopenia virus (FPV; Courchamp and Pontier 1994; Courchamp et al. 1997; Pontier et al.

1998; Cave et al. 2002). The most significant disease affecting dogs is rabies.

FIV and FeLV are retroviruses found globally in cat populations, often at low prevalence because of their low transmissibility (Fromont et al. 1997). FIV is less pathogenic than FeLV and is transmitted through bites; FeLV is transmitted by biting, licking, grooming, and sharing food (Fromont et al. 1997). Consequently, FIV infection is most common in males (Courchamp and Pontier 1994; Fromont et al. 1997; Lee et al. 2002; Natoli et al. 2005) and in more aggressive breeds (Pontier et al. 1998). FeLV tends to be most common in free-roaming, sexually intact cats (Fromont et al. 1997). Spatial organization and social behavior significantly affect the epidemiology of these viruses, with colonies in urban areas exhibiting higher prevalence rates or increased temporal variability in prevalence rates (Fromont et al. 1996, 1997, 1998). It has also been speculated that such infections may have contributed to the evolution of sexual size dimorphism in cats (Pontier et al. 1998).

FPV is widely distributed (Rypula et al. 2004) and is closely related to canine parvoviruses; hence, it is also known as feline parvovirus. As a group, parvoviruses cause gastrointestinal damage leading to vomiting, bleeding, and diarrhea and are particularly serious in kittens and puppies, where they are frequently lethal (Cave et al. 2002). Given its infectivity and mortality rate, FPV has been used to eradicate an introduced island population of cats (Bester et al. 2000, 2002); this is one of the few successful examples of biological control of mammals.

The principal routes of transmission of zoonotic infections from cats and dogs to humans are bites (Chomel and Trotignon 1992; Kato et al. 2003), scratches, ingestion of fecal material, and arthropod vectors. Approximately 400,000 and 800,000 people seek medical attention following bites from cats and dogs, respectively, in the United States each year, of which 28%–80% and 15%–20% require treatment (Presutti 2001; Kravetz and Federman 2002). Most dog bites occur in private residences rather than in public parks (i.e., are from pets rather than semiferal dogs; Macpherson et al. 2000; Miller and Howell 2008). In addition, approximately 300 human fatalities arising directly from dog attacks were recorded during 1979–1996 (Sacks et al. 2000). Most cat, but not dog, bites appear to be the result of human provocation (Patrick and O'Rourke 1998).

The most significant disease spread to humans through cat and dog bites is rabies, a viral infection that is fatal without treatment. In developed countries, the major cost of rabies is economic; postexposure prophylaxis costs $700–4,500 per person (Meltzer and Rupprecht 1998a; Patronek 1998). In one exceptional case, contact with a single pet-store kitten resulted in more than $1.1 million of postexposure treatment (Noah et al. 1996). Only 22 human deaths from rabies were recorded in the United States and Puerto Rico during 2000–2005 (Blanton et al. 2006). Although cats are the domestic animal most commonly recorded with rabies in the United States, the incidence rate is low (4.2% of cases in nonhuman animals in 2005 vs. 1.2% for dogs; Blanton et al. 2006); no human has been recorded to have contracted rabies from cats in the United States (Patronek 1998).

Rabies is a much more serious problem in developing countries, where dogs are often the principal vector (Meltzer and Rupprecht 1998b; Cleaveland et al. 2006; El-Yuguda et al. 2007). Rabies has been estimated to cause 55,000 human deaths annually in Africa and Asia, with an additional 0.04 million disability-adjusted life-years lost to nonfatal cases (Knobel et al. 2005); children and young men are particularly at risk from dog bites (Fèvre et al. 2005; Nogalski et al. 2007; Rezaeinasab et al. 2007). Many factors are known to limit the successful management of dog rabies in urban areas in developing countries, including the financial burden on owners of vaccinating pets and working dogs (Meltzer and Rupprecht 1998b). Free-roaming dog populations in these areas were also thought to contain many semiferal individuals, which were viewed as a substantial obstacle to achieving necessary vaccination levels; more recent programs have, however, indicated that many such dogs are actually owned and can be vaccinated effectively (Perry et al. 1995; Cleaveland et al. 2006). For those dogs that cannot be vaccinated parenterally, oral vaccines may be suitable (Matter et al. 1998).

Cat-scratch disease is a bacterial infection commonly found in pet and semiferal cats (Chomel et al. 2002; Boulouis et al. 2005) and which can be transmitted to humans by bites, scratches, fleas, and lice. Approximately 22,000 cases are recorded in the United States annually, of which about 2,000 result in hospital admission (Maguiña and Gotuzzo 2000). Surprisingly, cat owners do not appear to be at greater risk of infection (Skerget et al. 2003). The disease is particularly common in people with poor personal hygiene, such as the homeless, but is usually benign. It is a more serious infection in individuals who are immunosuppressed (Boulouis et al. 2005); the recent trend for such people to have increased contact with cats (and dogs) because of their therapeutic value poses a particular management issue (Robinson and Pugh 2002; see also Macpherson 2005).

Toxoplasma gondii is an intracellular protozoan parasite transmitted primarily between rodents and felids but which is capable of infecting all mammals (Tenter et al. 2000). Nationally, prevalence rates in humans can be >65% (Tenter et al. 2000; Lafferty 2006); prevalence in the United States in 1999–2000 was approximately 12% (Jones et al. 2003) but has apparently increased to >30% in recent years (McQuillan et al. 2004). Humans typically acquire oocysts through the ingestion of undercooked meat, transplacentally, or through the accidental ingestion of fecal material. For example, the single largest outbreak of toxoplasmosis occurred in British Colombia, Canada, following the contamination of drinking water with cat feces (Mullens 1996). Low prevalence rates in some urban cat populations may be attributable to reduced rodent prey availability because of pest control operations (Afonso et al. 2006). *Toxoplasma gondii* is known to cause changes in the behavior of infected rodents that increases their likelihood of depredation (Webster et al. 1994, Berdoy et al. 2000), and has also been implicated in some human mental health disorders (Torrey and Yolken 2003). It has even been suggested that latent infections of *T. gondii* may have influenced the development of national cultures by way of its effects on human behavior (Lafferty 2006). The greatest risk to humans is when infection occurs in the womb or when the immune system is depressed. More recently, *T. gondii* has been discovered in marine mammals (Conrad et al. 2005), raising speculation that the parasite may be entering marine environments through contaminated cat feces flushed into urban sewage systems that empty into the sea.

Other diseases affecting humans that can be transmitted by cats and dogs include echinococcosis, leptospirosis, toxocariasis, and visceral leishmaniasis (Kalz et al. 2000; Barutzki and Schaper 2003; Rembiesa and Richardson 2003; Suzán and Ceballos 2005). In addition, 23 cases of bubonic plague linked to cats were recorded in the United States during 1977–1998 (Carlson 1996; Patronek 1998; Gage et al. 2000).

CONFLICT AND MANAGEMENT

The principal management concerns caused by urban cats and dogs are (1) the impact on biodiversity; (2) the transmission of wildlife and zoonotic diseases

(see "Diseases and Parasitic Infections"); (3) the risk of physical attack; (4) the risk posed by collisions with vehicles (5) the nuisance value caused by rifling through garbage; (6) noise disturbance (Beck 1973; Meyer et al. 2003); and (7) the deposition of feces. For example, Dabritz et al. (2006) estimated that 9,330 free-roaming cats in Morro Bay, California, produced 107.6 tons of waste annually, and 213 kg of dog feces were deposited in composting bins in one park in Montreal, Canada, in just 2 months (Nemiroff and Patterson 2007). Similarly, it is conservatively estimated that 100 tons of dog feces and considerable amounts of urine are deposited in Richmond Park, London, annually; this waste has a considerable effect on the park's flora, especially around parking lots (McDowall 2006).

The deposition of feces (and urine) in public areas also creates a potential source of infection for people. Cats, however, also frequently defecate on private property, which may create a particular health concern, as many non-pet-owning householders may be unfamiliar with the associated zoonotic risks, although many pet owners also appear naïve of these risks (Bugg et al. 1999). The actual magnitude of risk of zoonotic infection from cat and dog feces is, however, poorly quantified. Although toxocariasis is frequently cited as a significant issue, only seven human cases were reported in England and Wales in 2005–2007 (Anonymous 2008a), although serological surveys, especially in undeveloped countries, can suggest annual incidence rates in excess of 10% of people surveyed (Filho et al. 2003). Furthermore, with the notable exception of rabies, few studies have quantified the effectiveness of cat and dog management programs in reducing zoonotic infection rates in humans.

Diet and Predation

Cats are obligate carnivores and are adapted to a predatory lifestyle typified by several small meals a day, whereas the dog's lifestyle is characterized by large infrequent meals (Bradshaw et al. 1996; Bradshaw 2006). The digestive tract of cats is much shorter than that of dogs, which limits their ability to extract nutriment from plant matter. Consequently, unlike dogs, the diet of cats tends to contain very little plant matter. Furthermore, cats are not able to synthesize or regulate effectively several key amino acids and vitamins (Morris 2002). It is essential, therefore, that these nutrients are present in their diet, and this is one possible reason why even well-fed pet cats may hunt.

The composition of the diet of feral cats in nonurban habitats has generally been studied through the analyses of feces and stomach contents (Pearre and Maass 1998; Fitzgerald and Turner 2000). These studies indicate that feral cats are opportunistic predators, generally taking prey in relation to its availability and that trophic niche breadth, a measure of the diversity of prey taken, tends to increase with decreasing latitude (Pearre and Maass 1998). Overall, cats tend to take prey weighing ~1% of their own body mass (Pearre and Maass 1998) but are capable of taking much larger species. In urban areas, semiferal cats show an increased reliance on human scraps (Meyer 1999).

The predatory behavior of urban pet cats has most commonly been studied by quantifying the number of prey animals brought home to their owner's house. One study indicated that cats return approximately 30% of the prey animals they kill (Kays and DeWan 2004). If this is typical, it is necessary to multiply the number of prey returned by a factor of 3.3 to estimate the actual number killed. Using this approach, estimates of the number of animals killed by cats have ranged from 11 to 217 prey/cat/year (table 12.3; see also George 1974; Borkenhagen 1978; Mead 1982; Dunn and Tessaglia 1994; Carss 1995; Flux 2007), although there are numerous problems associated with deriving these estimates (Baker et al. 2005, 2008). In addition to dead prey, a significant proportion of animals may be returned alive. For example, 32% of 358 prey animals returned home by 131 cats in Bristol, United Kingdom, were retrieved alive by their owners (Baker et al. 2005). The proportion of such animals that survive after release is not known, and this mortality is typically excluded in calculations of the impact of cats.

The tendency to kill prey has been observed even in well-fed pet cats, and food-deprived individuals in laboratory experiments may actually forgo eating food given to them in deference to killing live prey (Adamec 1976). Coupled with their high densities (table 12.1), and given that the size of their population is not directly regulated by prey availability, a long-running debate has arisen concerning the effect of free-ranging pet cats on urban prey populations, particularly birds (Barratt 1998; Weggler and Leu 2001; Anonymous 2003; Brickner 2003; Woods et al. 2003; Hawkins et al. 2004; Baker et al. 2005, 2008; Winter and Wallace 2006), and the consequent need for urban cat control (May 1988; Proulx 1988; Fitzgerald 1990; Jarvis 1990; Coleman et al. 1996; Patronek 1998). Perhaps the most infamous tale of the effect that even one cat can have is "Tibbles," who was reputedly responsible for the

Table 12.3. Summary of reported predation rates of pet cats, based on prey returned home to the animal's owner

Source[a]	Location	Method[b]	Habitat	Reported return rate	Standardized predation rate
A	England	A	Mixed	5.5 prey/cat/month	217
B	United States	A	Urban	56 prey/cat/year	184
C	England	A	Mixed	~3.2 prey/cat/month	126
D	United States	B	Urban	0.7–1.7 birds/cat/week	>120
			Suburban	0.6–1.1 birds/cat/week	>102
			Rural	0.8–1.5 birds/cat/week	>137
E	England	A	Mixed	2.9 prey/cat/month	114
F	New Zealand	A	Urban	21.6 prey/cat/year	71
			Urban	20.4 prey/cat/year	67
G	United States	A	Suburban	1.7 prey/cat/month	67
H	England	A	Village	14 prey/cat/year	46
I	Australia	A	Urban	10.2 prey/cat/year	33
J	England	A	Urban	7.9 prey/cat/year	26
			Suburban	26.4 prey/cat/year	87
			Rural	37.5 prey/cat/year	124
K	England	A	Urban	21 prey/cat/year	21
L	Australia	A	Village	5 prey/cat/year	16
M	United States	A	Urban	3.4 birds/cat/year	11

Note: For comparative purposes, recorded prey return rates have been converted to a standardized predation rate (prey killed per cat per annum) by multiplying returns by 3.3 to indicate that only 30% of prey killed are returned home (after Kays and DeWan 2004).

[a] A: Ruxton et al 2002; B: Crooks and Soulé 1999; C: Nelson et al. 2005; D: Lepczyk et al. 2003; E: Woods et al. 2003; F: Gillies and Clout 2003; G: Kays and DeWan 2004; H: Churcher and Lawton 1987; I: Barratt 1998; J: Howes 2002; K: Baker et al. 2005; L: Meek 1998; M: Fiore and Sullivan (undated).

[b] A = owners asked to record prey during a study period; B = owners asked to estimate amount of prey returned home retrospectively.

extinction of the Stephens Island wren, an endemic, flightless passerine found on Stephens Island, New Zealand. This supposedly is the only occasion in which a single individual has been responsible for the loss of a whole species. In reality, however, several factors may have contributed to the wren's demise (Galbreath and Brown 2004; Medway 2004).

The effect of cat predation on urban bird populations is equivocal, in part because no manipulative experiments of cat density have been undertaken thus far. However, Crooks and Soulé (1999) recorded that bird diversity in urban woodland fragments in California was highest where cats were least abundant; cat abundance was negatively related to the presence of coyotes. Instead, the potential influence of urban cat predation has been estimated by comparing the number of prey killed with the density and productivity of prey populations. Such comparisons suggest that, under some circumstances, cats may be limiting local populations of some prey species (Barratt 1998; Weggler and Leu 2001; Baker et al. 2005). Solitary, ground-feeding birds appear particularly vulnerable. In contrast, however, long-term monitoring studies of birds typically do not indicate the declines that would be expected from a pattern of increasing cat ownership (Cannon et al. 2005), nor is there a negative correlation between cat and bird density (Sims et al. 2008).

The impact of cats on commensal rodent populations is also unclear but is probably not substantial, given the reproductive capacity of these species and that they occupy habitats such as sewers or building cavities where they are not readily accessible. Cats may, however, have a localized effect on mammalian prey species, most likely through a combination of predation and predator avoidance. For example, native rodents were less abundant in one California park where semiferal cats were not fed relative to a second area where they were fed (Hawkins et al. 2004). Similarly, Flux (2007) reported that his pet cat caused the local extinction of rabbits from his suburban garden in New Zealand.

In evaluating the possible impact of urban cat predation, it is also necessary to consider whether it is likely to be an additive or compensatory form of mortality (i.e., would those animals killed by cats have survived had they not be taken?). Analyses of the physical condition of birds killed by cats versus those killed by factors such as window strikes have indicated that cat-killed birds generally have lower levels of fat reserves (Møller and Erritzøe 2000; Dierschke 2003; Baker et al. 2008),

suggesting that they may have been likely to die anyway. If so, cat predation would be compensatory and their impact on prey populations would be reduced. However, body mass regulation in passerines is complex. Although fat is an important source of energy, the costs associated with its transport mean that higher fat reserves may increase predation risk by affecting take-off velocity and maneuverability (Witter and Cuthill 1993). Consequently, dominant or better-quality individuals, which may be expected to have better access to more reliable food sources, may be more likely to reduce fat reserves to maximize predator evasion performance. Under this scenario, the lower fat reserves seen on cat-killed birds may actually indicate "better-quality" individuals in the population and that cat predation is an additive form of mortality. The impact of urban cats is, therefore, contentious.

It is clear, however, that urban cats do kill large numbers of prey animals. It has even been postulated that the large number of prey animals taken by cats in urban areas could reduce the prey base available to other carnivores and predatory birds, thereby leading to competition (George 1974), although this is equivocal. It may be sensible, therefore, to adopt a precautionary approach and attempt to limit losses even if it is not clear that prey populations are being reduced. Potential management methods include fitting pets with a bell (Ruxton et al. 2002; Nelson et al. 2005) or an "antipounce" bib (Calver et al. 2007), using ultrasonic deterrents to deter cats from key locations (Nelson et al. 2006; but see Mills et al. 2000), restricting ownership of cats near ecologically sensitive habitats, and reducing the amount of time that pets are allowed to roam (see Lilith et al. 2006), particularly at dawn when predation on birds is highest (Barratt 1997a). Declawing of cats is not recommended because of welfare concerns.

In contrast to cats, urban dogs are typically more reliant on anthropogenic waste (Romanowski 1985; Meyer et al. 2003) so do not appear to pose a substantial risk to wildlife but can be active predators in some instances (Gentry 1983; Campos et al. 2007). Semiferal dogs on urban fringes can, however, pose significant risks to livestock and wildlife through predation, competition, and disease transmission (Coman and Robinson 1989; Matter and Daniels 2000; Butler and du Toit 2002; Butler et al. 2004; Manor and Saltz 2004). For example, one female dog reputedly killed 13 of 23 radio-collared North Island brown kiwis during one scientific study (Taborsky 1988). Predation by domestic and feral dogs poses a significant risk to the endangered

Key deer (Anonymous 2008b), and chasing urban deer can increase levels of traffic mortality (Bender et al. 2004). In Britain, there is also direct predation on adults of smaller deer species such as Reeves' muntjac and roe deer.

Interactions with Other Urban Species

Cats and dogs interact with other urban carnivores in a number of ways. At one extreme, larger carnivores influence their distribution through mesopredation. For example, coyotes regularly kill cats and dogs (Quinn 1997a), such that the latter species avoid areas where coyotes are present (Crooks and Soulé 1999). In a similar manner, dogs predate red foxes (Harris and Smith 1987a), and foxes avoid areas where free-roaming dogs are present (Harris 1981b). Conversely, although cats, particularly kittens, may be killed by red foxes (Harris 1981b), this occurs much less frequently than is popularly supposed, and there is no evidence that cats and foxes actively avoid each other. In many instances, foxes and cats have been observed eating amicably alongside one another and even playing together. Dogs, of course, frequently chase cats, and may even kill them. The frequency with which urban cats and dogs fight with and injure (or are injured by) other carnivores is not known.

Increasing evidence suggests that walking pet dogs in natural or seminatural habitats may significantly affect the behavior of birds and mammals (Miller et al. 2001b; George and Crooks 2006; Randler 2006; Antos et al. 2007; Banks and Bryant 2007; Langston et al. 2007), although the impact of the presence of humans in addition to their dogs is likely to be a significant factor (Riffell et al. 1996; Miller et al. 2001b; Fernández-Juricic et al. 2003; Blumstein et al. 2005). However, there are no data on the impacts of walking dogs in more residential urban habitats nor on the impact of disturbance by pet dogs on the breeding patterns of birds within residential gardens. Conversely, it has been suggested that the presence of high numbers of cats and the constant elevated risk of predation might exert a chronic stress response in urban birds that could lower their reproductive capacity (Beckerman et al. 2007).

Similarly, there are few data to suggest that pet dogs significantly affect the movement patterns of other urban carnivores within their home range. This is most likely because the majority of urban carnivores are nocturnal, and pet dogs tend to be locked in the house at night. Where dogs are allowed access to their owner's garden, it is easy for wild carnivores to avoid these individual gardens with minimum disruption to their foraging patterns. For example, we have regularly observed red foxes foraging in gardens in Bristol, United Kingdom, separated from an irate neighbor's dog by the width of a wooden fence.

Cats and dogs living in areas in proximity to other carnivore species, most notably in villages in agricultural landscapes, may also act as a reservoir of infectious diseases and have been known to cause significant declines of these other species because of diseases such as rabies and canine distemper. Such risks have been effectively managed using programs that include vaccinating cats and dogs (Haydon et al. 2006). The importance of disease transmission from cats and dogs on the population dynamics of other urban carnivores is not known.

Management: The Practical Issues

The management of cats and dogs is frequently controversial and emotive and has tended to polarize into a debate of the ethics of lethal versus nonlethal control (Wulff 2003, 2005; Jessup 2004a–2004e, 2005, and references therein). Ethical concerns aside, the issues surrounding the management of companion animal species are the same as for other urban carnivores: Is it warranted? Does it resolve conflict(s) cost-effectively? Is it undertaken to acceptable welfare standards? Unfortunately, in most cases, data to support or refute the need for human intervention are unavailable (Stoskopf and Nutter 2004). Furthermore, within an urban context, it is rarely possible to disengage discussion of these ethical concerns, as human activities significantly affect the success or failure of cat and dog control strategies more than for any other carnivore species. For example, human caretakers feeding semiferal individuals derive a great deal of pleasure from doing so and may rigorously oppose any attempt to interfere with "their" charges. Similarly, eliminating the problem of abandoned pets will require substantial changes in human behavior (Fournier and Geller 2004).

Although population reduction is frequently viewed as the panacea of urban cat and dog management, many of the problems listed can be reduced using alternative approaches. For example, rabies has been managed effectively using vaccination programs and the postexposure treatment of bite victims, and cat predation could be reduced through the use of collars, deterrents, curfews, and banning ownership of cats in ecologically sensitive areas. Similarly, nuisance problems could be rectified by ensuring that, for example, garbage is deposited securely, dog walkers collect feces, and pet dogs are not allowed to range freely, including in wildlife habitats. Whether pet cats should be allowed to range freely is more contentious, as is the right and legal status of householders to control (including killing) other people's cats on their own property.

Despite these alternatives, semiferal cat and dog management is increasingly viewed in terms of reducing numbers using a range of strategies, such as culling, removing animals from the street that may be rehomed or euthanized, and, in the case of cats, especially trap-neuter-release (TNR) programs, where the goal is a long-term reduction in numbers by halting or lowering the rate of reproduction and recruitment. Sterility may be achieved by ovariohysterectomy or tubal ligation of females or ocrhiectomy or vasectomy of males (Gunther and Terkel 2002). Culling can be effective in reducing or eliminating populations in urban areas (Natoli 1985) and is potentially more desirable when a rapid resolution is required. The deliberate introduction of a disease has also been used to eradicate cats (Bester et al. 2002).

In developed countries, most pet animals tend to be sterilized, in part to prevent the production of kittens and puppies that subsequently may be abandoned. However, a significant proportion of these are only neutered after they have produced a litter (Manning and Rowan 1992), possibly due to a misapprehension that females need to first go through one estrus cycle or have a litter (New et al. 2000). In addition to the free-roaming cats and dogs collected by animal control agents, it is estimated that 6–10 million pet cats and dogs are relinquished to animal shelters in the United States each year (www.hsus.org), indicating a significant inability of prospective owners to choose suitable pets (Fournier and Geller 2004). Many of these animals are euthanized (Sturla 1993).

Compared with culling, TNR programs are more expensive and take much longer and more effort (Andersen et al. 2004; Nutter et al. 2004b) to reduce population size (e.g., 66% reduction in 11 years: Levy et al. 2003a; 22% decline in 10 years: Natoli et al. 2006; see also Levy and Crawford 2004) and are not always successful (Foley et al. 2005). The major obstacle to the rapidity and effectiveness of TNR programs is immigration from other semiferal populations and the abandonment of pets or their offspring. Consequently, the education of pet owners is required to ensure that they neuter and do not abandon their pets (see Levy and Crawford 2004).

Although TNR programs are considered ethically more acceptable (Slater et al. 2008), animals involved in these programs still have to endure the stress of capture, handling, veterinary intervention, and subsequent survival in the urban environment, the latter considered hostile relative to the life experienced by a companion animal (Jessup 2004a–2004e). Some authorities have expressed concern over the welfare standards associated with TNR programs and that "release" is actually tantamount to (illegal) abandonment (Jessup 2004b). Such viewpoints would appear at odds with those studies of urban carnivores described in this volume, in which animals are routinely trapped for reasons such as fitting radio-collars and then released. We would argue that few people would consider such activities to constitute abandonment, even though these animals experience the same dangers as free-roaming cats and dogs. Such dichotomy in the ethical considerations as applied to cats and dogs versus other carnivores almost certainly stems from our tendency to view the former as companions even when living relatively free of humans (Wilford 2004). Paradoxically, however, cats and dogs are often given markedly different legal status from one another. For example, in the United Kingdom, motorists are not obliged to stop and report an incident if they hit and injure or kill a cat, but they are if they hit a dog.

In conclusion, many methods are available for managing conflicts with cats and dogs in urban areas, and applying any method should be considered in terms of its need and the form of resolution required. In many cases, the need for management and its benefits in alleviating defined sources of conflict is equivocal. However, we believe that, where immediate benefits of control are desired and substantial, such as in conservation or the control of major zoonoses, killing individuals or whole groups, or colonies, is acceptable if no other solution is available. Of course, all such activities should be conducted to the highest welfare standards. Alternatively, where a gradual reduction in population size is acceptable, effectively conducted TNR programs may achieve satisfactory outcomes. The practical obstacles to such programs are, however, substantial, and further research into their efficacy and effectiveness is warranted. Pet owners need to be engaged and educated to increase the number of pets that are neutered and to reduce the number of abandoned animals (see Fournier and Geller 2004). The eradication of these species from urban areas, a call that is heard from some stakeholders, is unachievable in the absence of substantial legislative reform and enforcement, and for millions of people it would also be undesirable. The management of urban cats and dogs will, therefore, continue to be a source of contention and colorful debate.

CONCLUSION

Because of their role as companion animals, cats and dogs are by far the commonest urban carnivores. Furthermore, at a local scale, their densities are typically orders of magnitude greater than the density of other urban carnivores. Potentially, they have the most significant ecological role and societal impact of any of the species discussed in this volume. Despite their potential importance, however, relatively little information is currently available on the ecology and behavior of urban cats and dogs: even basic information, such as the home range area of pet cats and free-roaming dogs, is lacking in many instances (see tables 12.1, 12.2). In particular, there is a paucity of information on the social behavior of free-roaming dogs, probably because these are relatively uncommon in developed countries. Consequently, more research on both species is needed. Of particular interest in developed countries are their interactions with, and impacts on, other urban-dwelling species through processes such as predation, competition, and disturbance and how these processes affect ecosystem functioning.

These issues are also increasingly important from a societal perspective. As more people reside in urban areas as the human population grows, focus has increased on the beneficial role that exposure to wildlife and green spaces may have on human well-being and health (Fuller et al. 2007), as well as their function in educating urban residents about biodiversity and conservation (Miller and Hobbs 2002). Such benefits theoretically could have significant financial implications for national health systems. Although the therapeutic role of pets to owners and hospital patients has been well established (Beck 2000; Filan and Llewellyn-Jones 2006; Souter and Miller 2007), their impact in the wider urban ecosystem is likely to be mixed. For example, defecation by free-roaming cats and dogs is a significant nuisance (and possibly health) problem, and one that enrages a large number of people. In addition, if cats and dogs are indeed affecting other urban species, they are likely to be negatively affecting people's everyday experiences of urban wildlife by reducing the abundance or distribution of those positively viewed species, such as birds (sensu Fuller et al. 2008).

Therefore, we consider the following research questions to be particularly significant in terms of the ecology of urban ecosystems: (1) Does predation by domestic cats limit songbird numbers, in turn limiting the abundance of other species, such as avian predators and mammalian carnivores? (2) Does disturbance by cats and dogs influence the distribution of other urban-dwelling species? (3) Does disturbance by cats and dogs influence the reproductive success of other urban-dwelling species, and, if so, what is the underlying mechanism? (4) What proportion of primary productivity is removed by urban cats and dogs in comparison with other urban carnivores? (5) Do cats and dogs have a net positive or negative effect in terms of their effects on human well-being? From the perspective of a global society, however, the most significant research issues relate directly to formulating effective management programs to limit the spread of parasitic infections and diseases, most notably rabies.

13

A Taxonomic Analysis of Urban Carnivore Ecology

GRAZIELLA IOSSA

CARL D. SOULSBURY

PHILIP J. BAKER

STEPHEN HARRIS

ROSS-SPECIES comparisons can reveal important ecological correlates in patterns of behavior and have been used previously to examine patterns of terrestrial carnivore ecology and life history (Gittleman 1986; Sandell 1989). Applying such comparisons specifically to urban carnivores has a number of potential benefits. These can include (1) understanding which parameters within urban ecosystems are important for survival and (2) allowing the development of a framework to predict which species can potentially colonize urban areas in the future, as well as patterns of further colonization or extinction for current urban-dwelling species.

To understand the ecological requirements of terrestrial urban carnivores, a clear definition is needed of what constitutes an "urban carnivore." Three categories can be recognized (table 13.1): (1) populations that live entirely within the urban environment (resident urban carnivores); (2) individual home ranges that encompass some urban environment (transient urban carnivores); (3) individuals reported within the urban environment but for which we do lack detailed information to make a conclusive judgment on their ecology (urban carnivores of unknown status).

The number of species recorded in urban areas is extensive (table 13.1). Of 250 carnivores species (excluding the marine families Otariidae, Odobenidae, and Phocidae), 14% can be classified as urban carnivores; unfortunately, too little information is available to define many species properly. Even within these groupings, a further distinction between "colonizing" urban carnivores such as red foxes and raccoons that actively expand into

Table 13.1. Species of terrestrial carnivore recorded in urban areas and their interaction with the urban environment

Terms and definitions	Species	Scientific name	Distribution	Mass (kg) ♂	Mass (kg) ♀
Resident urban carnivore	American red fox	*Vulpes vulpes*	North America	5.1–4.1	4.3–3.4
Self-sustaining populations	Beech marten	*Martes foina*	Asia, Europe (North America: introduced)	1.8	1.5
with individual home ranges	Bobcat	*Lynx rufus*	North America	10.1	8.9
entirely within the boundary	Coyote	*Canis latrans*	Central and North America	17.9–13.3	16.0–10.2
of urban areas	Eurasian badger	*Meles meles*	Asia, Europe, Japan	9.9–8.0	9.1–6.0
	Eurasian red fox	*Vulpes vulpes*	Asia, Europe, Japan, North Africa (Australia: introduced)	6.4–6.0	5.2–5.5
	Gray fox	*Urocyon cineroargentus*	Central, North, and South America	4.0	3.3
	Kit fox	*Vulpes macrotis*	North America	2.2	1.9
	Raccoon	*Procyon lotor*	North America (Europe, Japan: introduced)	8.1	6.5
	Raccoon dog	*Nyctereutes procyonoides*	Asia, Japan (Europe: introduced)	4.5	4.5
	Striped skunk	*Mephitis mephitis*	North America	4.5–2.9	3.3–1.5
Transient urban carnivores	Asiatic black bear	*Ursus thibetanus*	Asia, Japan	200–60	140–40
Populations in which	Black bear	*Ursus americanus*	North America	225–60	150–40
individual home ranges may	Brown bear	*Ursus arctos*	Asia, Europe, Japan, North America, North Africa	550–150	250–80
include some urban area	Gray wolf	*Canis lupus*	Asia, Europe, Japan, North America	80–20	45–18
or may use urban areas for	Leopard[a]	*Panthera pardus*	Africa, Asia	37–90	
foraging but do not exist ex-	Mountain lion	*Felis concolor*	Central, North, and South America	67–103	
clusively within urban areas	Polar bear	*Ursus maritimus*	Asia, Europe, North America	800–300	300–150
Unknown status	American mink[b]	*Mustela vison*	North America (Asia, Europe, South America: introduced)	1.8–0.8	0.8–0.4
Species recorded in urban	Brown hyena[c]	*Parahyaena brunnea*	Africa		
areas but with insufficient	Brown mongoose[d]	*Herpestes fuscus*	Asia		
information to accurately	Brown-nosed coati[e]	*Nasua nasua*	South America	4.6–3.8	4.1–3.0
describe their ecology	Common genet[f]	*Genetta genetta*	Africa (Europe: introduced)	2.5–1.5	2.0–1.4

Crab-eating fox[g]	Cerdocyon thous	South America	5.7	5.7
Eurasian otter[h]	Lutra lutra	Asia, Europe, Japan	10–7	7–5
Fishing cat[d]	Prionailurus viverrinus	Asia	6.6	5.8
Golden jackal[i]	Canis aureus	Africa, Asia, Europe		
Indian mongoose[j]	Herpestes auropunctus	Asia (Japan, West Indies, and many islands: introduced)	0.65	0.43
Lesser Oriental civet[d]	Viverricula indica	Asia		
Masked palm civet[k]	Paguma larvata	Asia (Japan: introduced)	4.4	5.0
Ringtail[l]	Bassariscus astutus	North America		
River otter[b]	Lontra canadensis	North America	9.2–7.7	8.4–7.3
Ruddy mongoose[d]	Herpestes smithii	Asia		
Side-striped jackal[m]	Canis adustus	Africa	9.4	8.3
Spotted hyena[n]	Crocuta crocuta	Africa	62–45	83–55
Striped hyena[o]	Hyaena hyaena	Africa, Asia, Europe	23–35	23–25

Note: References for species with unknown status are provided. The American red fox and Eurasian red fox are not a separate species, but for the purposes of analyses they are treated so as they differ in their ecology.

[a] Generally transient but there is a resident population in Bombay because of unusual circumstances (the city surrounds a piece of forest where the leopards live).

[b] Mech (2003).

[c] Mills and Hofer (1998).

[d] E. Wikramanayake, personal communications.

[e] Alves-Costa et al. (2004).

[f] Gaubert et al. (2008).

[g] Pedó et al. (2006).

[h] Mason et al. (1992).

[i] Jhala and Moehlman (2004).

[j] Gorman (1975).

[k] Ninomiya et al. (2003).

[l] Barja and List (2006).

[m] Atkinson and Loveridge (2004).

[n] Gade (2006).

[o] Abi-Said and Abi-Said (2007).

urban environments and "remnant" urban carnivores, such as Eurasian badgers whose populations have been surrounded by increasing urbanization, could be made. This is an important between-species differentiation, as it may be expected that remnant species have more specific ecological requirements, while colonizing species are widely adaptable.

In this chapter, we describe the morphological and ecological characteristics shared by the majority of resident urban carnivores. Then, we examine three key aspects of resident urban carnivore ecology: (1) diet and habitat use, (2) home range, and (3) life history parameters (survival rates, litter sizes). The aim of these cross-species comparisons is to highlight which factors are important in determining urban carnivore ecology and to what extent urban carnivore populations differ from rural populations of the same species. Throughout the chapter, we refer to American and Eurasian red foxes, which is not a taxonomical distinction but a way to draw a comparison between two populations living in separate continents with different urban environments.

URBAN CARNIVORES: WHO, WHAT, AND WHERE?

There are differences between resident urban carnivore and transient carnivores. All resident urban carnivores can be classified as mesocarnivores (1–20 kg sensu Buskirk 1999; table 13.1), ranging in size from the 1.5-kg stone marten to a maximal 18-kg coyote. Many of the transient urban carnivores are significantly larger, up to 500 kg for the polar bear (table 13.1). Many factors may cause this, including human tolerance. Negative perceptions of carnivores typically increase with body size (see Røskaft et al. 2007), and larger urban carnivores, such as coyotes, are less tolerated than smaller ones, such as red or kit foxes (Gosselink et al. 2007; chapters 5 and 6). Clearly, human tolerance is a limiting factor for some species.

Dietary requirements and foraging behavior appear to be key. The majority of resident urban carnivores are classified as omnivorous; the sole exception is the bobcat (table 13.2). Omnivorous species appear to be less sensitive to habitat fragmentation and patch isolation as they can use a various resources. Crooks (2002) indicated that among resident urban carnivores in Southern California, bobcats were the most sensitive to habitat fragmentation caused by urbanization (see

also Tigas et al. 2003). In addition, bobcats have the lowest usage of urban habitats (table 13.2).

Foraging behavior is also a major characteristic distinguishing resident from transient urban carnivores. All resident urban carnivores are solitary foragers and none are group hunters. For example, in certain environments, coyotes may hunt cooperatively (Gese and Grother 1995), but in urban environments, they are only reported to hunt solitarily. Urban environments do not preclude the formation of groups, and some species, including Eurasian badgers, red foxes, and coyotes, form or maintain social groups in urban areas (Way 2003; Baker and Harris 2004; chapter 9). Urban landscapes appear to be incompatible with group-hunting due to lack of open areas of suitable size, human disturbance, and other factors that hinder the coordination and communication for group-hunting species.

That urban areas are predominantly colonized by omnivorous species is interesting, as urban areas also contain potentially useable resources for obligatory carnivorous species. Indeed, rodents, lagomorphs, and deer are important and sometimes primary constituents of the diet of resident urban carnivores (Harrison 1997; Lavin et al. 2003; Morey et al. 2007). As the local abundance of these larger prey items is not generally known, it is not possible to estimate whether enough food is available for obligately carnivorous species.

Most resident urban carnivores are found in developed countries. Of those listed in table 13.1, six are found in North America, four in Europe, four in Japan, and one in Australia (Eurasian red fox and raccoon counted multiple times). In contrast, none are from developing countries. While this almost certainly reflects, in part, a lack of research in these areas (see carnivores with unknown status; table 13.1), another key factor may be the high prevalence of populations of stray and feral dogs in developing countries. Stray dogs cause significant mortality to some carnivore species in the periurban vicinity (Courtenay and Maffei 2004). Even in developed countries, stray dogs can limit the distribution of resident urban carnivores (Harris 1981b). In addition, intraguild competition is known to affect the distribution of carnivores within urban areas (e.g., coyotes vs. other mesocarnivores; Crooks and Soulé 1999; red foxes vs. stone martens; Le Lay and Lóde 2004; chapter 14), so it would seem likely that the presence of a larger, high-density population of predators or competitors may limit the propensity of some

Table 13.2. Dietary components, dietary breadth, and mean (±SE) habitat use of resident urban carnivores

	% anthropogenic food	Mean dietary breadth	Mean % natural habitat used
American red fox[a]	62	0.89	79.6 ± 20.4
Bobcat[b]	0	0.57	86.6 ± 6.4
Coyote[c]	0–14	0.89	56.6 ± 13.1
Crab-eating fox[d]	0	0.81	—
Eurasian badger[e]	25–48	1.27	17.3
Eurasian red fox[f]	42–61	1.53	19.6 ± 9.77
Gray fox[g]	2–19	1.44	—
Kit fox[h]	12	1.11	—
Raccoon[i]	14–43	1.24	79.1 ± 3.5
Raccoon dog[j]	0–72	1.11	—
Stone marten[k]	4–12	1.43	—

Note: Where multiple items were present, the one with the greatest frequency was included. Dietary breath was calculated using Shannon's diversity index (Magurran 1988).

[a] Adkins and Stott 1998, Lewis et al. 1993, Roundtree 1999.

[b] Riley 2001, Riley et al. 2003.

[c] MacCracken 1982, Quinn 1997a, 1997b, Fedriani et al. 2001, Cepek 2004, Morey et al. 2007.

[d] Facure and Monterio-Filho 1996, Pedó et al. 2006.

[e] Harris 1982, 1984, Kaneko et al. 2006.

[f] Harris 1981a, Doncaster et al. 1990, Kolb 1985, Saunders 1992, Ansell 2004, Contesse et al. 2004.

[g] Harrison 1997, Riley 2001.

[h] Cypher and Warrick 1993.

[i] Bozek et al. 2007, Hadidian et al., chapter 4.

[j] Takatsuki et al. 2007.

[k] Lucherini and Crema 1993, Prigioni and Sommariva 1997, Apáthy 1998, Lanszki 2003.

carnivore species to colonize urban areas, particularly in developing countries.

Urban Carnivore Diet and Habitat Use

Although the majority of resident urban carnivores are classified as omnivorous, there is considerable variation in the relative use of resources at both species and population levels. To make direct comparisons between species, we collated dietary data from urban carnivore studies across the published literature. Foods were grouped into five broad categories: (1) anthropogenic food, (2) mammalian prey, (3) avian prey, (4) fruit, and (5) invertebrates. Broad groupings were selected because many subcategories affect the calculation of dietary breadth and does not allow for direct comparison. Frequency of occurrence data were used, as this was the most commonly reported measure; from this, dietary breadth was calculated using the Shannon index (Magurran 1988). We collated information on body mass and mean percent natural habitat used (the mean time spent by carnivores in natural habitat; table 13.2).

Dietary breadth was unrelated to body size (males, $n = 9$ species, 14 studies, $r_p = -0.32$, $P = 0.26$; females, $n = 9$ species, 15 studies, $r_p = -0.33$, $P = 0.26$), although there was a significant negative relationship between use of natural habitats by each species and dietary breadth ($n = 5$ species, 9 studies, $r_p = -0.69$, $P = 0.04$; table 13.2); species with lower dietary breadth exhibited greater use of natural habitats. In parallel, higher use of anthropogenic resources was associated with lower use of natural habitats ($n = 4$ species, 9 studies $r_p = -0.84$, $P = 0.02$). However, studies of habitat preference indicate that most urban carnivores preferentially select natural habitats when available, such as badgers (Kaneko et al. 2006), bobcats (Riley et al. 2003), and coyotes (Gibeau 1998). Even species with low usage of natural habitats may actually show a preference for it. For instance, at high population density Eurasian red foxes preferentially selected for residential gardens (Saunders et al. 1997); yet, when density was lowered, more natural habitats were used to the same extent (Newman et al. 2003). Such a pattern suggests that increased density may act as a driver for increased usage of urban habitat or even altered diet composition rather than vice versa.

Urban Carnivore Home Ranges

Home range size in carnivores is generally thought to increase with increasing body size, as greater metabolic needs require greater food resources (Gittleman and Harvey 1982). However, within this limited mesocarnivore guild, home range size did not correlate with body size for either males ($n = 11$ species, 43 studies, $r_p = 0.24$, $P = 0.13$) or females ($n = 11$ species, 40 studies, $r_p = 0.21$, $P = 0.17$). Instead, a significant negative relationship between dietary breadth and home range size was apparent for males ($n = 7$ species, 14 studies, $r_p = -0.62$, $P = 0.02$) but not females ($n = 8$ species, 15 studies, $r_p = -0.48$, $P = 0.07$) although the pattern was similar (figure 13.1A, 13.1B). Although the pattern is similar, the possibility of sex differences in diet and its effect on home range size has been little studied and warrants further investigation. Concurrently, the reliance on anthropogenic resources also negatively correlates with male ($n = 7$ species, 12 studies, $r_p = -0.58$, $P = 0.05$) but not female ($n = 7$ species, 12 studies, $r_p = -0.54$, $P = 0.07$) home range size. The strongest relationship was between the percent use of natural habitat and male ($n = 5$ species, 7 studies, $r_p = 0.87$, $P = 0.01$) and female ($n = 5$ species, 7 studies, $r_p = -0.81$, $P = 0.03$) home range size, but, as shown earlier, this pattern is likely to be linked to diet.

It is unclear whether home ranges are larger or smaller in urban areas compared with natural habitats. Certainly, they are larger for some species (e.g., coyotes), similar for others (gray foxes, American red foxes), or smaller (Eurasian red foxes). Patterns may also vary within a species. Where species do not rely heavily on anthropogenic resources, such as the coyote, home range size tends to increase with the degree of urbanization (Riley et al. 2003). Conversely, Atwood et al. (2004) demonstrated that coyote range size decreased in urban areas because key habitat patches were larger in rural areas. Such a pattern would not normally be expected in urban environments, but it serves to demonstrate that the size and distribution of patches can be key in determining home range size for some species (Crooks 2002; Tigas et al. 2003).

For some species, the reverse pattern appears true (e.g., raccoons: Prange et al. 2004; Bozek et al. 2007). In these species, diet is dominated by anthropogenic foods; this pattern is enhanced with increasing urban development, such that home ranges are smaller as the

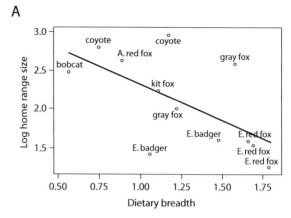

A

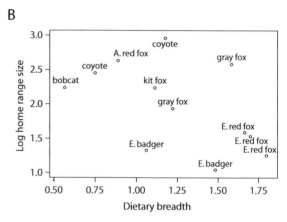

B

Figure 13.1. The relationship between dietary breadth and log home range size in (A) male and (B) female urban carnivores. A. = American, E. = Eurasian. Studies used for home range analysis: Bozek et al. (2007), Boydston (2005), Frost (2005), Gehrt (2004b), Gehrt and Prange (2007), Grinder and Krausman (2001b), Harrison (1997), Herr and Roper (2007), Lewis et al. (1993), Lewis (1994), Prange et al. (2004), Prigioni and Sommariva (1997), Riley et al. (2003), Rosatte et al. (1991), Roundtree (2004), Shargo (1988), Soulsbury et al. (chapter 6), Tanaka et al. (2002), Tigas et al. (2002), Way et al. (2002), Way (2007).

degree of urbanization increases. However, as Harris and Rayner (1986b) indicated, the type of development is important in affecting number and distribution of foxes in British cities. Thus, for all carnivore species, the type of urban development may have varying impacts on home range size; these differences are linked to species' dietary breadth.

Urban Carnivore Life History Parameters

Prange et al. (2003) have suggested that urban environments can alter the life history parameters of resident urban carnivores, such as by increasing litter size. In turn, these can lead to higher population densities in urban than in rural areas, as is found for some (Riley et al. 1998), but not all, species. We collated data from the published literature on litter sizes and percentage adult females breeding in urban and rural populations and annual survival rates in urban and unexploited rural populations. A cross-species analysis did not indicate that urban populations have larger litter sizes; indeed, the converse trend is evident ($t_5 = -4.07$, $P = 0.01$; table 13.3). However, differences in litter sizes are small, with mean differences of more than one offspring per litter. There was no compensation in the proportion of adult females in a population breeding ($t_5 = -0.08$, $P = 0.94$; table 13.3). Increased productivity would be achieved if a greater number of juveniles survived to adulthood; yet, although annual survival rates of juveniles and adults are high, neither juvenile ($t_4 = 0.12$, $P = 0.91$) nor adult ($t_8 = 0.79$, $P = 0.46$) survival was significantly higher than in rural populations (table 13.3). Again, some species-specific trends are evident. The only species with higher mortality rates in urban areas for both juveniles and adults is the Eurasian badger (Harris and Cresswell 1987), a remnant species and one that is not well adapted to the urban environment (chapter 9). Conversely, some species have higher survival rates in urban areas, presumably because they are sheltered from persecution (e.g., coyotes) or from predation (e.g., kit foxes). Coyotes in particular appear to have higher survival in urban areas, but densities do not appear much higher than in rural populations. Indeed,

the two species with the greatest differences between urban and rural densities (table 13.4), raccoons and red foxes in Europe, appear to have limited differences in litter size and mortality (Table 13.3). Thus, across urban carnivore species, urban environments do not appear to enhance population size or productivity by increasing litter size or by decreasing mortality.

URBAN CARNIVORE DENSITY: IMPLICATIONS FOR THE FUTURE?

Clear differences exist among carnivores in how individual species interact with urban areas. By examining density (table 13.4), a sliding scale of success for urban carnivore species is apparent, with positive, neutral, and negative influences on carnivore populations. Diet appears to be the primary determinant of species success. Fedriani et al. (2001) suggested that increased usage of anthropogenic resources increases the density of urban coyotes, but this does not demonstrate causation. Increased densities may alter diet (Ansell 2004) in the same way higher densities have been shown to alter habitat preferences (Newman et al. 2003). Densities may be increased through a reduction in other variables, such as mortality (table 13.3). However, this is the case only for some, but not all, species. Having carried out cross-species comparisons of several ecological parameters, we found no single, unified pattern

Table 13.3. Litter sizes, % parous adult females, and survival rates of juveniles and adults of urban and rural populations of carnivores

	Urban populations				Rural populations			
	Litter size	Juvenile survival rate	Adult survival rate	% parous females	Litter size	Juvenile survival rate	Adult survival rate	% parous females
American red fox[a]	4.0	0.33	0.32	—	4.7	0.20	0.28	—
Bobcat[b]	—	—	0.83	—	—	—	0.82	—
Coyote[c]	5.0	0.84	0.80		5.6	0.21	0.56	—
Eurasian badger[d]	—	0.36	0.64	35	—	0.71	0.80	44
Eurasian red fox[e]	4.5	0.40	0.48	82	4.7	—	0.41	97
Kit fox[f]	3.8	—	0.92	79	3.8	—	0.51	61
Raccoon[g]	3.2	0.54	0.74	73	3.5	0.66	0.70	94
Striped skunk[h]	—	—	0.62	87	—	—	0.67	75

[a] Storm et al. 1976, Lewis et al. 1993, Gosselink et al. 2007.

[b] Riley et al., chapter 10.

[c] Kennelly 1978, Windberg et al. 1985, Atkinson and Shackleton 1991, Grinder and Krausman 2001a, Way et al. 2001, Riley et al. 2003.

[d] Harris and Cresswell 1987, Cresswell et al. 1992, Macdonald and Newman 2002.

[e] Lloyd 1980, Harris and Smith 1987a, Heydon and Reynolds 2000, Baker et al. 2006, Soulsbury et al., chapter 6.

[f] Ralls and White 1995, Cypher et al. 2000, Frost (2005), Cypher, chapter 5.

[g] Rosatte et al. 1991, Gehrt and Fritzell 1999, Prange et al. 2003, O'Donnell and DeNicola 2006.

[h] Gehrt 2005, Rosatte et al., chapter 8.

Table 13.4. Minimum and maximum recorded densities of urban and rural populations of carnivores

Species	Urban (individuals/km²)	Rural (individuals/km²)
American red fox[a]	1.5	0.4–2.7
Bobcat[b]	0.6–0.8	0.04–1.5
Coyote[c]	0.6–3.1	0.2–2.5
Eurasian badger[d]	0.8–33	0.2–44
Eurasian red fox[e]	2–37	0.2–2.5
Raccoon[f]	25–333	1–56
Stone marten[g]	6–8	0.7–2
Striped skunk[h]	0.6–6.5	0.5–26

[a] Storm et al. 1976, Adkins and Stott 1998.

[b] Riley et al., chapter 10.

[c] Gehrt 2004b.

[d] Kowalczyk et al. 2000, Macdonald and Newman 2002, Harris et al., chapter 9, Huck et al. 2008a.

[e] Webbon et al. 2004, Soulsbury et al., chapter 6.

[f] Riley et al. 1998, Gehrt 2004b.

[g] Herr 2008.

[h] Gehrt 2004a, 2004b.

examined in the future, detailed studies on basic parameters of urban carnivore biology are needed (i.e., density, litter size, survival rates). By studying urban populations for longer periods, subtle interactions between carnivores and their environments can be understood. Last, the greatest advancement in understanding urban carnivore ecology would be to study the species classified here as "unknown status urban carnivores," for which we currently have little or no information.

ACKNOWLEDGMENTS

We thank Stanley Gehrt, John Hadidian, Seth Riley, and Suzie Prange for providing hard-to-find manuscripts and information and Brian Cypher, Steve DeStefano, Stanley Gehrt, and one anonymous referee for comments on an earlier draft of this chapter.

for urban carnivores. Species clearly respond to urban environments in different ways. Furthermore, urban environments differ greatly in their suitability for each species.

Our analyses have shown that certain species (e.g., Eurasian red foxes, kit foxes, raccoon dogs, stone martens) are likely to make greater use of urban habitats, as is evident from their diet and home ranges and can be expected to colonize additional areas in the future. Others (e.g., coyotes) also appear to be positively influenced by urban areas but to a lesser extent because they are limited by diet and habitat requirements. Some species (e.g., American red foxes, striped skunks) appear to find urban areas neutral, perhaps gaining some limited benefits, but would not be expected to colonize urban areas rapidly. Finally, the Eurasian badger appears to be a remnant species not well adapted to the urban environment. Although urban badger diet is diverse (table 13.2), their requirement for suitable dens is the greatest limiting factor to the colonization of urban environments. Indeed, the smaller sett sizes in urban versus rural areas may reflect this (Davison et al. 2008).

CONCLUSION

The ecology of urban carnivore species is diverse; central to species differences is diet flexibility. Although many areas of urban carnivore ecology should be

Stone Martens (*Martes foina*) in Urban Environments

Jan Herr

The stone marten is common in much of continental Europe and, of the eight marten species that occur worldwide, is the only one that has successfully entered urbanized habitats. The species has been recorded in urban areas, ranging from small rural villages to the centers of large cities across Central and Eastern Europe (Herr 2008 and references therein). Information on when stone martens began to exploit urban environments is largely missing, but records dating back to the 1940s and 1960s suggest that this was a gradual process that had already started by the mid-twentieth century (Nicht 1969). In urban areas, the diet is usually dominated by fruit, birds, and small mammals. However, although stone martens will feed on just about anything edible they come across, they have not been reported to specialize on scavenging from refuse or to profit from deliberate feeding by human residents, as do other urban carnivores (Holišová and Obrtel 1982; Tester 1986).

Conflicts between stone martens and people are relatively common. Martens have a tendency to den in attics or roof spaces, where the resulting noise, damaged roof insulation, odors, or staining from feces and urine often cause them to be considered a nuisance. They have also earned a reputation for climbing into car engine compartments, where they cause damage by tearing up noise and heat insulation mats and chewing through ignition leads, coolant hoses, or electrical wiring. This behavior was first observed in Switzerland in the late 1970s and subsequently spread across most of Central and into Eastern Europe (Kugelschafter et al. 1984/1985). A yearly peak in damage occurrences coincides with a heightened interest for parked cars, which, in spring and early summer, the martens systematically patrol and scent-mark. This suggests that territorial behavior plays an important role in this phenomenon (Herr 2008).

A telemetry study on 13 stone martens in two towns in southern Luxembourg revealed a very strong association with urban habitat (Herr 2008). Almost all the martens' daytime denning and just over 90% of their nightly activities took place in residential areas within the urban perimeter of the relevant towns, while neighboring forested and rural areas were rarely visited (see figures). Buildings were strongly favored as daytime den sites, with 42% of days being spent in inhabited and 54% of days in uninhabited buildings. Urban martens, like their rural counterparts, adhered strictly to a pattern of intrasexual territoriality (see figures). This ensured that, despite home ranges being somewhat smaller in urban areas (Bissonette and Broekhuizen 1995; Herr 2008), population densities remained limited to a moderate 4.7 to 5.8 adult martens/km^2.

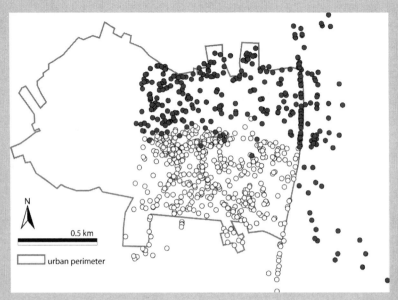

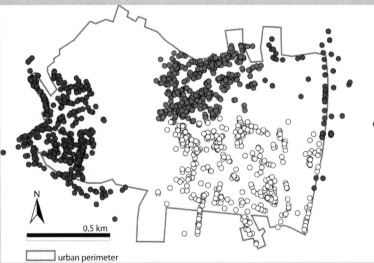

Nocturnal telemetry fixes for (*above*) two male and (*below*) four female stone martens in Bettembourg in southern Luxembourg. Although there is very little overlap between ranges within either sex, male ranges overlap with those of one or two females.

14

Community Ecology of Urban Carnivores

KEVIN R. CROOKS

SETH P. D. RILEY

STANLEY D. GEHRT

TODD E. GOSSELINK

TIMOTHY R. VAN DEELEN

A FUNDAMENTAL AVENUE of ecological research involves the study of communities—assemblages of species that occur together in space and in time. Indeed, questions such as the relative importance of top-down versus bottom-up forces in structuring ecological communities often have dominated research and debate in the ecological literature for decades (Hairston et al. 1960; Fretwell 1987; Power 1992). As arenas for their research, community ecologists have often avoided urban areas in favor of more pristine natural habitat. However, understanding the structure and function of ecological communities in urban systems is becoming progressively important (Faeth et al. 2005). Biological diversity is increasingly affected by, and found within, urban landscapes (McKinney 2002), and urbanization is the leading cause of species endangerment in the United States (Czech et al. 2000).

Similarly, for decades, carnivore biologists have examined the community ecology of carnivores, including competitive interactions among predators and predation impacts on prey. Although much of this research has occurred in rural areas, there has been increasing interest in the food web dynamics of predators in urban systems, particularly because mammalian carnivores can differ considerably in their responses to urban fragmentation, with some species markedly sensitive to urbanization and other species enhanced by it (Crooks 2002; Faeth et al. 2005). Herein, we review the community ecology of urban carnivores, focusing on interactions among carnivore species and impacts on prey. First, we will define and describe potential community-level interactions of carnivores to illustrate the complex role of predators in the dynamics of food webs. We then review a series of case studies that have explored some of these pathways for carnivores in a variety of urban settings. In light of these and other studies, we then discuss how community-level interactions of carnivores might change along a gradient of urbanization and provide directions for future research and conservation efforts regarding the community ecology of urban carnivores.

COMMUNITY-LEVEL INTERACTIONS

We can categorize several major types of interactions of carnivores with other species in the ecological community, following from standard definitions in the ecological literature. *Competition* is where two (or more) species use the same limited resource, or seek that resource, to the detriment of both (Krebs 2001). Two different types of competition have been defined (Birch 1957): (1) *resource* (or *exploitative*) *competition*: when organisms use common resources that are in short supply and (2) *interference* (or *contest*) *competition*: when organisms seeking a resource harm one another in the process, even if the resource is not in short supply. Schoener (1983) further subdivides resource competition into (1) *consumptive*: consuming a renewable resource, specifically food items and (2) *preemptive*: competition for a finite, permanent resource, usually space (e.g., among plants or other sessile organisms). Schoener (1983) also subdivides interference competi-

tion into (1) *chemical*: competition through chemical mechanisms (e.g., allelopathy), (2) *overgrowth*: competition through overgrowth (e.g., among more sessile organisms), (3) *encounter*: direct interactions that may involve harming or killing other individuals, and (4) *territorial*: one species excludes the other from specific areas. Both resource and interference competition can occur among carnivore species, specifically consumptive, encounter, and territorial competition. *Intraguild predation* might be considered an extreme form of encounter competition, where the subordinate carnivore is not only killed but eaten by a dominant species in the same ecological guild (Polis et al. 1989; Polis and Holt 1992). More generally, *predation* occurs when members of one species consume as food members of another species (Krebs 2001).

Figure 14.1 presents a simple stylized diagram of community-level direct and indirect interactions of urban carnivores. On the left is a pathway from apex carnivores to vegetation, with top predators interacting with consumers (e.g., ungulates) through predation and consumers interacting with vegetative communities through herbivory. The arrows go both ways, in that we envision both top-down and bottom-up forces to be important drivers of the structure, function, and composition of ecological communities, often interacting simultaneously and interactively (Estes et al. 2001). The identity of apex predators will certainly vary among systems, depending on which mammalian carnivores still persist in the area and which have

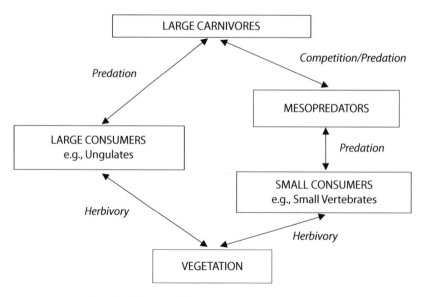

Figure 14.1. A simple stylized diagram of direct and indirect interaction pathways among urban carnivores and the ecological community.

been extirpated. For example, after their reintroduction to Yellowstone National Park, wolves are once again considered the apex predator (Smith et al. 2003), whereas coyotes might be the top predator in urban systems in which larger predators, such as wolves, mountain lions, or bears, have been extirpated (Crooks and Soulé 1999).

The right side of figure 14.1 shows an indirect and perhaps less intuitive pathway. Here, large carnivores influence the distribution, activity, or numbers of smaller-bodied "mesopredators" (typically 1–15 kg mammalian carnivores; sensu Soulé et al. 1988; Buskirk 1999) through competition and predation. Although top carnivores will kill and sometimes eat small predators (Palomares and Caro 1999), spatial and temporal avoidance, territorial shifting, behavioral exclusion, and resource partitioning are primary mechanisms by which dominant predators can limit mesopredator populations (e.g., coyote-induced fox declines through avoidance mechanisms; Sargeant et al. 1987). Apex predators are particularly vulnerable in urbanizing landscapes because of large home ranges and resource requirements, low densities and reproductive rates, and direct persecution by humans (Noss et al. 1996; Gittleman et al. 2001; Crooks 2002). Consequently, as the dominant carnivores decline or disappear from urban areas, relaxed competition and predation is hypothesized to facilitate an ecological release of subordinate mesopredators. In turn, some mesopredators may be predation threats on a variety of prey populations (e.g., small vertebrates), which themselves interact directly with vegetation through herbivory. This process of "mesopredator release" therefore assumes strong top-down interactions along the pathway and that mesopredators have a greater impact on smaller prey populations than do apex predators. The mesopredator release pathway has been implicated in affecting both carnivore and prey populations and community structure generally in a variety of systems worldwide (Sargeant et al. 1987, Soulé et al. 1988; Palomares et al. 1995; Rogers and Caro 1998; Courchamp et al. 1999; Crooks and Soulé 1999; Henke and Bryant 1999; Estes et al. 2001; Terborgh et al. 2001; Mezquida et al. 2006; Johnson et al. 2007).

Although studies suggesting top-down effects of carnivores in general, and mesopredator release specifically, are increasingly widespread, the strength of top-down pathways certainly varies across taxa, systems, and time (Faeth et al. 2005). For certain combinations of species in certain areas, top-down effects may be weak or nonexistent, as in the case study regarding responses of raccoons and striped skunks to coyotes in urban forests in Chicago. Indeed, predator guilds are diverse and the relationships among species are often complex and poorly understood (Gehrt and Clark 2003). We would expect those combinations of carnivores that are most similar ecologically to exhibit the most pronounced agonistic interspecific interactions. This is exemplified within the family Canidae, where at times strong dominance relations have been recorded as a linear hierarchy based on body size (Johnson et al. 1996; Cohn 1998; Gehrt and Prange 2007), with widespread evidence of displacement and mortality between wolves and coyotes (Mech 1970; Peterson 1977; Fuller and Keith 1981; Carbyn 1982; Smith et al. 2003) and between coyotes and red foxes (Dekker 1983; Voight and Earle 1983; Sargeant et al. 1987; Harrison et al. 1989; Theberge and Wedeles 1989; Sargeant and Allen 1989; Gosselink et al. 2003, 2007), gray foxes (Crooks and Soulé 1999; Fedriani et al. 2000a; Farias et al. 2005), swift foxes (Sovada et al. 1998; Kitchen et al. 1999; Kamler et al. 2003), and kit foxes (Ralls and White 1995; Cypher et al. 2000). For those species that are less similar ecologically, we might expect weaker dominance interactions. We would also expect weaker top-down effects emanating from rare predator species (e.g., felids such as mountain lions or bobcats) compared with those that occur in higher densities (e.g., canids such as wolves or coyotes).

Further, the existence of top-down effects certainly should not preclude or minimize the importance of bottom-up effects in anthropogenically disturbed systems. For example, Elmhagen and Rushton (2007) analyzed the response of a mesopredator (red fox) to declines in top predators (wolf and Eurasian lynx) and agricultural expansion in Sweden, accounting for the bottom-up impact of productivity along a landscape-level bioclimatic gradient. They recorded a distinct mesopredator release effect in relation to declines of top predators, but ecosystem productivity determined the relative strength of the trophic interactions. The release effect for foxes was more pronounced in more productive areas, whereas top-down forces were less important than bottom-up constraints on foxes in less productive sites. That both top-down and bottom-up processes are important determinants of community structure should be expected, and the challenge therefore is to evaluate the relative importance of each in urban systems (Patten and Bolger 2003). We would expect bottom-up constraints to be most important for those carnivores without a sympatric ecologically similar top predator.

The complexities in relationships among urban carnivores are illustrated in the following case studies. We synthesize information from these and other studies to highlight patterns and predictions regarding the potential mechanisms structuring carnivore communities and how these might be altered with urbanization.

CASE STUDIES

Mesopredator Release in Urban Habitat Islands in San Diego, California (K. R. Crooks)

In coastal Southern California, human population growth and widespread urban sprawl has created one of the world's major endangerment and extinction hot spots (Myers 1990; Dobson et al. 1997). In San Diego and neighboring cities, much of the remaining natural shrub habitat occurs as undeveloped canyons dissecting coastal mesas; many of these fragments essentially persist as habitat "islands" immersed within a vast urban sea. In this system, Soulé et al. (1988) formally proposed the mesopredator release hypothesis as a potential mechanism to help explain the rapid disappearance of scrub-breeding birds from urban habitat fragments. They predicted that the decline of the most common large predator in these fragments, the coyote, would result in the ecological release of urban mesopredators and increased predation rates on birds and other small prey.

To test this prediction, we (KRC and colleagues) exploited spatial and temporal variation in the distribution and abundance of the coyote among small (<100 ha) urban fragments to investigate the direct and indirect effects of this apex predator on the ecological community. In accordance with the mesopredator release hypothesis, Crooks and Soulé (1999) found negative relationships between visitation rates by coyotes and mesopredators to track stations within fragments, even after statistically accounting for the confounding effects of fragment size, age, and isolation on carnivore activity. Simply the presence of coyotes in fragments (i.e., if they visited the fragment at all during the study) was also an important predictor of overall mesopredator activity. Mesopredators appeared to temporally avoid coyotes as well. Specifically (1) in fragments that coyotes visited episodically during the course of the study, overall mesopredator activity was higher during sampling sessions when coyotes were absent than when coyotes were present; (2) temporal variation in

mesopredator visitation rate was higher in fragments in which coyotes came and went compared with those in which they were either constantly present or absent; and (3) coyotes and mesopredators visited the same track station on the same night less than would be expected based on random visitations.

Of the suite of mesopredator species detected in these fragments, gray foxes, followed by domestic cats and opossums, appeared to have the most consistently negative relationship with coyotes, with most analyses revealing statistically significant spatial and temporal avoidance of coyotes. Raccoons and striped skunks generally had the least consistent relationships with coyote activity in the fragments, particularly when controlling for confounding effects of fragment size, age, and isolation.

As predicted by the mesopredator release hypothesis, species richness of scrub-specialist birds was lower in fragments with fewer coyotes and more mesopredators, after accounting for the positive effect of fragment area and the negative effect of fragment age on bird persistence. Crooks and Soulé (1999) suggested that the top-down effect of coyotes on cats likely had the strongest impact on scrub-breeding birds in the urban fragments. Coyotes kill domestic cats in the fragments, as evidenced by cat remains found in fragments and coyote scats, as well as direct mortality of radio-collared cats by coyotes. Moreover, the top-down impact of coyotes on cats was evident by the behavior of not just cats, but also of their owners. In questionnaire surveys distributed to residents bordering fragments, residents generally understood the threat of coyotes to cats, and consequently, almost 50% of cat owners somehow restricted their cat's outdoor activity because of coyotes in the fragment. This represents an uncommon example of mesopredator release at least partially mediated by the behavior of humans.

Reduced cat activity in these fragments likely benefits populations of smaller vertebrate prey. Unlike wild predators, domestic cats are recreational hunters maintained far above carrying capacity by nutritional subsidies from their owners; they continue to kill prey species even when populations of that species are low (Churcher and Lawton 1987; Coleman and Temple 1993; see chapter 12). From questionnaire surveys, nearly one-third of residents bordering San Diego fragments owned cats, on average each owner owned 1.7 cats, three-fourths of cat owners let their cats outdoors, and 84% of outdoor cats brought back kills to their owner's house. Using data on cat densities and predation rates, Crooks and Soulé (1999) estimated that cats surrounding a moderately sized fragment return approximately 840 rodents, 525 birds, and 595 lizards to residences per year. Collection of prey items returned to houses bordering fragments indicated that most were native species, some of which were relatively rare within these fragments. Because most cats in these fragments were either owned or at least fed by surrounding residents, cat predation is largely an edge effect that spills over from the urban matrix. Predation impacts by cats, therefore, should be expected to be higher in smaller fragments, which have a higher edge to interior ratio. Indeed, cat visitations to track stations were more common in smaller fragments, with cat activity relatively high within 50 m of the urban edge but then tending to decline into the interior of habitat fragments (Crooks 2002).

Extinctions of some birds appear to be frequent and rapid within some of these urban fragments. Scrub-breeding bird species richness decreases with fragment age (Soulé et al. 1988), and local extinctions exceed colonizations across the urban matrix (Crooks et al. 2001), suggesting ongoing avifaunal collapse. At least 75 local extinctions may have occurred in these fragments over the past century (Bolger et al. 1991). Existing population sizes of some birds do not exceed 10 individuals in small to moderately sized fragments (Bolger et al. 1991), so even modest increases in predation pressure from mesopredators may quickly help drive native prey species to extinction.

Although top-down constraints appear to be operating in this system, bottom-up forces clearly help structure the ecological community as well. Trophic interactions between coyotes, mesopredators, and prey were embedded within a network of complex fragmentation effects that exerted considerable bottom-up constraints on both predator and prey populations. For instance, landscape geographic variables, such as fragment size, age, and isolation, and local habitat variables, such as vegetative cover and distance to the urban edge, were important predictors of distribution and abundance of populations of carnivores (Crooks 2002) and potential prey, including birds (Soulé et al. 1988; Crooks et al. 2001; Morrison and Bolger 2002; Patten and Bolger 2003; Crooks et al. 2004), rodents (Bolger et al. 1997), and invertebrates (Suarez et al. 1998; Bolger et al. 2000).

An overarching trend is that many of these taxa require native vegetative shrub cover to persist within urban fragments (Soulé et al. 1992), indicating the importance of bottom-up limitations, such as resources available for movement, foraging, cover, and shelter. Indeed, in a study of reproductive success of four scrub-breeding bird species along a fragmentation gradient in San Diego, Patten and Bolger (2003) found that the strength of top-down forces varied with characteristics of predator and prey and their species-specific responses along the urban gradient. Although this study did not specifically quantify abundance of mammalian mesopredators or monitor sites without coyotes, they did not find evidence that mesopredators were important nest predators, instead they found that snakes and avian predators reduced scrub-bird nest success. Given these results, Patten and Bolger (2003) suggested that if top-down trophic cascades through mesopredator release occurred, they were more likely caused by increased mortality on fledglings and adults rather than predation on eggs and nestlings. Overall, this case study again exemplifies how top-down and bottom-up constraints work in concert to shape communities of mammalian predators and their prey in urban systems.

Competition between Bobcats and Gray Foxes in a National Park Bordering Urban Development in the San Francisco Bay Area, California (S. P. D. Riley)

From 1992 to 1995, I (SPDR) studied the behavior and ecology of bobcats and gray foxes in urban and rural zones of Golden Gate National Recreation Area, a national park just north of San Francisco, California. Studying the two species in sympatry allowed an analysis of interspecific interactions and how they might vary between urban and rural areas (Riley 2001).

Consumptive competition is difficult to establish because detailed knowledge of the population dynamics of both predator and prey is necessary. Consumptive competition requires that prey resources are limiting for the predator populations, which may often be the case, but also that competing predators limit access to the resource, which may be less common and certainly more challenging to determine. In this study, even though gray foxes were omnivorous and used a variety of prey, including fruit, invertebrates, trash, and pet food, there was extensive overlap between the diets of bobcats and foxes. Both species relied heavily on meadow voles, especially in the urban zone, where 90% of bobcat scats contained vole remains and voles comprised two-third of the estimated fresh weight of prey in the diet (% FWP; Kelly and Garton 1997; Riley

2001). Dietary overlap values (Pianka 1973) between bobcats and foxes were 0.62 (% occurrence) and 0.92 (% FWP) in the urban zone, and 0.56 (% occurrence) and 0.90 (% FWP) in the rural zone. Overlap was highest during the wet season (November–April) in the urban zone, when it was 0.75 for % occurrence and 0.95 for % FWP (Riley 2001).

There is certainly the potential for consumptive competition between gray foxes and bobcats in this case, particularly during the wet season when fruit and nuts are less important in the fox diet. Consumptive competition, however, would also require one species limiting the supply of voles for the other. Although voles appeared to be dense in this area, vole populations can fluctuate and even cycle, so during a vole population low, competition could be more severe. Consumptive competition did not appear more likely in the urban zone because dietary overlap was high in both areas. In fact, if interspecific spatial overlap is reduced in the urban zone, consumptive competition may be reduced. Unlike bobcats, foxes used developed areas outside the park, and female bobcats, which have smaller and therefore more intensively used home ranges, were not found even at the edge of development (Riley 2006). This increased flexibility in landscape use by foxes, as well as their ability to take advantage of anthropogenic food resources, may decrease the likelihood of consumptive competition.

Interference competition also was evident between bobcats and gray foxes in this study area, particularly in the urban zone (Riley 2001). Bobcat and fox home ranges overlapped extensively in both zones, and in the rural zone, there was also significant interspecific overlap of core areas (50% adaptive kernels). In contrast, in the urban zone, fox core areas were almost entirely outside those of bobcats; mean overlap of fox core areas by those of bobcats was only 4%. This spatial separation was potentially related to a focus on different habitat. The core areas of female bobcats, in particular, centered on grassy willow riparian valley bottoms, whereas fox core areas were located higher on slopes in dense coastal sage scrub and chaparral. However, there was also direct evidence of bobcat-fox interactions, which could further explain spatial separation of the two species in the urban zone. Three of eight foxes that died in the urban zone were killed, although not eaten, by another predator, likely bobcats (coyotes had been extirpated from the urban zone and were just beginning to return during the study), and I witnessed a bobcat chasing a fox. No radio-collared foxes in the rural zone were killed by predators, even though bobcats were also common in the area and coyotes were still present there.

Why would interference competition potentially be stronger in the urban zone? The available natural habitat in the urban zone was constrained on three sides by development or open ocean, while in the rural zone, there were fewer such constraints. The potential restriction of female bobcats to areas away from the urban edge could increase the intensity of carnivore use within the urban zone of the park. Although foxes may avoid contact with bobcats by inhabiting areas in and near development, avoiding core areas of intense bobcat use in interior natural areas of the urban zone may lead to increased fox survival as well. In the final section of the chapter, we discuss in more detail the hypothesis that resource limitations in habitat bounded by urban development may heighten agonistic encounters among urban carnivores.

Carnivore Community Interactions in the Santa Monica Mountains and Simi Hills North of Los Angeles, California (S. P. D. Riley)

At Santa Monica Mountains National Recreation Area, a national park north of the Los Angeles Basin in Southern California, we (SPDR and colleagues) have been studying carnivore behavior, ecology, and conservation since 1996. These studies have involved monitoring, through radio-tracking and other methods, bobcats from 1996 to the present, mountain lions from 2002 to the present, coyotes during 1996–2004, and gray foxes during 1997–1999. The simultaneous study of multiple species has also allowed investigations into the interactions among carnivores and how these interactions may be affected by the urban landscape.

Fedriani et al. (2000b) studied potential interactions, including dietary overlap, for sympatric gray foxes, coyotes, and bobcats at three sites in the Simi Hills and Santa Monica Mountains. These sites consisted predominantly of core areas of natural habitat but also some suburban/residential development. The diets of the three species overlapped significantly (overlap values of 0.6–0.8 based on dry weight of prey items in scats), with small mammals and lagomorphs important prey for all three. Although both gray foxes and coyotes are omnivorous and seasonally include significant amounts of fruit and invertebrates in their diets, diet overlap was highest between gray foxes and bobcats. This may have been because gray foxes were more restricted to denser chaparral and so focused more on specific small mammal species, such as woodrats, which were also heavily preyed on by bobcats (Fedriani et al. 2000b).

Recently, we have analyzed the diets of bobcats and coyotes in a much more fragmented part of the landscape, collecting scats from some of the same core natural areas in Fedriani et al. (2000b) as well as in small and large habitat fragments in the Simi Hills (National Park Service, unpublished data). Gray foxes were rare or absent in these habitat fragments. Dietary overlap between bobcats and coyotes was still high in this analysis (overall overlap of 0.51 based on percent occurrence in scats). In habitat fragments, overlap was consistent across seasons (0.49 in the dry season, 0.54 in the wet season), and lagomorphs (desert cottontails) represented a consistent portion of the diets (~60% occurrence for bobcats, ~30% occurrence for coyotes). In the core natural areas, dietary diversity was generally higher for both species, with increased use of birds, reptiles, and other mammals (e.g., voles, woodrats), and in the dry season dietary overlap dropped to 0.33. Overall, carnivores in this urban landscape are certainly using similar food resources, but whether consumptive competition is occurring is unknown. In habitat fragments, the greater focus on a smaller number of species, specifically rabbits, may make resource competition more likely, particularly in the wet season when the proportion of fruit in the coyote diet is reduced. However, higher rabbit densities in these fragmented areas could negate the effects of increased diet overlap.

Evidence is stronger for potential effects on carnivore individuals or populations through direct interactions. Other carnivores, especially coyotes, were the major source of mortality for radio-collared gray foxes in the Simi Hills (Farias et al. 2005) and have also been a significant source of mortality for bobcats (Fedriani et al. 2000a; Riley et al. 2003), particularly females (National Park Service, unpublished data). Coyote predation was also the main source of mortality for young (<12 weeks) bobcat kittens in habitat fragments (Moriarty 2007). In turn, other carnivores have killed coyotes, including other coyotes and mountain lions (see also chapter 11). Despite these mortalities, which obviously have severe implications for individuals, bobcats are broadly sympatric with coyotes in the region, as are coyotes with mountain lions. Gray foxes and coyotes also co-occur in core natural areas in the area (Farias et al. 2005; and based on scat, sightings, and remote camera photos).

However, recent work indicates that gray foxes are apparently largely or entirely absent from the Simi Hills habitat fragments. Intensive trapping over 8 years (1998–2006) in these areas, using methods that are effective for gray foxes (Riley 2006), resulted in no fox captures. In addition, we have recorded no recent sightings or remote photos of foxes in these fragments, and specific intensive surveys of carnivore diversity (Tigas et al. 2003) produced no evidence of foxes. In contrast, bobcats and coyotes have maintained a continuous presence in these same patches. Similar methods in nearby areas have documented foxes, and longtime residents of the area report observing foxes in the past when development and roads had not yet isolated some of the fragments.

Because it is known that (1) coyotes can be a significant source of mortality for populations of smaller foxes, including gray foxes in this area (Farias et al. 2005), (2) coyotes may spatially exclude smaller canids through territorial competition, (3) bobcats have the potential to both kill (Riley 2001; Farias et al. 2005) and possibly spatially restrict gray foxes (Riley 2001), and (4) gray foxes can survive and even thrive in urban landscapes (Riley 2006), we speculate that gray foxes have been extirpated from habitat fragments by encounter and territorial competition with bobcats and perhaps especially coyotes. This hypothesis is also consistent with results from the San Diego case study, where gray foxes were often rare or absent in urban habitat fragments supporting coyotes. In 2007, we captured a gray fox in a Simi Hills fragment where bobcat activity has been significantly reduced because of a disease epizootic (Riley et al. 2007; see chapter 10). In 2008, we radio-collared a fox in another fragment, although the collared fox moved to a core natural area before dying from anticoagulant poisoning. It will be interesting to see whether fox populations recolonize any of these habitat fragments over the long term.

Weak Interactions among Raccoons, Striped Skunks, and Coyotes in Urban Forest Parks of Chicago, Illinois (S. D. Gehrt)

We (SDG and colleagues) used a combination of observational and manipulative approaches to investigate relationships between raccoons, striped skunks, and co-occurring coyotes in urban forest parks in Chicago, Illinois. We tested the prediction (prediction 1: predation effect) that coyotes directly limited raccoon/skunk populations through predation and a second nonexclusive prediction (prediction 2: avoidance effect) that the smaller carnivores would avoid areas used by coyotes.

We monitored raccoons ($n = 27$) and coyotes ($n = 13$) with radiotelemetry during 2000–2002 at a study site in Kane County, Illinois (Gehrt and Prange 2007). This study area was a 546-ha private reserve with limited human visitation but substantial human use outside the reserve. Surrounding land use included 39% residential or commercial use, a human density of 2,488 individuals/km², and traffic volume of 118,850 vehicles/24 h (Prange et al. 2003). Similarly, we concurrently monitored striped skunks ($n = 106$) and coyotes ($n = 9$) in an urban park in Cook County, Illinois, during 1999–2004 (Prange and Gehrt 2007). The park had between 1.5 and 3 million human visitors annually, which is typical of many urban parks in Cook County. Human use adjacent to the park also was substantial, with 78% residential and commercial land use, a human density of 3,420/km², and traffic volume of 302,550 vehicles/24 h (Prange et al. 2003). Both study sites occurred within the greater Chicago metropolitan area and had high-density coyote populations (we were not able to radio-collar all resident coyotes in either area but recorded visual observation of group sizes).

We found little support for prediction 1 for either species. During the raccoon study, no radio-collared raccoons were killed by coyotes, and overall, only 1% of 90 radio-collared raccoons died from predation during a long-term study of this population over a cumulative 8 years (1995–2002; Prange et al. 2003; S. D. Gehrt, unpublished data). Likewise, less than 5% of skunks ($n = 106$) died from predators over 6 years of radio-tracking (Prange and Gehrt 2007; S. D. Gehrt, unpublished data). These results were consistent with previous estimates of mortality rates from other studies of raccoons in primarily rural areas, in which mortality rates from predation have been consistently <5% (see Gehrt and Clark 2003 for a review). Previous studies reporting cause-specific mortality of striped skunks, largely from rural areas, also typically reported <5% predation rates (Prange and Gehrt 2007).

To test prediction 2, we compared the spatial patterns of raccoon and striped skunk home ranges relative to coyotes for both species. We observed no apparent avoidance of coyotes at the scale of the home range for either species (for details, see Gehrt and Prange 2007; Prange and Gehrt 2007). We observed no avoidance of coyote core areas for those raccoons with home ranges overlapping those core areas. Indeed, there was a nonsignificant trend for greater use of coyote core areas than would occur with random use of the home range.

As a more rigorous test of prediction 2, we used manipulative experiments to assess possible avoidance by raccoons and striped skunks at the microscale. For the raccoon study, we recorded raccoon visitation at scent stations marked with lure (fish oil), and compared that visitation to randomly selected stations for which we liberally applied coyote urine. We predicted that raccoon visitation would decline with the coyote urine treatment but instead recorded no significant difference before and after treatments. We repeated this test on another study site, in which we designed scent stations with lure as controls and had paired stations with a combination of lure and coyote urine. Visitation rates of raccoons to treatment stations in these trials were actually higher than for control stations following the application of coyote urine (Gehrt and Prange 2007).

The skunk experiment entailed overlaying the portion of the study area used by most of the skunks with grid cells and artificially increasing simulated coyote activity (i.e., auditory and olfactory cues) within treatment grid cells. For the first test, we partitioned the area into treatment and control areas and monitored the movements of 21 radio-collared skunks residing in the area. In the treatment area, we randomly selected grid cells for treatments, including howling and urine tests. Howling tests involved playing a tape of coyote howls at specific times during the night, and the urine tests were similar to those in the raccoon study. We observed no difference in skunk use of treatment and control cells. Following this, we repeated the experiment, but we shifted from a control-treatment design to a before-treatment-after design. Again, we detected no differences in use of cells by skunks before and after howling or urine treatments. These experimental results were consistent with visual observations we recorded during radio-tracking sessions at night, in which we observed skunks ignoring or repelling coyotes, but in no case did a skunk retreat from a coyote (Prange and Gehrt 2007; S. D. Gehrt, unpublished data).

Our results at the local level within an urban landscape are consistent with statewide population trends. In Illinois, coyotes apparently increased across the state during the 1990s (Gosselink et al. 2003), raccoons also exhibited a similar increase during the same time period (Gehrt et al. 2002), and striped skunks exhibited little trend during this period (Gehrt et al. 2006). However, red and gray foxes experienced negative trends, which is consistent with intraguild predation.

Overall, this case study emphasizes that we should not expect to see universally strong top-down effects emanating from all apex predators to all smaller carnivores and throughout the food web. Rather, the intensity of top-down interactions will vary, depending on the specific situation, and the possibility exists that larger predators do not govern populations of some smaller mesopredators. The interesting research question then becomes: In which situations do we expect particularly strong dominance interactions? We return to this question at the end of this chapter.

Red Fox Refugia from Coyotes along a Rural Urban Gradient in East Central Illinois (T. E. Gosselink and T. R. Van Deelen)

As suggested at the end of this chapter, carnivore communities vary along urban-rural gradients, with interior urban areas possibly serving as refugia from top predators, which may be reluctant to use the more urbanized areas because of human interactions. Coyotes have expanded their range into eastern North America, and, as the top mammalian carnivore in much of the region (Parker 1995), may instigate top-down effects on prey and smaller sympatric predators, including foxes. The habitat partitioning, spatial segregation, and direct killing of red foxes by coyotes documented in rural areas may also occur in urban system, but research is lacking.

In central Illinois, we (TEG and TRV) examined interactions between sympatric coyotes and red foxes along an urban-rural gradient using radiotelemetry on 64 coyotes (rural and urban-rural interface) and 335 red foxes (113 urban and 222 rural). We hypothesized interior urban areas would serve as red fox refugia from coyotes, and red foxes along the urban-rural interface and rural areas would experience antagonistic interactions with coyotes. The cities of Champaign and Urbana were the focal study areas, but we also included four other cities (pop. = 2,500–104,000) and 22 surrounding counties in central Illinois from 1997 to 2001 (Gosselink 2002). We classified foxes as urban if their home ranges included any portion of urban development, ranging from foxes that used entirely interior urban habitat to others that lived on the urban-rural interface. We captured coyotes along the urban-rural interface but did not capture coyotes in interior urban areas, nor did we observe radio-marked coyotes using the interior urban areas, or unmarked coyotes in the interior urban areas (Gosselink 1999). Human tolerance levels likely influence the lack of coyotes in interior

urban areas. Coyotes in the urban-rural interface were often reported to be shot by local law enforcement throughout the year (Gosselink 1999), and being shot opportunistically by hunters was the principal mortality source of rural coyotes (Van Deelen and Gosselink 2006).

In the interior urban areas, the lack of coyote presence was evident because we recorded no coyote-induced mortality of foxes, which was a main mortality source for rural foxes (juveniles 40%, adults 29%; Gosselink et al. 2007). In the urban-rural interface, we recorded fox mortalities due to coyote predation but significantly less often than among their rural counterparts (juveniles 14%, adults 4%; Gosselink et al. 2007). Sarcoptic mange in urban foxes (45% of all mortalities) appeared to be compensatory for predation in rural foxes (only 2% recorded mange mortalities) because both urban and rural foxes had similar annual survival rates (rural 38%, urban 32%). We did not detect a difference in mange incidence among the interior urban and urban/rural interface foxes. During the study, we did not record any coyotes with mange. Foxes that succumbed to coyote predation also never had mange (that we could visually detect). This may be due to coyote avoidance of mange-infected foxes, which also may further explain fewer foxes predated by coyotes along the urban/rural interface. Because of the annually cyclic pattern of mange of urban foxes (a repeated pattern of a high-incidence year followed by a low-incidence year), coyote predation was also cyclic (higher in years of low mange presence), which may be related to the possible avoidance of mange-infested foxes by coyotes combined with the low survival rates of foxes with mange.

Spatial distribution of coyote and fox home ranges in the urban/rural interface clearly showed a distinct separation of the two canids, with foxes residing in the interior urban areas as well as the small undeveloped portions along the urban/rural interface. Coyote ranges were directly adjacent to the fox ranges along the edge of developed urban areas, with overlap of home ranges nearly absent (Gosselink 1999). Urban foxes shifted habitat use from urban-adjacent row crop fields to interior urban habitats during the winter when crop fields were devoid of vegetation due to harvest (Gosselink et al. 2003). Similarly, rural foxes avoided coyote-rich habitats (e.g., grassland, drainage ditches) and gravitated toward human establishments in rural areas (e.g., farmsteads, rural housing developments) in central Illinois (Gosselink et al. 2003).

In both urban and rural areas, drainage culverts along roadsides (consistently dry and large enough for foxes but not for coyotes) served as fox dens and refugia from predators and humans (Gosselink et al. 2003). In contrast, coyotes selected den sites away from human establishments (e.g., center of mile section). Culverts under driveways in urban areas were often used as daytime resting sites for foxes as well as other urban carnivores (e.g., skunks, raccoons, opossums). This man-made microhabitat feature was critical in providing havens for red foxes from human activity in urban areas as well as from coyotes in rural areas.

This study suggests that in a coyote-dominated landscape, red foxes seek out refugia from coyotes, which translated into more intense urban use, especially during the winter months when coyote predation on foxes was most intense (Gosselink et al. 2007). Coyote persecution of red foxes is well documented and was also evident in this study in both the rural areas and the urban-rural interface (Gosselink 2002; Gosselink et al. 2007). The reestablishment of coyote populations in the Midwest landscape has likely resulted in foxes using urban areas more than historically, when coyotes were absent. Previous research in the Midwest on red foxes before coyote establishment never mentioned red fox use of urban areas (nor did wildlife biologists during that time period; R. Andrews, personal communication), and foxes in rural areas used den sites away from human establishments (Storm et al. 1976; Pils and Martin 1978). The shift of red foxes into human-developed areas (urban and rural) is likely due to structural changes in the carnivore community, with foxes using the small rural pockets free of coyotes and interior urban areas where coyotes are typically excluded by humans.

CARNIVORE INTERACTIONS ALONG THE URBANIZATION GRADIENT

What lessons might we derive from these case studies regarding interactions among urban carnivores? How might community-level interactions among carnivores differ in urbanizing landscapes? If we simply dichotomize the landscape into urban and natural, we might posit two alternative hypotheses regarding the intensity of agonistic interactions among dominant and subordinate carnivores (figure 14.2). In urban areas, limited space and resources surrounding natural habitat might create "pile-up" of home ranges along the hard boundaries of urban edges and roadways (Riley et al. 2006),

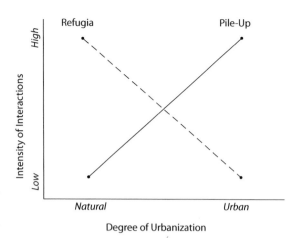

Figure 14.2. Simplified conceptual model of the intensity of dominance interactions between apex predators and subordinate mesopredators in urban versus natural areas. The "pile-up" hypothesis predicts limited space and resources surrounding circumscribed urban habitat patches will force agonistic interactions in urban compared to natural landscapes. Alternatively, the "refugia" hypothesis predicts that the decline of apex predators in urban areas will reduce agonistic interactions.

thereby forcing increased spatial and temporal overlap and hence agonistic interactions among carnivore species compared with natural areas. We label this the pile up hypothesis in figure 14.2. Alternatively, we might predict waning of agonistic interactions between dominant and subordinate predators in urban areas if apex carnivores decline with urbanization. We label this the "refugia" hypothesis in figure 14.2.

These contrasting hypotheses can perhaps be reconciled by viewing the landscape not as an urban-natural dichotomy but rather as a complex gradient of urbanization along which both the landscape and predator community changes (figure 14.3). Along the gradient, progressive urbanization creates landscapes characterized by high spatial heterogeneity, with remaining natural areas bordering, and often surrounded by, an extensive human-dominated matrix (Pickett et al. 2001; Faeth et al. 2005). The urban extreme of the gradient consists of highly urbanized areas with small, isolated habitat islands immersed within an urban matrix. Larger patches of natural habitat, often with at least some connections among them through the urban matrix, characterize moderate urbanization. The natural extreme of the gradient consists of a much-less-urbanized landscape with relatively few to no residential developments.

Such dramatic transformation of the landscape alters the relative abundance and composition of the

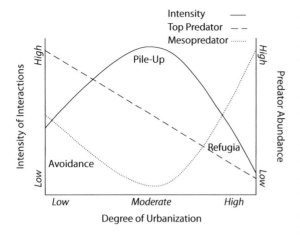

Figure 14.3. A conceptual model of the intensity of dominance interactions between apex predators and subordinate meso-predators as a function of changes in the landscape and predator community along an urban-natural gradient. The left axis represents the intensity of dominance interactions (solid line), and the right axis represents the relative abundance of dominant (dashed line) and subordinate (dotted line) predators (both standardized to the same high-low scale). Top predators are expected to decline with urbanization, potentially relaxing dominance interactions and creating refugia for mesopredators in highly urban areas. In less urbanized systems where both top predators and mesopredators occur, spatial and temporal avoidance of dominant carnivores within areas of sympatry may be common. The intensity of dominance interactions might be most pronounced at intermediate levels of urbanization, where top predators still persist but home range pile-ups along urban edges force agonistic encounters with mesopredators. Throughout, more intense interactions are expected among ecologically similar species with high niche overlap.

mesopredators can move through and reside within developed areas and perceive the urban matrix as somewhat permeable, enabling them to persist where some top predators cannot.

Given alterations in both the landscape and predator community with increasing urbanization, where along the urbanization gradient might interspecific interactions among dominant and subordinate carnivores be most pronounced? We suggest that the intensity of top-down effects and the mechanisms generating them may depend on the degree of urban fragmentation and the responses of carnivores along that portion of the gradient. At the natural end of the gradient, in landscapes with little to no urban fragmentation, we hypothesize that spatial and temporal avoidance may be a primary mechanism by which large predators impact mesopredators (figure 14.3). That is, given sufficient area, mesopredators may be able to coexist broadly with larger predators but may exhibit substantial avoidance of top predators within areas of sympatry. Such spatiotemporal avoidance has been shown repeatedly, often among canids.

At the opposite extreme of the gradient, in habitat islands where large dominant predators are rare or absent, we would predict a decrease in agonistic interactions among dominant carnivores and mesopredators according to the refugia hypothesis described previously. For example, in the case studies reviewed here, isolated urban patches potentially serve as fox refugia from coyotes in Southern California (Crooks and Soulé 1999), Northern California (Riley 2001), and Illinois (Gosselink et al. 2003, 2007). Although dominance interactions between apex and smaller predators decrease in this scenario, as mesopredator populations expand in highly urban areas, competitive interactions among mesopredator species may consequently increase. If so, we would then expect increased control of mesopredators through bottom-up constraints, as well as an increased chance of top-down control of their prey (Faeth et al. 2005). Predation impacts via mesopredators may be particularly evident for those prey species that are rare or are habitat specialists restricted to urban fragments completely circumscribed by urban development.

We hypothesize that perhaps the most intense dominance interactions among carnivores, however, may be at moderate degrees of urban fragmentation (figure 14.3), where top predators can still persist, but habitat and resource limitations heighten agonistic encounters with subordinate carnivores, as envisioned under the

predator community (figure 14.3), with some species responding negatively and some positively to urban fragmentation (Crooks 2002; Prange and Gehrt 2004; Faeth et al. 2005). Larger, apex predators may not be able to persist at the urban extreme end of the gradient, where urban habitat patches are too small, too isolated, and too disturbed by humans. In contrast, populations of some mesopredators, including native and nonnative carnivores, often thrive at the extreme end of the urban gradient because of a combination of factors (e.g., raccoons; Riley et al. 1998; Prange et al. 2003; Prange and Gehrt 2004). Ecological release from dominant carnivores may contribute to increases in mesopredators in urban areas. Habitat suitability, however, also plays an important role. As highlighted in the case studies, residential development often represents suitable habitat for generalist, omnivorous mesopredators, where they readily exploit human-generated food such as trash and cultivated fruits. As such, generalist pile-up hypothesis. In this scenario, habitat fragments that support dominant carnivores are at least partially bounded by a relatively impermeable matrix of urban development and roadways. If so, depressive competition and intraguild predation may be particularly severe, potentially resulting in top-down population impacts on mesopredators, at times to the point of their complete exclusion from urban fragments. We would expect the most pronounced interactions between those species most similar ecologically (e.g., canids) and in situations where the top predator is relatively abundant. Such may be the case in urban natural areas in Southern California where gray fox populations are uncommon or absent altogether in habitats supporting coyotes (Crooks and Soulé 1999; Fedriani et al. 2000). As with mesopredators in highly urban areas, we would also expect large predators in moderately urbanized landscapes to exert the strongest top-down effect on their prey, particularly for rare or specialized prey species within patches largely bounded by urban development.

The framework we provide here, and indeed the chapter as a whole, has focused specifically on interactions among different species of carnivores rather than between carnivores and their prey, largely because the available studies have largely addressed competition and intraguild predation. However, predator-prey interactions involving mammalian carnivores could well also be altered in urban landscapes. Fifty years ago, Huffaker (1958) elegantly showed that spatial structure can significantly alter predator-prey dynamics in mites, with both species going extinct on single oranges, or "patches," but both species persisting in complex, connected systems of patches. Although there are few studies of mammalian carnivore-prey interactions in urban landscapes or on urban "islands," many of the oft-cited cases of top-down effects by mammalian carnivores are on true islands, such as the wolves and moose of Isle Royale (Peterson 1977) or the predator-free islands of Lago Guri reservoir in Venezuela (Terborgh et al. 2001). Aside from potential spatial constraints on predator-prey interactions in urban areas, other human actions may alter these relationships as well. For example, in the desert Southwest, Faeth et al. (2005) argue that increased water availability in urban areas can have both bottom-up and top-down effects on bird-arthropod predator-prey interactions. Further examination of predator-prey relationships, and perhaps other types of interspecific interactions,

is a largely untapped and rich area for study in urban carnivore ecology.

Although the framework we present provides a context in which to predict the community-level interactions of carnivores in urbanizing landscapes, it represents one of many possible scenarios for how the community ecology of carnivores might change with urbanization. Community interactions will certainly differ among species, systems, and over time; such interactions are difficult to predict without detailed knowledge of the response of both predator and prey to urban fragmentation (Faeth et al. 2005). Like many conceptual models, deviations from these predictions might turn out as interesting and useful as examples that support them. We hope, however, that such a framework will stimulate continued thought about and research into the community ecology of urban carnivores.

Gray Foxes (*Urocyon cinereoargenteus*) in Urban Areas

Seth P. D. Riley and Paula A. White

Gray foxes are small canids (around 5 kg) that are distributed widely across North America and south into Central and South America. They are less closely related than the other foxes described in this volume (kit, red, and arctic foxes; Geffen et al. 1992) and are distinct from other foxes because of their relatively shorter snout and legs, gray body with red highlights and black tip and top of tail, and the ability to climb trees. Gray foxes favor thick habitat generally, specifically woodlands in the east and shrublands in the west. Gray foxes certainly overlap distributionally with many urban areas throughout the United States and elsewhere (e.g., Mexico City), but to date, there are very few studies of the species in urban landscapes. The most detailed studies of gray foxes in urbanized or developed areas were in Northern California (Riley 2001, 2006; Riley et al. 2004) and New Mexico (Harrison 1997). Both studies compared fox ecology between more and less developed landscapes: in California, foxes were monitored in urban and rural zones of Golden Gate National Recreation Area, a national park north of San Francisco (see figures), and in New Mexico, and Harrison compared foxes between an undeveloped area and a nearby rural residential area. Gray foxes have also been studied in the overall urban landscape of Santa Monica Mountains National Recreation Area near Los Angeles (Farias et al. 2005), but these foxes were not using the developed or fragmented areas (see also chapter 14).

Both studies found that gray foxes could adapt to and use developed areas to a degree, but less so than some other carnivores such as Eurasian red foxes (chapter 6). In California, 15 of 18 collared foxes either used (10) or lived adjacent to (5) developed areas, either outside or inside (park residences and offices) the park. The use of developed areas outside the park was generally at night, however, and no fox core areas overlapped adjacent development. In New Mexico, gray foxes avoided large housing developments but were able to navigate roads and development to use the developed landscape, exhibiting more home range complexity (e.g., satellite and peninsular regions of the range) than foxes in the undeveloped area. New Mexico foxes also used residential areas more at night, avoiding them during the day but selecting them at night. A large road formed home range boundaries for a number of foxes in the residential area in New Mexico, while in California, foxes appeared to cross roads in both the urban and rural zones of the study more readily than bobcats (Riley 2006). Vehicles were a cause of mortality in foxes in California, although not a major one. Anthropogenic food items did not reflect a major portion of the diet in either area (10% of less in both studies) and consisted mostly of ornamental fruits—pet food was not detected in fox scats in either study. Small mammals were the most important item in the diets of foxes in developed areas in both New Mexico and California.

Neither study showed much evidence of smaller home ranges for foxes in more developed landscapes. In both studies, there was no significant difference in home range size between the more and less developed areas, although in New Mexico, the residential home ranges were smaller (384 ha vs. 549 ha, on average) and the sample size was small ($n = 6$ in both cases, $p = 0.109$; Harrison 1997). In examining home range size in nonurban studies, there is some evidence of large home ranges in more eastern states such as Alabama, Florida, Missouri, and Mississippi, where gray fox home range sizes have been 500–600 ha (Cypher 2003). Home range size in the Northern California study (in both urban and rural zones) was similar to the 100-ha size seen in small previous studies in Utah and California (Cypher 2003). However, home range size in the New Mexico study was more similar to those in some eastern states.

Gray foxes can be greatly affected by disease, and disease exposure was an interesting aspect of the Northern California study. Foxes in the urban zone had significantly higher seroprevalence for canine parvovirus (CPV), a disease present in domestic dog populations. Moreover, the few foxes in the rural zone that were seropositive for CPV, were also the same ones that visited the small towns in the rural zone. Interestingly, all foxes (in both zones) were seronegative for canine distemper, and just after the study ended, an epizootic of canine distemper swept through the population, killing many animals. Overall, gray foxes are able to use developed areas, perhaps to a certain housing density (Harrison [1997] suggested that 1 house/3.1 acres was perhaps the limit), but threats such as disease and larger carnivores (see also chapter 14) may threaten their ability to persist in developed landscapes. Clearly, there is much more to learn about gray fox ecology in urban areas.

Gray fox eating marshmallows at Golden Gate National Recreation Area.

15

Responding to Human-Carnivore Conflicts in Urban Areas

PAUL D. CURTIS

JOHN HADIDIAN

I N NORTH AMERICA, raccoons, striped skunks, and red foxes are widely distributed in urban habitats, while black and brown bears, mountain lions, gray foxes, fisher, weasels, river otter, mink, and kit foxes are present across a limited range of urbanized zones or have recently appeared on the urban scene. Coyotes have been present in and around Los Angeles for decades, while recently completing a documented expansion into urban areas of the eastern United States. These carnivores, and hence their potential for conflicts with humans, are extremely diverse. With body sizes ranging from 0.05 kg in the least weasel to more than 680 kg for the brown bear, it would be difficult to detail comprehensively the many potential impacts of carnivores. General categories of conflict, however, range from the hard to objectify annoyance and intolerance of carnivores as "nuisances" to measurable property damage and impacts on the health and safety of people or their companion animals (De Almeida 1987). In some cases, urban zones seem to confer partial protection to carnivores, as is the case with the San Joaquin kit fox (Cypher 2003). That aside, we will focus on generalizations involving a few species to highlight general principles for reducing negative impacts.

Many factors can influence the potential for conflict between people and urban carnivores. Although specifics and complexity differ across circumstances, most conflict situations seem to involve animals attempting to satisfy their needs for food, shelter, or breeding opportunities, rather than animals practicing predation. If the landowner or wildlife professional keeps this in mind, many potential problems can be prevented. Proper design and maintenance of structures and removal of food attractants will help minimize the potential for problems.

Not all carnivore impacts are predictable or preventable. Urban coyotes may defend their territories by attacking domestic dogs that intrude, even when humans are walking them. Usually the pet owner can intervene, but the attack may be so swift that the pet is severely injured or killed. Carnivore attacks typically garner much media attention, particularly if a person is injured or killed (Gore et al. 2005). Although human injuries from carnivores are rare, the high level of media involvement when they do occur may create elevated perceptions of wildlife-associated risks (Riley and Decker 2000a).

PERCEPTION OF RISKS ASSOCIATED WITH WILDLIFE

There are many types of wildlife-associated risks, including disease transmission, physical injury, and the potential for property damage. Tolerance of wildlife depends in part on how people perceive these risks (Knuth et al. 1992; Decker et al. 2002; Gore et al. 2007). Two elements of risk perceptions are important: (1) perceptions of the probability of an undesirable outcome and (2) the worry or fear associated with that outcome (Slovic 1987). It is helpful to distinguish between these aspects of risk perceptions (Slovic 1993).

Some risks—such as the risk of a bear attack—may be perceived to have low probability but are highly feared because of the potential for injury or death (Herrero and Fleck 1990). Other risks, such as bears damaging bird feeders, may be perceived as highly probable but create little dread (e.g., causing terror or fear). Both aspects influence how people respond to risks and their tolerance for a given wildlife conflict. In general, studies examining risk perception related to carnivore species indicate that if a wildlife-related risk is unfamiliar, uncertain, novel, or uncontrollable, it will likely lead to a higher perceived risk (chapter 3).

Nine factors apparently may influence how people perceive bear-related risks (Gore et al. 2006), including control, dread, environment, trust, responsiveness, agents, seriousness, frequency, and volition. These constructs are discussed in-depth elsewhere, so we will provide only a summary here. It is likely that these characteristics influence how people react to possible carnivore conflicts.

People's risk tolerance decreases as their perception of the probability or frequency of the risk increases (see chapter 3 for an in-depth discussion). In studies of two different kinds of wildlife problems, the risks of cougar attacks in Montana and the risks of deer-vehicle collisions in New York state, people were more likely to favor wildlife population reductions as their perception of risks increased (Riley and Decker 2000a; Stout et al. 1993).

Risk perceptions motivate people to take action, whether or not such perceptions are accurate. Often a discrepancy will exist between the technical assessment of risk and perceived risk among individuals (Slovic 1993). Riley and Decker (2000a) reported that many Montana residents perceived the risks of cougar attacks to be orders of magnitude higher than any reasonable objective assessment of risk. Risks with low probability, but serious or dreaded consequences, tend to increase fear and elevate perceived risks. Inaccurate perceptions provide a focus for educational communication because people's perception of risk motivates management action.

People are more willing to accept risks assumed voluntarily, which they believe they can control (Kleiven et al. 2004). For example, a homeowner who feeds birds may tolerate elevated risks of attracting bears or coyotes into their yard. A neighbor may protest the same level of risk because he or she did not assume these risks by choice.

Risks to children are less tolerable than risks to adults (Decker et al. 2002). Concern about children will be expressed in any wildlife issue that involves a threat to human health or safety. Risks to children may be a greater concern in part because adults recognize that children have less ability to evaluate risks accurately. Tolerance for coyotes in suburbia may be low because children are most likely to be attacked (Timm et al. 2004). However, children are much more likely to be bitten by a domestic dog than a coyote. The Centers for Disease Control and Prevention (2003) indicated that in 1994, approximately 4.7 million people were bitten by dogs in the United States, and 799,700 persons required medical care. Injury rates were highest among children ages 5 to 9 years. To put this in some perspective, children under 10 are twice as likely to die from drowning in a 5-gallon bucket, than from a dog bite (Bradley 2006). Although rare, coyotes have injured people near sites where food conditioning has occurred, often attacking young children (Carbyn 1989). Beier (1991) documented 9 fatal mountain lion attacks on 10 humans and 44 nonfatal attacks on 48 humans in the United States and Canada during 1890–1990. Most victims (66%) were unsupervised children or lone adults.

Perceptions of risk decrease if the benefits associated with carnivores are tangible. Many people enjoy viewing wildlife, especially carnivores, and are willing to tolerate some level of risk as a trade-off for the presence of these charismatic mammals. As noted in chapter 11 on the mountain lion, most Americans have a positive, protective attitude toward mountain lions, wolves, and bears. Thus, mountain lions and other charismatic carnivores can help build public support for wildlife conservation, but much of this support is dependent on the public perception of these animals as nonthreatening, a perception that will require proactive outreach and educational efforts to sustain.

INTEGRATING ECOLOGICAL AND HUMAN DIMENSIONS OF CONFLICTS

The integration of social and ecological information is critical for management of viable wildlife populations in human-dominated landscapes (Loker et al. 1999). Riley et al. (2002) emphasized that managers should focus on desired impacts, the recognized important effects resulting from interactions between people and wildlife. Levels of positive or negative impacts, rather than animal abundance, are the primary determinants of wildlife acceptance capacity (Riley et al. 2002; chapter 3). Decker and Purdy (1988) defined acceptance capacity as a range of population sizes, including an upper limit of tolerance for the negative effects of wildlife on people. However, the cumulative outcome of perceived positive and negative effects associated with a wildlife population will influence overall acceptance capacity (Carpenter et al. 2000).

Recent research has shown that individuals' acceptance of management actions was influenced by their perceptions of wildlife-related impacts (Decker, Jacobson, and Brown 2006). Impact dependency suggests that tolerance of carnivores is less about acceptance of the species or its population density and more about acceptance of the impacts that result from living with the species (chapter 3). For example, acceptance capacity for cougars (Riley and Decker 2000a) and black bears (Gore et al. 2007) was associated with both real and perceived risks associated with these carnivores. In New York, both wildlife managers and the public agree that food attractants caused most black bear

conflicts and that bear problems were common (Gore et al. 2007). These stakeholders also agreed that wildlife managers were trusted to reduce the risk associated with bears. Setting management objectives in terms of impacts expands options available to managers beyond manipulation of habitat or species abundance, to include management of the frequency of interactions and evaluating the effects of those interactions (Lischka et al. 2007).

IMPACTS ASSOCIATED WITH MANAGEMENT ACTIONS

Management of wildlife conflicts often presents a challenge because people associated with conflict situations may disagree about how to resolve problems (Decker et al. 2002). Proposed management actions may create controversy associated with the cost and safety of implementing a program or the pain and suffering experienced by animals removed. It is essential to distinguish public agreement with management objectives from the acceptance of management actions. For example, a community may generally agree that it is desirable to reduce the risk of coyote attacks on pets, but stakeholders within the community may disagree strongly about how to achieve this goal (Fox and Papouchis 2005; Gehrt 2006). Some may favor stricter leash laws or keeping pets indoors at night. Others may favor direct reduction of the coyote population. Often debate arises over the relative merits of lethal versus nonlethal control methods. A similar scenario can exist for bears, mountain lions, or other large carnivores.

Among stakeholders for whom reducing conflicts is most important, the solutions with the greatest cost-effectiveness may be preferred. Some people will favor shooting or trapping in such cases. For example, in some communities where pets have been repeatedly killed by coyotes, professional wildlife control operators have removed coyotes by trapping, or calling and shooting specific problem animals. Those who place higher importance on other impacts, such as minimizing the pain and suffering of wildlife, may seek different management strategies (figure 15.1). Therefore, understanding the "collateral" impacts of management actions can aid in developing acceptable management strategies (Organ et al. 2006).

Changing public values and sentiments may contrast with conventional wildlife policies and management approaches derived from historic species propagation, wildlife conservation, and sustainable harvest

Figure 15.1. Sign warning residents of coyotes near a playground in the Chicago metropolitan area. Max McGraw Wildlife Foundation.

perspectives. Residents of urban areas tend to express humane or moralistic concerns for wild animals, including those with whom they may be experiencing conflicts (Kellert 1984; Braband and Clark 1992). This raises questions with respect to whether ethical guidelines exist, or need to exist, for how "problem" wildlife is controlled (Hadidian et al. 2006). It is widely, but not universally, recognized that simply trapping and killing or translocation of problem animals does not permanently resolve human-wildlife conflicts because other animals fill in the gap created by those that were removed. However, strategies to inform, address, or mandate more holistic forms of conflict resolutions are still wanting (De Almeida 1987; Prange et al. 2003; Gates et al. 2006; Gehrt 2006).

Economic Impacts

The costs for reducing negative impacts may be borne by individuals (e.g., fencing or scare devices for a specific property) or by communities (e.g., selective culling programs carried out by paid shooters or trappers). High costs may deter landowners from taking steps to reduce conflicts on their properties, even if benefits will greatly exceed expenditures. Opposition to some community-wide management strategies, such as carnivore removal, often is based partially on costs to taxpayers and perceived effectiveness. However, people may be willing to pay for expensive programs to protect public safety or even reduce pet losses. For example, by 2003, more than 50 million baits for oral rabies vaccination of carnivores were distributed in 13 states (Slate et al. 2005). The cost to conduct oral rabies

vaccination (ORV) campaigns in the United States by the USDA, APHIS, Wildlife Services program varies annually but is currently about $15 million per year. Smaller amounts of resources are expended annually for ORV in four states and a few counties. About 75% of the ORV costs are for baits containing oral rabies vaccine. Air distribution contracts and fuel for airplanes make up much of the remaining costs. The target species for ORV in the United States are the raccoon in the eastern United States and the gray fox and coyote in Texas (USDA 2006). Sometimes such situations result in co-management, with state and federal government agencies, local communities, and stakeholder groups sharing the cost and responsibility for management actions (Schusler et al. 2000).

Health and Safety Impacts

Potential disease transmission by carnivores may also raise health concerns. Diseases transmitted by contact are more likely to impact dense wildlife populations. Carnivores are often rabies-vector species (Rosatte et al. 1992a), and this disease is highly dreaded. Suburban landscapes often support elevated raccoon densities (Prange et al. 2003), which may increase transmission rates of rabies (Riley et al. 1998; Wolf et al. 2003) and a perceived, if not actual, threat to pets and domestic livestock in areas with high raccoon densities. The challenges of oral rabies vaccine distribution across suburban landscapes differ from those for rural areas because the potential for human exposure to vaccine baits is high, there is competition for baits from nontarget species, and there is limited access to private property (Boulanger et al. 2008). As another example, coyotes frequently harbor canine heartworms and could serve as a reservoir to transfer these parasites to domestic dogs. In human dimensions surveys, disease transmission associated with wildlife is a concern frequently expressed by stakeholders (Decker et al. 2006; Leong et al. 2007).

The safety of shooting and trapping strategies (Andelt et al. 1999) for managing wildlife is a major concern to some stakeholders. Landowners in rural areas who believe that lethal control is necessary for reducing impacts may nevertheless restrict these activities on their property because of perceived safety risks. One recent survey associated with deer hunting in New York, reported that 28% of landowners were "very concerned" about "the fear of being shot by hunters" who may gain access to their property (Enck and Brown 2004). Similar results were observed in southern Michigan where landowners expressed "worry about being hurt by stray shots" (Lischka et al. 2007). High housing density in suburban areas may influence people to believe that lethal management strategies are unsafe, even if they feel wildlife populations must be reduced. In New York, environmental conservation laws prohibit the discharge of a bow or firearm within 155 m (500 ft) of a dwelling without landowner permission. Many other states and communities have similar discharge regulations to protect public safety.

MANAGEMENT OF CARNIVORE CONFLICTS

Much information exists about management options for reducing human-carnivore conflicts. We will focus on management techniques in urban areas that can be used for a variety of species, and we will provide specific examples for several common carnivores. The full range of management options includes both lethal and nonlethal techniques (Knowlton et al. 1999; Mitchell et al. 2004) because individual carnivores that attack and injure, or kill, humans will need to be selectively removed from the population. Predation is a learned behavior, and a carnivore that has successfully practiced this on humans would be expected to turn to them as prey again. Management of urban carnivores should be focused on individual animals. There are few urban carnivores managed at the population level. With many small land parcels under private control, it is difficult to gain access to a sufficient number of animals to change overall population growth and survival rates. Also, it is a challenge to apply conventional wildlife management tools (e.g., shooting and trapping) to urban wildlife populations.

Given the tremendous diversity of carnivore conflicts, various management tools are needed. No single technique can be expected to work in all situations. The methods discussed here have limitations and often yield mixed results. Researchers and managers continue to test and refine management techniques, attempting to develop more consistent and reliable approaches for reducing negative impacts.

Tenets of Management

The professional practice of wildlife damage management is a relatively new field, even though humans have had to manage conflicts involving wild animals since time immemorial. One necessity of a growing profession has been the development of principles on which management decisions are made and acted on. Taking a compilation of these from available sources (Fisher and Marks 1996; Marks 1999; Littin et al. 2004), some basic tenets of any management intervention might include: (1) Is the proposed action justified by defensible needs, economic considerations, and other reasonable criteria? (2) Does the management option selected have a reasonable probability of success for reducing negative impacts? (3) Will the technique be species specific, or better yet, address a problem-causing individual? (4) Have the human-behavioral aspects of the problem been identified and considered? (5) Are the most advanced and appropriate methods used? (6) Are the solutions matched to the scale of the conflict (e.g., not conducting widespread programs for highly localized problems)? (7) Are the methods considered humane (e.g., preventing reasonable and foreseeable harm and eliminating animal suffering)?

Food Attractants

People often intentionally feed wildlife because they enjoy viewing animals at close distances. For example, millions of households feed birds; yet, bird feeders are one of the primary food attractants for black bears during spring and early summer. Spilled birdseed may attract rodents and consequently draw coyotes into a suburban backyard. Proper management of food attractants is an important tool for reducing human-carnivore conflicts. In addition, the deliberate feeding of wild carnivores, for entertainment purposes or out of some desire to be "kind," must be discouraged. Although the temptation to observe wildlife close-up may be compelling, the dangers of acclimating such animals to human presence can only lead to adverse consequences, mostly for the animals, but sometimes for people.

Food conditioning and habituation of carnivores (McNay 2002; Herrero 2005) may exacerbate human-wildlife conflicts, particularly in urban areas (chapter 3). The availability of anthropogenic food sources may attract carnivores to residential areas (figure 15.2), thereby increasing the potential for conflicts (Bounds and Shaw 1994). If carnivores become comfortable foraging in urban habitats, their habituation to humans in such environments may increase the likelihood of a negative encounter (Kitchen et al. 2000; McNay 2002; Lambert et al. 2006).

Educational programs often focus on changing human behavior and reducing food attractants for bears, coyotes, raccoons, and other wildlife (figure 15.3).

Figure 15.2. People feeding a raccoon. Photos by J. Bruskotter.

Figure 15.3. Bear sign in New York. Photo by Meredith L. Gore.

Evaluating outreach interventions is important for rating them relative to other management approaches (Zint et al. 2002). Objective assessment of effectiveness of education efforts can be a key element in an adaptive management approach. For example, despite an intensive educational program to reduce human-black bear conflicts in New York, there was no short-term evidence of an environmentally responsible change in human behavior (Gore at al. 2008). It is critical that evaluation be conducted as part of outreach efforts so that the associated effects on mitigating human-wildlife conflicts can be ascertained.

Examples of Management Interventions

Coyote interactions with pets and people appear to occur infrequently and are usually not predictable. When pets are killed or people are attacked, the community response is often to eliminate as many coyotes as possible near the location where the incidents occurred. Humane shooting or culling of specific coyotes causing damage may be acceptable when excessive losses occur in agricultural settings (Arthur 1981), but these practices are controversial in urban contexts (Fox and Papouchis 2005). Anecdotal information and recent studies (Bogan et al. 2008) indicate that few coyotes may be responsible for these negative impacts. Yet, how does a management agency identify the individual problem animal? Indiscriminate culling may have little effect on a specific problem because of behavioral variation based on animal age or sex. For example, young coyotes are more easily captured during late summer and fall, yet it may be older, established adult coyotes that have habituated to people and are responsible for the conflicts. Trapping without focus may capture animals but not resolve the concerns.

Intensive coyote removal may resolve an immediate problem, but this approach is unlikely to reduce impacts over the long term. Voids created by culling coyotes are rapidly filled by other coyotes, and this may occur in as little as 2 to 6 weeks (Blejwas et al. 2002; Bogan 2004). Coyotes may respond to removal by breeding at an earlier age and increasing their average litter size (Beckoff and Gese 2003). Consequently, reductions of coyote numbers are difficult to maintain and unlikely to have long-term effects on conflict management.

Unintended consequences of actions need to be considered. For example, there is a possibility that broad-scale coyote removal may destabilize existing territories, a situation that could create more problem individuals. In suburban landscapes, coyotes appear to use patches of natural habitat as core areas in their territories (Riley et al. 2003; chapter 7). Removing a breeding animal may cause territory abandonment and create transient individuals that have a greater propensity to present a problem.

Many suburbanites who reside in areas with mountain lions hike, run, bicycle, and enjoy outdoor activities in adjacent natural habitats. These activities create opportunities to encounter a mountain lion. Recent research by Beier et al. (chapter 11, this volume) suggests that a near-urban mountain lion is not inherently more dangerous than a nonurban one. Radio-collared cats were rarely observed by people and behaved similarly to mountain lions studied in nonurban areas. The near-urban lions primarily hunted and consumed deer, used natural habitats with dense cover, and remained out of sight of humans. These near-urban animals were not attracted to areas of human activity and were not

treating people as prey. Similar results were observed for 10 radio-collared lions using a popular state park in California (Sweanor et al. 2008).

When should lethal control of coyotes, mountain lions, or other urban carnivores be implemented? What level of risk or tolerance should be acceptable? If a coyote is watching children get on a bus every morning, and showing no fear, should that individual animal be removed? These are difficult questions that affect individuals, communities, elected officials, and public agencies. We encourage wildlife agencies to develop and vet their policy and operation plans before an unfortunate incident occurs. It is much better to be proactive in gaining stakeholder understanding and cooperation rather than to be reactive to rising public outcry.

Aversive Conditioning

Aversive conditioning is often promoted as a nonlethal approach to reduce carnivore conflicts. Aversive conditioning modifies animal behavior through association and negative conditioning (Gustavson et al. 1976; Shivik and Martin 2001). The premise behind aversive conditioning is to impose a negative association (i.e., creating pain or distress) for a specific undesirable action. It is assumed that a cognitive link will develop between the action and the negative consequences for the action, thus creating avoidance if the relative gains are not sufficient to risk the consequences.

Conditioned taste aversion is induced by the ingestion of a food that causes gastrointestinal discomfort, creating a desire to avoid that food. Gustavson et al. (1982) and Cornell and Cornley (1979) established conditioned taste aversion in wild coyotes to decrease sheep depredations. Later testing showed poor efficacy as a practical taste-aversion treatment (Burns 1983). Burns (1983) also noted that, if lithium-chloride (Li-Cl) baits were effective for teaching coyotes not to kill sheep, then averted coyotes would be valuable assets and require protection. Although aversive conditioning studies for suburban coyotes have been proposed, it has been impossible to conduct reliable field trials. Coyote use of anthropogenic foods appears to be infrequent and unpredictable (Bogan 2004; Morey et al. 2007; chapter 7).

Human-bear interactions increase as black bear ranges expand into residential landscapes and development encroaches into bear habitat (figure 15.4). Bears often display bold behavior to exploit food resources, damaging property and possibly threatening human

Figure 15.4. Wildlife agency staff and volunteers check a black bear that was found in a winter den under a deck of a suburban New Jersey home. Photo by Meredith L. Gore.

safety (Weaver et al. 2003). Stakeholders expect wildlife managers to reduce bear conflicts in nonlethal ways (Beckmann and Berger 2003a), thus aversive conditioning has been used to discourage nuisance bears (Beckmann et al. 2004).

Woolridge (1980) reported conditioned taste aversion in both free-ranging and captive black bears and polar bears using Li-Cl, emetine hydrochloride, and alpha-naphthyl-thiourea. Although captive bears showed a greater aversion to previously treated foods, free-ranging animals' visit to garbage dumps decreased after treatment, displaying the potential for site aversion.

Ternent and Garshelis (1999) aversively conditioned free-ranging black bears in Minnesota with thiabendazole. Thiabendazole is an anthelmintic drug, used by veterinarians to de-worm mammals (Brown et al. 1961). When ingested, it causes gastrointestinal discomfort, nausea, diarrhea, and vomiting (Rollo 1980). Unlike Li-Cl, which has a strong salty flavor that may reduce its effectiveness (Burns 1980), thiabendazole is odorless and tasteless.

Wegan (2007) evaluated thiabendazole for reducing food consumption by black bears in northern New York, but results were mixed and inconclusive because of small sample sizes (6 bears). He recorded 175 visits by black bears to bait stations in 82 consecutive nights. From digital photographs, Wegan (2007) was able to identify the individual tagged bear in 75 of these vis-

its. Six marked bears visited a bait station at least three times during 10 days and met the minimum standards for thiabendazole treatment. Two bears were randomly selected as control animals, while the remaining four were given treated baits. No treated bears rejected the thiabendazole-laced bait. Although the mean visitation rate for the treated bears (5.2%) was less than the mean rate for the control group (28.9%), the confidence intervals overlapped and the difference was not significant. However, there was a significant reduction in the likelihood of a treated bear returning to the specific bait station where it consumed thiabendazole when compared with the control group. Therefore, thiabendazole treatment may create a site-specific aversion for black bears, but additional research should be conducted to verify these results.

Rauer et al. (2003) used rubber slugs and buckshot to aversively condition European brown bears but found effectiveness was variable, and most attempts were unsuccessful. Beckmann et al. (2004) aversively conditioned black bears trapped in urban areas with rubber buckshot and slugs, pepper spray, cracker shells, yelling, and dogs. These techniques did not reduce the likelihood that a treated bear would return to the site and repeat problem behaviors.

A critical review of aversive conditioning theory, the design of interventions, methods of implementation, and expectations from applications is needed. Although much behavior theory has been used to support this technique, practical field applications have been very limited. To date, there is little convincing evidence that conditioned aversion will provide long-term management of urban carnivore conflicts. At best, it seems that this method will provide short-term relief from problems caused by specific animals.

Improved Livestock Management

Poor animal husbandry practices on farms adjacent to developed areas may lead to conflicts with carnivores. For example, discarding livestock carcasses aboveground, and keeping vulnerable mother and offspring livestock away from secure buildings, may lead to conflicts with coyotes or bears (Robel et al. 1981; Atkinson and Shackleton 1991). This behavioral conditioning caused by anthropogenic food sources sets the stage for heightened negative interactions. Mitchell et al. (2008) documented that carcass dumps influenced the home range and movements of suburban coyotes in Rhode Island. Landowners should focus on preventive steps to eliminate carcasses and break the habituation cycle. Despite such management efforts, however, one can expect coyotes to continue to kill sheep and other livestock in rural areas (Timm and Connolly 2001).

Relocation

Relocation of nuisance carnivores has shown varying degrees of success. Problem black bears with repeated offenses in suburban areas are often captured in culvert traps and sometimes transported more than 100 km. Experience in the northeastern North America indicates that if a relocated bear is a juvenile male, chances are about 70% that the bear will not return, or not continue to cause problems. However, the success rate for juvenile females is only about 40% and is even lower (20% or less) for adult bears of either sex (Landriault et al. 2000). Adult bears may travel 200 km or more to return to their established home range. It is difficult to document, but between 10% and 60% of relocated bears may repeat their nuisance behaviors (Landriault et al. 2000).

Dense raccoon populations living in urban areas frequently occupy attics and chimneys (O'Donnell and DeNicola 2006; chapter 4). Raccoons using buildings have often been live captured and released elsewhere, and relocated raccoons may show high survival rates (Mossillo et al. 1999) while other problem animals, such as squirrels, might not (Adams et al. 2004). Surprisingly little information exists about the behavior of problem-causing raccoons and their responses to different control strategies. O'Donnell and DeNicola (2006) documented that most of the nursing raccoons they removed from suburban houses relocated to other homes or outbuildings during nursing, then to tree dens during weaning, returning to houses the following spring. Hatten (2000) recommended that effective management of nuisance raccoons included capturing and euthanizing the offending animal that had entered a home. Nuisance raccoon situations should be prevented over the long term by: (1) physical exclusion (chimney caps, one-way doors, etc.), (2) removing food attractants (securing garbage cans, feeding pets indoors, etc.), and (3) proper home maintenance to reduce access points associated with vents and soffits. Control efforts should focus on these long-term options because such proactive management is more likely to be effective, economical, and consistent with the public interests (see chapter 4).

More research is needed to determine the fate of other relocated carnivores. Relatively few studies have documented the fate and behavior of nuisance animals that are trapped and moved elsewhere to reduce conflicts.

INTERNATIONAL CARNIVORE CONFLICTS

Up to this point, we have confined discussion primarily to carnivore conflicts in North America, but this issue is international in scope, as conflicts between wildlife and humans are a significant problem in many parts of the world. As long as people have been raising livestock and producing crops, they have had to deal with wildlife-related safety concerns and animal losses. Factors associated with carnivore impacts include expanding human populations, habitat losses, and growing carnivore populations in some areas resulting from successful conservation programs (Saberwal et al. 1994).

Predation losses may be particularly severe where pastoral farmers live in proximity to protected areas and parks (Mishra 2001; Conforti and de Azevedo 2003; Wang and Macdonald 2006). Lax herding practices, inadequate fencing and guarding, and overgrazing may contribute to livestock losses from predators (Wang and Macdonald 2006). Countries in Southeast Asia reported predation losses ranging from one-quarter to almost half of a family's annual cash income (Oli et al. 1994; Mishra 1997; Stahl et al. 2001). In Bhutan, tigers and leopards attacked cattle, mostly in areas with low human activity, and dhole often killed sheep (Wang and Macdonald 2006). Where livestock is abundant, losses increase, resulting in predation hotspots (Nass et al. 1984; Yom-Tov et al. 1995; Wang and Macdonald 2006).

Livestock predation attracts carnivores near human settlements, resulting in occasional direct attacks to people. In Sumatra, Indonesia, between 1978 and 1997, tigers reportedly killed 146 people and injured 30 and killed at least 870 livestock (Nyhus and Tilson 2004). Conflict was more common in intermediate disturbance areas such as multiple-use forests, where tigers and people coexist.

Tigers are considered to be one of the most dangerous species of large carnivores (McDougal 1987). From 1970 to 2001, there were increasing numbers of people attacked or killed by Amur tigers in the Russian Far East (Miquelle et al. 2005). This trend parallels increasing attacks by mountain lions in western North America (Beier 1991) during the end of the twentieth century. With conservation efforts and protection, populations of large cats appear to be increasing near populated areas. In 1999, a Tiger Response Team was created to address conflict situations in the Russian Far East, and this team responded to 73 conflicts during 1999–2002 (Miquelle et al. 2005). Often (50% of responses) the team investigated complaints and provided security to people who felt threatened by tigers. Although the perceived risk of an attack was greater than the actual risk, the team provided government acknowledgment of local concerns and reduced the antagonistic relationship between people and tigers. The goal of education programs was to reduce fear and change local attitudes toward tigers, but the impacts of this program have not been evaluated. Impacts to people must be reduced to an acceptable level in the communities that directly incur the cost of living with tigers. The most important contribution of the Tiger Response Team may be to reduce the perceived risk of tiger attacks.

Databases documenting the number of carnivore attacks on humans around the world are few, often fragmented, and difficult to find (Loe and Roskraft 2004). Apparently, resources and personnel to develop reliable filing systems are usually lacking. In India, sometimes the family of a person killed by a leopard or tiger is compensated (Veeramani et al. 1996). Compensation is an incentive to report claims but may result in false information. Although the actual number of carnivore attacks on people is unknown, available evidence indicates that worldwide, large carnivores have killed at least 150 people annually throughout the twentieth century (Loe and Roskraft 2004). Management strategies should focus on avoiding encounters between carnivores and people, including (1) creating protected areas where carnivores are undisturbed by humans, (2) removing problem animals, and (3) providing people with information on how to behave in the presence of carnivores.

CONCLUSION AND MANAGEMENT IMPLICATIONS

The increasing abundance of carnivores, one of many wildlife management success stories, has led to the recent appearance of unexpected species in urban and suburban habitats. Many suburban residents have few experiences with carnivores and are uncertain how to respond to perceived risks and potential conflicts. Suc-

cessful management of carnivore impacts will require modifying both human and wildlife behaviors. Greater long-term success will likely be attained by changing human behavior, especially peoples' attitudes toward, and tolerance of, conflict situations.

As urban sprawl and development continue around the world, concomitant with the recovery and expansion of carnivore populations, it can be expected that human-carnivore conflicts will increase. State and federal wildlife management agencies should proactively plan to address these issues, rather than respond to a tragic event after it has occurred. Although it is unrealistic to think that the risks posed by carnivores can be completely avoided, steps can be taken to reduce the likelihood of problems.

Eliminating intentional feeding of carnivores through aggressive enforcement, removing unintended food attractants, and greater public awareness of the negative consequences of both actions, can probably lower the risk of a negative encounter. Changing human behaviors is more likely to provide greater long-term reduction in conflicts with carnivores than attempting to modify animal behavior. Despite many attempts with wildlife species, aversive conditioning has shown mixed results for lowering negative impacts. At best, this method has provided short-term relief from damage for individual problem animals.

Many cases of community-based wildlife management indicate that public involvement will be critical for addressing human-carnivore conflicts. Education is often mentioned as an approach for reducing perceived risks of carnivore attacks and for reducing food attractants associated with negative encounters. Although we support such education programs, they need to be evaluated to determine which components have the greatest likelihood of changing human behavior. Education programs and outreach materials are seldom critically examined (very little rigorous evaluation of such programs exists in the literature), and in the few cases such evaluation has occurred, minimal positive change in people's behavior has been documented to date (Gore et al. 2008). Consequently, better education programs and delivery methods may be required.

Given the diversity of carnivore species and potential conflict situations, no single management tool or solution will resolve most problems. Management agencies need the policy and capacity in place to apply a variety of techniques, including nonlethal and lethal methods, to address the array of potential concerns. Particularly with large carnivores in suburban areas, protecting public health and safety must be paramount. Agencies should develop policies and protocols for determining when lethal removal of a high-risk animal is appropriate. Actions should be justified by defensible criteria, using the most appropriate and humane methods. The many stakeholders involved in wildlife management decisions will demand this from public officials, agency staff, and wildlife management professionals.

16

Conservation of Urban Carnivores

BRIAN L. CYPHER

SETH P. D. RILEY

RAYMOND M. SAUVAJOT

MANY CARNIVORE species are permanent or occasional residents of urban environments; therefore, conflicts between urban carnivores and humans are not uncommon. Consequently, most management efforts are directed toward the mitigation and amelioration of conflicts, including occasionally reducing the abundance of urban carnivores, or at least removing specific individuals. However, increasingly, it may be desirable to encourage and conserve populations of carnivores in urban environments. In this chapter, we will explore situations in which such conservation efforts may be desirable, important factors to consider in these efforts, and some potential strategies for conserving urban carnivores.

WHY CONSERVE URBAN CARNIVORES

Reasons for conserving carnivore species in urban environments include benefits for the species of interest as well as benefits that extend beyond the species. Species benefits can include enhancing local and regional abundance, increasing population stability and security, and possibly elevating public awareness and appreciation. Numerous species, carnivores as well as others, have declined in abundance because of profound loss, degradation, and fragmentation of their natural habitats, resulting in a significant risk of extinction for some species. The endangered San Joaquin kit fox discussed in chapter 5 is a prime example. As human populations grow and conversion of natural habitats continues, additional species will likely suffer a similar fate. However, as this book documents, some carnivore species are sufficiently adaptable to use urban environments, which expands the habitats available to these species and increases local and regional abundance. Indeed, densities of some carnivores are higher in urban areas compared with nearby nonurban areas (Harris 1981b ; Rosatte et al. 1991; Beckmann and Berger 2003b).

Urban populations also might enhance a species' regional stability and security. The presence of urban populations increases the overall number of individuals and populations, which reduces the vulnerability of species to catastrophic events, such as disease epidemics. Furthermore, urban populations also may be subject to different environmental conditions compared with nonurban populations, and this may further enhance stability and security. Food may be more consistently abundant in urban habitats compared with nonurban habitats. For example, prey availability for endangered kit foxes can fluctuate markedly in natural habitats due to annual variation in environmental conditions, particularly precipitation, and this produces significant fluctuations in kit fox abundance (White and Garrott 1997; Cypher et al. 2000). However, such fluctuations have not been evident in urban habitats where irrigated landscaping may bolster prey abundance and anthropogenic foods (e.g., food discarded by humans, pet food) are abundant (Cypher and Frost 1999). Greater food availability in urban areas may increase reproductive rates, as observed in red foxes (Gosselink et al. 2007), black bears (Beckmann and Berger 2003b), and kit foxes (B. Cypher, unpublished data). Smaller carnivores may experience higher survival rates in urban environments because of lower abundance of larger predators, as observed among kit foxes (B. Cypher, unpublished data) and red foxes (Gosselink et al. 2007). Thus, urban populations may experience less variation in abundance or may fluctuate asynchronously with nonurban populations, resulting in greater regional population stability.

In some situations, conditions in urban habitats may be sufficiently favorable that "surplus" animals are produced that disperse into adjacent nonurban habitats. In these cases, the urban populations could serve as source populations and help to bolster regional abundance. Gosselink et al. (2007) provided evidence that this was indeed the case for some urban red fox populations in the Midwestern United States, and the same may be true for urban red foxes in the United Kingdom (Baker et al. 2000) and possibly for kit foxes in the San Joaquin Valley, California (B. Cypher, unpublished data).

Urban carnivores may serve as "ambassadors" for their species and for wildlife in general. People's observations of animals commonly stimulate interest in the species and may generate an appreciation and compassion for it. This interest and concern can extend beyond urban populations to nonurban populations, thereby benefiting the entire species. People who had observed urban kit foxes were more likely to favor conservation not only of urban kit foxes (figure 16.1) but also of kit foxes in natural lands (Bjurlin and Cypher 2005). Thus, conservation of urban populations can provide benefits for entire species.

In addition to benefits for specific species, conservation of carnivores in urban areas may be important for ecological, aesthetic, or ethical reasons. Conserving biodiversity in general and in specific species in particular contributes to the maintenance of ecosystem structure and function. The removal of any species from an ecosystem is likely to result in significant adverse impacts to that system. The removal of carnivores in particular is thought by many to have potentially significant effects on other ecosystem components (e.g., Crooks and Soulé 1999; Ray et al. 2005; see also chapter 14). However, because ecological processes are complex, identifying and quantifying these impacts is usually difficult if not impossible. This problem likely is even more pronounced for carnivores, which because of their generally higher trophic positions, typically occur at lower densities and range over larger areas. Thus, collecting information on these species and their ecological interactions is more difficult, and yet their reduction or removal can affect ecological processes over a broad area.

From an ethical perspective, many people believe that humans have a fundamental obligation to attempt to prevent human-caused extinction, including the local extirpation of species. This belief extends to wildlife populations inhabiting urban environments. Furthermore, conservation of urban carnivores, or any wild species in urban areas, may help to maintain a link between human urbanites and nature. Observing wild animals in urban environments may be a pleasurable and rewarding experience, but it also helps to remind people of wild creatures and places outside of urban areas. This could be important psychologically for some people, especially those infrequently able to visit wild places and may help to garner support for conservation of wild places and the species that inhabit them.

Finally, conservation of carnivores as well as other

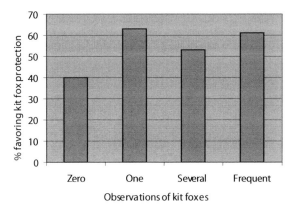

Figure 16.1. Willingness of citizens in Bakersfield, California, to protect urban kit foxes relative to the number of observations of kit foxes.

species in urban areas may be a goal of local, state, or federal agencies. Land management and conservation agencies commonly have a responsibility or at least a goal of conserving biodiversity on lands under their ownership or management. For example, under the U.S. Organic Act of 1916 the U.S. National Park Service (NPS) has been mandated to "to conserve the scenery and the natural and historic objects and the wildlife therein . . . in such manner and by such means as will leave them unimpaired for the enjoyment of future generations." This mandate applies to NPS units in urban environments such as Rock Creek Park in Washington, D.C.; Gateway National Recreation Area in New York City, Saguaro National Monument in Tucson, Arizona; Golden Gate National Recreation Area in the San Francisco Bay Area; and Santa Monica Mountains National Recreation Area next to Los Angeles, as well as for parks such as Yellowstone, Yosemite, or Grand Canyon, which are far from urban development. Carnivores, such as raccoons, red foxes, and recently coyotes in the eastern United States, and coyotes, bobcats, and mountain lions in the western United States, are present in these as well as other urban parks. Similar situations occur in state and local parks across the United States and in nature reserves worldwide, where maintaining a natural and complete an ecosystem is the goal.

SPECIES ATTRIBUTES FACILITATING CONSERVATION

Depending on the circumstances, it may be desirable to attempt to conserve almost any carnivore species in ur-

ban environments. Attributes of both the species and the specific urban area will dictate the level of challenge in conserving a given urban population. Certain species attributes may significantly facilitate conservation efforts.

The species that will be least challenging to conserve will likely be those species that are highly tolerant of human disturbances, can satisfy all life requirements in urban areas, are least conspicuous, and pose few or no threats to humans, domestic animals, or property. Some species are highly tolerant of disturbance, including striped skunks, raccoons, red foxes, kit foxes, and coyotes. These species are commonly observed by people in urban areas, may take up residence within (e.g., under houses) or very close to human habitations, and in some instances seem unfazed by high levels of noise and nearby activity.

Raccoons and skunks in North America and red foxes in Europe are notorious for dwelling close to people. Raccoons and skunks have been reported to enter dwellings through pet doors in search of food. Red foxes and coyotes have been observed foraging adjacent to extremely busy highways. Kit foxes have been reported to chase and steal golf balls on golf courses in Bakersfield, California, and in one instance, a kit fox made several appearances on the field during a high school baseball game thus holding up play. Such tolerance of human activities will facilitate conservation efforts. On the other end of the spectrum, some species, particularly felids such as bobcats and mountain lions, tend to be less tolerant and more sensitive to disturbance. Conservation of these species will be challenging and likely will require additional and larger refugia away from disturbances.

Species able to find adequate food, water, cover, space, and mates in urban habitats generally will be easier to conserve. Foraging generalists, particularly those tending toward a higher degree of omnivory, are likely to have an easier time finding food. Species that readily try novel foods also may have success, although this also could be problematic (e.g., preying on pets and livestock, consuming garden fruits and vegetables, damaging ornamental plantings, raiding garbage cans). Water generally is easy to find in irrigated landscaping, pools, water features, and other sources. Cover also is probably rarely a limiting factor. Most animals can find cover in landscape plantings, in patches of native vegetation, and in anthropogenic structures such as buildings, culverts, pipes, and firewood piles (figure 16.2). However, space could be a limiting factor for some species. Larger species such as mountain lions, black

Figure 16.2. San Joaquin kit fox using portable buildings as daytime cover at a high school in Bakersfield, California. Photo by C. Van Horn Job.

bears, and coyotes obviously will require more space per individual, and conserving large urban populations of these species will be challenging. For example, mountain lion home ranges in urban landscapes can range from 93 km² to 476 km² (see chapter 11). Conversely, smaller species tend to have smaller space requirements; therefore, more individuals can be conserved per unit area.

Habitat preferences also may influence conservation efforts. Species that are readily able to use anthropogenic habitats will be easier to conserve in urban areas. Species like kit foxes and red foxes can readily use urban habitats such as residential, commercial, and industrial areas. These species are able to persist even if little or no natural habitat is present. Conversely, species such as mountain lions are able to use anthropogenically altered habitats to a very limited extent and primarily persist only in urban environments that include considerable quantities of natural habitats that support their ungulate prey. Thus, conserving this species will be difficult, if not impossible, in urban areas lacking such habitats.

Less conspicuous species may be easier to conserve in urban environments. Species that are nocturnal, secretive, generally silent, use inconspicuous hiding places, and leave little sign usually will attract less attention: "out of sight, out of mind" could facilitate conservation. Species that are visible (e.g., red foxes and coyotes), noisy (e.g., coyotes), malodorous (e.g., striped skunks), or leave obvious signs (e.g., raccoon feces, kit fox dens, skunk diggings) readily alert humans to their presence and may elicit concerns that could impede conservation efforts.

Finally, species that constitute a potential threat to humans, domestic animals, or property will be challenging to conserve in urban areas. Potential threats include direct attacks or disease transmission. For example, attacks by mountain lions and coyotes in urban areas (Baker and Timm 1998; Adams et al. 2006), or even the perceived threat of attack, generate trepidation regarding cohabitation with these species. Black bears, by virtue of their large size, are viewed as dangerous cohabitants even though few attacks on people have been recorded in urban areas. Arctic foxes in urban areas have been reported to threaten and attack people (P. White, personal communication). Coyotes are notorious for preying on cats and small dogs. All species have the potential to transmit pathogens to humans or domestic animals. However, some species are readily associated with actual or perceived disease transmission, which can generate opposition to their presence (e.g., rabies in striped skunks and arctic foxes, zoonotic roundworms in raccoons, distemper in coyotes and gray foxes, and bovine tuberculosis in Eurasian badgers). Finally, species that cause property damage may be less welcome in urban environments (e.g., black bears breaking into residences, skunks digging in lawns, raccoons and Eurasian badgers raiding vegetable gardens, coyotes killing pets, and kit foxes establishing dens in inconvenient locations such as golf courses; see chapter 5, figure 5.6).

CONSERVATION STRATEGIES

Potential conservation strategies for carnivores in urban environments might be broadly categorized as "passive tolerance," "conflict resolution," or "opportunity enhancement." At present, most urban carnivore conservation efforts generally fall into one of the first two categories. Considerable potential exists under the third category, and strategies from multiple categories can be implemented for a given species in a given urban environment.

Passive Tolerance

Passive tolerance consists of permitting a species to be present in an urban environment. No measures are implemented to discourage the presence of the species or reduce abundance, but neither are any measures implemented to encourage or benefit the species. Passive tolerance is the most common "conservation strategy," largely because it requires no actions or effort. Thus, this is a default strategy. Passive tolerance is commonly observed when species have opportunistically colonized an urban environment and are not engaged in any significant conflicts with humans (e.g., nuisance issues, threat to human or pet health). In essence, humans are content to allow the species to coexist with them.

With the possible exception of larger carnivores (e.g., bears, mountain lions), passive tolerance generally is initially practiced whenever a species is found to occur in an urban area. Larger carnivores usually are perceived as a threat to humans and domestic animals and often are removed or forced out of urban areas. Smaller carnivores are tolerated until they cause problems, and then conflict resolution is usually implemented, which can constitute either management or conservation. Some species for which passive tolerance tends to be the primary conservation strategy include bobcats and gray foxes.

Conflict Resolution

Conflict resolution consists of implementing measures to increase human tolerance of species in urban environments by mitigating conflicts. Conflict resolution overlaps considerably with conflict management (see chapter 15), and the actions implemented are essentially identical. The difference is of purpose or intent, and even that distinction is not sharply delineated. When the intent of mitigating conflicts is to benefit a species by facilitating coexistence with humans, then this constitutes a conservation strategy. When the intent is to mitigate conflicts for the convenience of humans with no regard for benefits or impacts to a species, then this constitutes a management strategy. The two strategies obviously overlap when management reduces conflicts but coincidentally and unintentionally facilitates coexistence between carnivores and humans.

Conflict resolution measures are thoroughly discussed in chapter 15 and elsewhere throughout this volume, and examples will only be briefly mentioned here. One example is reducing access to refuse, as is commonly implemented for raccoons and black bears (figure 16.3). Another example is reducing access to structures, as is occasionally implemented for raccoons and skunks. Reducing risky human-animal interactions, such as discouraging or prohibiting feeding, is a measure commonly implemented for many species. In certain situations, overly aggressive individuals are removed, as has been implemented with coyotes and arctic foxes. Another measure includes vaccination to reduce potential disease transmission, as has been

Figure 16.3. Bear-proof Dumpster in Incline Village, Nevada. Photo by P. Kelly.

implemented for raccoons (Rosatte et al. 1992a) and red foxes (Brochier et al. 1991). Finally, providing educational information to the public to ameliorate conflicts is another common measure. Adams et al. (2006) describe in detail excellent examples of conflict resolution efforts to conserve coyotes in Vancouver, British Columbia, and black bears in four U.S. cities.

Opportunity Enhancement

Opportunity enhancement consists of proactive actions implemented to encourage the presence of a species in an urban area and to enhance their potential persistence. This strategy has been implemented frequently for noncarnivore species; songbird feeding is an example. Only rarely has it been implemented for urban carnivores, but much potential exists to do so. Modest and inexpensive measures can contribute significantly to the conservation of urban carnivores. Obviously, a consideration when implementing measures to encourage urban carnivores is to do so in a manner that avoids or minimizes the potential for subsequent conflicts with humans, and this could be a significant challenge that likely will limit this strategy to certain species in selected urban areas.

Landscape Planning

Urban landscape planning is an extremely important opportunity enhancement strategy. Compared with other species, carnivores have relatively high spatial requirements (see chapter 2). In general, carnivores maintain relatively large home ranges, require considerable space for foraging, are often territorial, and can disperse long distances. Thus, ensuring adequate space in urban landscapes for carnivores is critical for their conservation.

Furthermore, the juxtaposition of suitable habitat patches and connectivity among these patches is critical. Some carnivores (e.g., mountain lions, coyotes) require patches of relatively natural habitat (e.g., remnant forests or shrub communities). Other species may not require natural habitat but may not be able to use all types of urban habitats. For example, kit foxes use most urban habitats but are rarely found using high-density residential areas. In urban landscapes, suitable habitat rarely occurs in large contiguous blocks but instead occurs as dispersed patches. Therefore, maintaining connectivity among patches is important. In highly urbanized and fragmented Southern California, the Missing Linkages project has been analyzing the importance of remaining and missing linkages between large natural areas and working both to conserve them and to highlight their importance to land managers, planners, and citizens (South Coast Wildlands 2008). This effort has been successful in drawing attention to connectivity, in conserving the land around linkages, and, in some cases, adding crossing structures to improve connectivity for wildlife. Conservation of urban carnivores will be significantly enhanced if urban planners include suitable habitat patches and movement corridors in their landscape designs. It will be important for species experts to provide planners with required habitat elements for target species.

Education and Outreach

Arguably the most important facet for carnivore conservation in urban landscapes is effective public education and outreach. For the carnivores that live in a particular urban area, it is critical that people understand their ecology, their behavior, how they may conflict with humans, and how to avoid those conflicts. Effective and widespread education about larger carnivores such as mountain lions or bears that are potentially dangerous to humans is especially important. However, many urban residents may be concerned about or express fear of smaller carnivores as well, and reducing conflicts over property damage or nuisance issues may be as important as safety concerns, given the potential frequency of the two types of conflicts. Outreach and education may be valuable in reducing the effects of anthropogenic mortality sources such as the exposure of nontarget carnivore species to lethal toxicants or vehicle collisions.

Mitigating Road Effects

Collisions with vehicles commonly are the primary cause of mortality among urban carnivores, which is detailed in the species-specific chapters. Strategies and measures have been developed to mitigate impacts to wildlife on roadways (Forman et al. 2003; Glista et al. 2008), and these same approaches can be implemented in urban landscapes to reduce adverse impacts to urban carnivores and other urban wildlife. Such approaches include providing crossing structures, implementing speed controls, or erecting caution signs in critical road-crossing locations (figure 16.4). In the United States, transportation agencies and universities are working together to review the need for and efficacy of techniques that help reduce wildlife impacts due to transportation infrastructure (Cramer and Bissonette 2007; Western Governors' Association 2008). These efforts and initiatives promise to provide many new opportunities to help enhance carnivore survival in urban environments where roadway impacts can be severe.

In urbanized areas of Japan, mortality of raccoon dogs, from collisions with vehicles, is considerable, and roads with high-traffic volume may serve as barriers to movements (Saeki and Macdonald 2004). Patterns of raccoon dog road kills suggest that mortality is highest at major intersections near woodland habitat and water (Saeki and Macdonald 2004), and ecological modeling indicated that the development of green corridors at these intersections might reduce vehicle-related mortalities as much as 40% (Shinobu et al. 1999).

Figure 16.4. Speed warning sign in Yosemite National Park, California. Photo by C. Kelly.

Given the low population densities and extensive movements of carnivores, the installation of road-crossing structures may be a particularly beneficial conservation strategy for these species in urban environments. Crossing structures, in combination with fencing to protect animals from roads, could be an important conservation strategy in urban areas where the risk of mortality from vehicle strikes is significantly enhanced (e.g., Bjurlin et al. 2005). Wildlife-friendly crossing structures have been used widely in various nonurban situations for species ranging from toads to Florida panthers, often with considerable success (Clevenger and Waltho 2000). In Europe, transportation planners have designed innovative solutions to assist wildlife across roads, which can substantially enhance the probability of persistence of carnivores in urban areas (Iuell 2003).

Control of Nonnative Competitors

Nonnative carnivores occasionally are able to colonize urban habitats, possibly because of the altered nature of these areas. These nonnative species may then compete with native carnivores, which is the case for nonnative raccoons and raccoon dogs in Japan (Ikeda et al. 2004). Initially imported into the country as pets, raccoons were released by individuals either intentionally or accidentally in various cities during the 1970s and 1980s, and raccoons are still primarily associated with cities. Some evidence indicates that raccoons may displace raccoon dogs in cities where both occur (Ikeda et al. 2004). Conservationists have argued for an organized removal program for raccoons and other alien species to protect native species in urban and nonurban areas in Japan (Ikeda et al. 2004).

Another example involves red foxes, kit foxes, and gray foxes in California. Kit foxes and gray foxes, both native species, are present in many cities in California. Nonnative red foxes are rapidly increasing in abundance and distribution in California (Lewis et al. 1993). Red foxes are rarely observed in natural habitats used by kit foxes and gray foxes but are commonly observed and appear to be increasing in altered environments, including urban areas. Red foxes potentially engage in interference and exploitative competition with kit foxes (Ralls and White 1995; Cypher et al. 2001; Clark et al. 2005). In Bakersfield, California, red foxes have been observed using known kit fox dens, and red foxes use many of the same foods as kit foxes. Interactions between gray foxes and red foxes are unclear, but Trapp

and Hallberg (1975) suggested that red foxes may be a limiting factor for gray foxes. Thus, efforts to limit the abundance of nonnative red foxes could benefit native fox species in urban areas.

Species-Specific Efforts

Relatively few examples of proactive conservation measures have been implemented for particular species of urban carnivores. One notable exception is the kit fox (see chapter 5). Enhancement measures for this species in California have included installing artificial dens, constructing road-crossing structures, erecting caution signs near natal dens, setting aside refugia areas, maintaining movement corridors, and conducting educational outreach.

Another example of proactive conservation efforts involves Eurasian badgers. In the Netherlands, badger abundance may have been reduced to only about 1,500 individuals (Wiertz and Vink 1986), a decline largely due to habitat conversion (much of this from urban development) and fragmentation associated with roads (van der Zee et al. 1992). However, badger numbers have been rebounding since the 1990s due in part to several proactive conservation strategies, including fencing to exclude badgers from roads, passage tunnels beneath roads, installation of artificial setts (dens), and relocation / reintroduction programs (Apeldoorn et al. 2006). Proactive habitat enhancements for badgers in the Netherlands have been coupled with extensive outreach efforts by badger-focused conservation organizations (e.g., Das and Boom). Artificial setts also have been installed for Eurasian badgers near some urban areas in the United Kingdom (British Broadcasting Company 2007).

CONCLUSIONS

As urbanization continues worldwide, the occurrence of various carnivore species in urban areas likely will increase. Carnivores may opportunistically colonize expanding urban habitats, or shrinking natural habitats will force carnivores into urban areas. Given the continuing loss of natural habitats and concomitant declines in populations of many carnivores, these urban populations present excellent opportunities for conserving species and, in some cases, avoiding local or even range-wide extinction. Indeed, for some species, such as the Eurasian badger, urban areas at the urban / rural interfaces may constitute the bulk of the remaining habitat, and therefore conservation of urban populations will be critical to the long-term persistence of these species.

This circumstance will certainly become common for other carnivores, too, as remaining habitats are urbanized or are juxtaposed with urban areas. Although the profound trends of land conversion and urbanization present major challenges for carnivore conservation, the ability of many carnivores to successfully adapt to urban situations, and the willingness of humans to tolerate or even encourage carnivores in urban areas, will be critical for their long-term survival. For some species, the ability to persist at least sometimes in urban environments may be crucial to their conservation (e.g., mountain lions in Southern California, wolves and brown bears in Europe). Carnivores possess a suite of adaptations that can contribute to urban survival and conservation, and management tools are available to help complement these adaptations. In general, effective landscape planning in urban areas, including maintaining sufficient amounts of effectively connected natural areas, will be critical for urban carnivore conservation. The efficacy of conserving populations of urban carnivores also will be inextricably linked with successfully mitigating conflicts and educating the public, both of which will enhance acceptance of carnivores in urban areas, generate support for their conservation, and ultimately enhance public appreciation of these fascinating animals, even amid continuing urbanization.

17

Urban Carnivores
Final Perspectives and
Future Directions

SETH P. D. RILEY

STANLEY D. GEHRT

BRIAN L. CYPHER

W E HAVE SEEN over the course of this book that the study of urban carnivores touches on an array of important and interesting topics across the fields of ecology, behavior, conservation, and wildlife management. In this final chapter, we highlight some of these topics, discuss the value of classifying urban carnivores and the landscapes they inhabit, and indicate some of the subjects not covered in this book that may be the object of fruitful future work.

FUNDAMENTAL CONCEPTS ADDRESSED BY URBAN CARNIVORES

Ecology and Behavior
Population Biology
Population biology, or the study of how and why populations vary over time and space, is fundamental to all other aspects of ecology. In urban carnivores, we have repeatedly seen how critical demographic parameters, particularly survival but also reproduction, may vary between urban and nonurban populations. Reduced reproductive output or survivorship likely prevents many carnivore species from thriving or even existing in urban areas. However, many of the species that persist in urban landscapes have survival rates that are as high or higher in urban areas than outside them (see also chapter 13), and urban kit foxes actually have inflated rates of reproduction (chapter 5). High survival and reproduction rates in turn can lead to greater population density, which has implications for management

and for interactions with other species. High population density in urban areas is something that many urban carnivores share.

Another aspect of population ecology that has received attention in the past decade or more, is the idea of outside, or "allochthonous" inputs moving from one ecosystem or habitat to another (sensu Polis et al. 1997). In the ecological literature, this has generally been investigated at hard boundaries between systems such as an aquatic-terrestrial boundary. However, it is also relevant for urban ecosystems, where there is often a stark transition between developed and natural habitats. Humans, or the effects of humans, may often provide extra resources such as food, water, or cover (dens, etc.). Even though carnivores may venture into developed areas to acquire these resources, the resources are certainly provided by an "outside" source. The effects of these extra inputs for urban carnivores remain to be determined, but it has been suggested that they can result in altered population parameters, including density (e.g., Fedriani et al. 2001).

Finally, urban carnivores, because of their ability to move large distances and to often cross relatively inhospitable, developed habitat, may exist in metapopulations (Levins 1969) more frequently than is common for other species or in other types of landscapes. True metapopulations are often hard to find in nature (Harrison and Taylor 1997), but they may be more common in fragmented, urban landscapes, and metapopulation dynamics have important implications for conservation and management strategies. One type of metapopulation model, the "source-sink" model (Pulliam 1988), may be particularly relevant for the conservation of carnivore populations in urban landscapes when reproduction does not keep pace with mortality, and dispersal from core natural areas is required to maintain species presence (as in "remnant" urban carnivores, see "Defining Types of Urban Carnivores").

Community Ecology

Interactions between carnivores and other species may be particularly strong in urban areas if population densities of one or both species are high and species distributions are limited by roads or development (see also chapter 14). In these situations, the strength of all kinds of interspecific interactions may be increased, including those between competitors, predators and prey, and parasites and their hosts. Urban "islands" can also help to demonstrate that some interspecific interactions may not be as strong or ubiquitous as previously supposed,

for example, mesopredator release (Crooks and Soulé 1999). Even in a confined space with dense populations, there was no evidence of any significant negative effects of coyotes on raccoons in forest parks in Chicago (Gehrt and Prange 2007). Finally, if there is a difference in the ability of different species to persist in the face of urbanization, urban areas may actually reduce or eliminate interspecific interactions. In Illinois, urban red foxes are released from the intense competition with coyotes, including interspecific killing, that occurs in nearby rural areas, because coyotes do not survive human persecution in the cities (Gosselink et al. 2003, 2007).

Urban landscapes include dense populations of human commensal species such as domestic cats (chapter 12), a situation that can result in important and potentially detrimental (for wildlife) interactions with other species. The interaction between pet cats, native prey species, and pet food supplied by humans represents a classic (if anomalous) case of "apparent competition" (Holt 1977), where one prey species is abundant, thereby supporting more predators and leading to increased predation on the other prey species. In many urban areas, the constant supply of cat food often leads to high predator (cat) densities with potentially negative consequences for native prey species, including birds, small mammals, and reptiles (chapter 12). Because domestic cats and dogs are carnivores, they also share many diseases with wild carnivores, with potential consequences for both groups. Rare carnivores, such as the endangered San Joaquin kit fox, may be at increased risk for exposure to disease in urban areas (chapter 5) because of the high densities that can be attained by both wild and domestic carnivores. The implications of increased disease exposure in urban carnivores can carry over to humans, when zoonotic diseases such as rabies (e.g., in raccoons in the United States [chapter 4] or red foxes in Europe [chapter 6]) are present in wild species.

Behavioral Ecology

Urban carnivores also raise questions about animal behavior and behavioral ecology, such as how their behavior changes when animals are faced with the most extreme human-altered environments and what the limits of behavioral flexibility may be for different species. In high-density urban carnivore populations, social organization and structure may be altered. In British red foxes (chapter 6) and Californian kit foxes (chapter 5), larger social groups with more nonreproductive individuals are present in cities, including the interesting and anomalous situation of older female kit foxes helping to rear the offspring of other females. Eurasian badgers, however, perhaps the most social of urban carnivores, have more solitary individuals, more intergroup movement, and reduced territorial behavior in cities (chapter 9). Many urban carnivores, including bobcats (chapter 10), coyotes (chapter 7), and Eurasian badgers (chapter 9), adjust their activity patterns when they are around human development. Bobcats and coyotes use residential or commercial areas and cross roads mostly later at night, and urban badgers delay their extensive movements, staying near their setts in the evening hours. Some classic animal behavior issues are relevant for conflict management in urban areas as well (see also chapter 15). At what point do wild carnivores become habituated to human residents of cities? How does it happen? What are the consequences? Is it an irreversible process? Is aversive conditioning potentially effective for avoiding, or reducing, this habituation? These are important questions, particularly for larger carnivores, such as mountain lions, bears, or even coyotes, as these species can be dangerous to pets and occasionally to humans.

Management and Conservation

From an applied perspective, carnivores in urban areas present great challenges for both management and conservation. As mentioned in chapter 1, carnivores hold a great fascination for people, perhaps as icons of wilderness, and perhaps also because of their relatively close relationship with domestic dogs and cats. For some people, the reaction to carnivores is extremely negative, resulting in fear and even hatred. In nonurban areas, these feeling are often related to the loss, or the threat of loss, of livestock to carnivores. However, in urban areas, with so many people and their pets living in proximity to carnivores, people's fear, whether justified or not, may be more for the safety of themselves, their children, or their pets. The feelings are not related to a threat to income and livelihood, but the emotional content of the fear may be even greater if it is related to one's own safety or that of loved ones (including pets). However, on average, city dwellers are often more liberal politically and socially than rural dwellers, the result of which can be that city residents have more of an animal welfare perspective and may be very concerned about attempts to manage urban carnivores, especially with lethal control. Finally, urban areas are frequently political and media centers, and conservation and management issues may unfold in a highly charged and often public arena. Nowhere are these issues more contentious and emotional than with domestic cats, particularly free-ranging feral cats and cat colonies. Cat management will continue to be a major challenge for those dealing with urban carnivores, and the results will have implications for the cats, for the wild species they prey on, and for wild carnivores (chapter 12).

Carnivores in urban landscapes can represent a major conservation challenge as well. While some species may thrive in urban areas, for many other species that exist in or near urban areas, their long-term persistence may be threatened by the loss and fragmentation of habitat from urbanization. This may be particularly true for larger carnivores, such as mountain lions or bears, and perhaps other less-well-studied (at least in urban areas, chapter 13) animals such as wolves, leopards, or spotted hyenas. Large carnivores represent perhaps the ultimate challenge for wildlife conservation in urban landscapes because they need a greater amount of suitable habitat than any other animal and therefore exist at lower densities. That they move over great distances also means that carnivores are vulnerable to the barrier effects of roads and development. While there are certainly significant threats to the preservation of large carnivores anywhere in the world, even in the national parks of Africa, conserving big cats next to a huge metropolis such as Los Angeles may prove particularly difficult. However, if even large carnivores persist over the long term in an urban landscape, perhaps it is a sign that conservation efforts are indeed effective, or at least that some larger carnivores may be more resilient to anthropogenic pressures than we had imagined.

Finally, perhaps because of the intense reactions (of all kinds) that people in cities can have to them, urban carnivores represent an interesting study for the human dimensions of wildlife, specifically the attitudes of people toward carnivores and the implications of those attitudes. A potentially rich research subject is the difference between urban and nonurban residents in their attitudes about carnivores, and given the large numbers of people in cities and the political weight they carry, those differences can have significant influence. One interesting phenomenon (see also chapter 3), along with carnivores that habituate to humans, is the opposite effect—humans that habituate to carnivores. It is interesting to see the difference in reaction to a particular urban carnivore such as coyotes, for example, between residents of newly colonized areas, such as the eastern United States, versus residents of Southern California, where coyotes have been present in the urban landscape for a century or more. In Chicago, where coyotes are a relatively recent addition but have now been present for a number of years, the change in attitude as people acclimate to their presence, going from fear and concern to acclimation and acceptance, may occur on a short time scale. Increased knowledge about human attitudes toward urban carnivores may prove instrumental in effective management and conservation policy and practice.

CLASSIFYING URBAN AREAS AND URBAN CARNIVORES

Measuring Landscape Use by Carnivores

As discussed in chapter 1, there is no universal convention on what defines "urban"; this issue comes up repeatedly in the study of urban ecology and thereby throughout this book in the study of urban carnivores. It is relevant for any study of the effects of urbanization (or any type of land-use change) on ecosystems, and sometimes it seems that there are almost as many ways to measure urbanization as there are urban studies. Certainly, the proper measurement depends on the goals of the particular project. For example, studies of urban streams often measure the amount of impervious surface in a watershed (e.g., Schueler 1994), since this determines the amount of urban runoff. For this book, the question is how best to measure how "urban" urban carnivores are, both as individuals and as species. There are four (at least) relevant issues: (1) How do we classify the landscape for the study area? (2) How do we classify the landscape for individual animals (generally for use with radio-tracking data)? (3) How do we then describe how these individual carnivores use the landscape? (4) How do we analyze and communicate this use, for management and conservation purposes and for comparisons with other studies or species?

At the broad scale of the region or study area, classification is generally accomplished with demographic information about human population size, roads, and housing. It is best if these are calculated for a specifically defined area relative to the study and reported as a density (per unit area). Throughout the book, we have tried to include as much of this information as possible for as many studies as possible (e.g., chapter 6, table 6.1; chapter 7, table 7.1; chapter 10, table 10.1). These statistics can facilitate comparisons between urban studies, which, in this book, span the length and

breadth of the United States and include cities in Canada, Britain, and Europe. These measures can also have specific relevance for carnivore populations. For example, red fox density in Britain is strongly correlated with the availability of medium-density housing (chapter 6); Eurasian badgers also seem to prefer areas of intermediate density housing (chapter 9); and kit foxes in Bakersfield thrive, despite the almost complete lack of natural habitat, because of the widespread availability of relatively open, nonresidential, anthropogenic habitat (chapter 5).

For describing land-use patterns for individual carnivores, the simplest way to classify the landscape is with just two categories—urbanized and not urbanized, or "natural." For some species or situations, this may be all that is needed to evaluate how urban their movements and home ranges appear. For example, for raccoons in Rock Creek Park in Washington, D.C. (chapter 4, figure 4.3), the residential areas bordering the park are old and stable, and the park is generally bordered by roads, so that the transition between urban and natural areas is pretty clear. However, even in this study area, variation between residential areas, for example, between lower-density housing with yards and trees and multistory apartment buildings, may have relevance for raccoon movements or populations (e.g., raccoon density was highest in the spurs of parkland surrounded by high housing densities; Riley et al. 1998). The next most complex classification is three categories, such as with bobcats in Southern California, where we defined (1) natural, (2) developed, and (3) "altered open" areas (e.g., chapter 10, figure 10.3). In this study area, we wanted to identify areas that were not protected open space (natural) but were also not as intensely developed as high-density residential or commercial / industrial areas (developed). Altered open areas include low-density residential areas where there is often significant landscaped or natural vegetation between houses; landscaped areas such as golf courses, parks, schools, or large office complexes; small pieces of natural or vegetated habitat surrounded by development, and landfills. These altered open areas likely vary from both intensely developed and natural areas in ways that are important to urban carnivores, such as having different amounts or schedules of human activity, affording different amounts of cover, and supporting different species and abundances of prey.

The different types of developed or altered areas may be different enough from one another to warrant explicit distinction, and it is possible to classify the landscape into as many categories as one wants, bearing in mind the accuracy of the land-use classification and the animal locations. For example, to describe landscape use for coyotes in Chicago (chapter 7, figure 7.3), we used eight different categories, some of which are similar to those mentioned earlier for the bobcat study area, but in this case, they are analyzed separately. For kit foxes in Bakersfield (chapter 5), categories included undeveloped lands (e.g., vacant lots, fallow crop fields), storm water catchment basins (sumps), industrial areas, commercial areas (e.g., office and retail facilities), manicured open space (e.g., parks, school campuses, golf courses), linear rights of way (e.g., canals, railroad corridors, power line corridors), and residential areas. Regardless of how many categories are used, or what they are called, detailed descriptions are needed to enable readers, land managers, and other researchers to understand them and potentially combine them for their own analyses or purposes. For example, in chapter 13, Iossa et al. (table 13.2) compare multiple species using the 1 category of "mean percent natural habitat used."

The last two questions, about how to describe and communicate landscape use by (often radio-collared) urban carnivores, may seem highly technical or academic, but they are nonetheless important. One relevant issue is whether one is interested in where specific animal locations are in the landscape, or in the percentage of an animal's home range that consists of different land-use types (or both). These measures are never exactly the same, they measure somewhat different things, and they can both be relevant for conservation and management. It is most important to be clear about why you are using a particular measure. Second, assuming the land-use composition of the home range is important, a decision must be made about the type of home range to construct. Urban landscapes are different from many nonurban landscapes, in that there are many hard, and often linear, boundaries between areas of different land uses. Such elements as roads (of various sizes) may also act as barriers to movement and may bisect otherwise suitable natural habitat. Over the last couple of decades, the wildlife literature has emphasized density-based home range methods, specifically kernel methods (Seaman and Powell 1996). While these methods certainly have their value, there are two problems (at least) with them in urban landscapes: (1) they assume some use of areas outside the points, which can result in areas included in a home range that are across roads or other boundaries that actually act as

barriers to movement, and (2) they often produce separate and unconnected home range polygons, which fail to capture movements between them that must be occurring (unless the animals are sprouting wings). These movements are often extremely relevant because they occur through developed areas. A number of potential solutions to the first problem include clipping out (using mapping software) portions of the home range that are across known or likely barriers for the species, or using older, non-density-based home range models such as minimum convex polygons (e.g., chapter 7; Riley et al. 2003). Recent home range models are being developed to deal specifically with some of these issues (Horne et al. 2007), and these models deserve careful attention, where hard, linear boundaries, such as rivers or roads, are significant, including in urban landscapes.

Finally, the last question, how to report the data once analyzed, is important. Another growing trend in wildlife research is to report habitat selection (or, in the case of urban animals, often "land-use" selection), as opposed to reporting and describing the actual use. Knowing which habitats or land-use types are "selected" by carnivores (i.e., used more than they are "available") and which are avoided (i.e., used less than their availability) can certainly be valuable and is often worth determining and reporting (e.g., in chapter 5 for kit foxes, chapter 7 for coyotes, chapter 11 for mountain lions). However, the basic descriptive data on the use of the landscape, regardless of what the availability may be, should be reported. From a management or conservation point of view, this information is more important than the "selection index." For example, a red fox in England using residential areas 35% of the time, if its home range is 40% residential, would be classified as "avoiding" residential areas (assuming sufficient sample size for statistical significance). However, from a management standpoint, this animal is clearly spending a lot of time around people. Conversely, if a GPS-collared mountain lion is located in developed areas 5% of the time, but only 2% of its home range is developed, this animal might be classified as "selecting" developed areas (again depending on the sample size, which with a GPS collar is likely to be large). Which animal is more ubiquitous across the urban landscape and more visible and relevant for the public and land managers: the one that "avoided" developed areas or the one that "selected" them? Whether or not "habitat selection" is computed, it is critical to report the actual use of different habitats or land-use types, particularly in urban studies. This importance is demonstrated by table 13.2 in chapter 13 in which the most basic class of information about land use by urban carnivores, namely, the percentage of natural area used, is only available for a handful of species.

Defining Types of Urban Carnivores

In this book, we have addressed a range of carnivore species across four families and across a large size range from kit foxes to mountain lions. These species also show a range of urban association and of responses to urbanization. Raccoons, coyotes, or British red foxes immediately come to mind when people think of urban carnivores. However, other species, such as mountain lions or endangered San Joaquin kit foxes, may be surprising inclusions in this book. In the broadest sense, any species that is even occasionally seen or reported in urban areas may be considered an urban carnivore, and the full list in chapter 13 (table 13.1) includes 35 different species, from 800-kg polar bears down to the tiny Indian mongoose (<1 kg). However, given that so many species may be at least occasionally in or around urban areas, classifying urban carnivores may be useful, for comparative or scientific purposes, but also because of management and conservation considerations.

Two attributes that may be useful for classifying urban carnivores are (1) their population or demographic status and (2) their urban association, or the degree to which they use or live in developed areas. Contributors to this book have used both attributes. In the kit fox chapter (chapter 5), Cypher separates urban carnivores from a demographic point of view into: those that cannot persist in urban areas because of disturbance or lack of habitat such that their populations would be *declining* in urban areas; those for which urban areas are not optimal, but they are able to continuously or at least intermittently occupy them—populations that we might call *persisting*; and others that are able to thrive in urban landscapes perhaps even more so than in nonurban areas, such that their populations may be *growing*. These distinctions have obvious implications for management, as growing urban carnivore populations are more likely to (although won't necessarily) result in various kinds of conflicts with humans. In the case of kit foxes in Bakersfield (and potentially other nearby cities), that they fit into the "growing" category means that they have significant potential to assist in the conservation of this endangered species.

In chapter 13 on taxonomic analysis, Iossa et al. use both demographic and landscape characteristics

to divide carnivores into five categories. For grouping the 35 species in table 13.1, they divide carnivores into *residents*, or self-sustaining populations with individual home ranges entirely "within the boundary of urban areas"; *transients*, or populations in which home ranges may include some urban area but do not exist entirely within them; and *unknown status* species, which have been recorded in urban areas, but too little information is available to effectively describe their urban ecology.

The demographic component is relevant because the resident carnivore populations are self-sustaining, but Iossa et al. further classify these species as *colonizing* carnivores that may actively expand into urban areas (e.g., raccoons and Eurasian red foxes) and *remnant* species, carnivores whose populations may have been present in an area before urbanization and have become "urban" by default (e.g., Eurasian badgers; see chapter 9). This classification is also useful. It indicates that a great deal remains to be learned about many species in the "transient" and, especially, "the unknown" categories. These species should be priorities for future study, especially if they may be adversely affected by urbanization or could produce significant conflict concerns (e.g., leopards, wolves, polar bears). The "remnant" classification may also, unfortunately, become increasingly relevant if additional carnivore populations and species inadvertently find themselves closer to urban development. Mountain lion populations across much of the western United States may soon be considered remnant urban carnivores, and in some places, such as Southern California, this may already be the case. Remnant species and populations may be in particular need of conservation attention.

In Iossa et al.'s classification (chapter 13), landscape use is addressed on the broad scale of populations, as opposed to on the scale of individual animals. Bobcats and gray foxes are included as "resident" urban carnivores because they exist as self-sustaining populations within an overall urban landscape, even though individual animals may live largely or entirely within natural areas present in that landscape. Another way of classifying urban carnivores is by the specific landscape use of individual animals. How carnivores use the urban landscape varies significantly, ranging from animals that don't approach the boundary with urbanization (these might also be the *declining* urban populations in the Cypher classification and *transient* or *unknown status* carnivores in the Iossa et al. classification) to animals whose entire home ranges exist within developed habitats. This range could be thought of as a measure of adaptability, or flexibility, with which animals are able to use fragmented landscapes (figure 17.1). By "flexibility," we mean the ability to use developed areas and thereby the ability to stray from the theoretical ancestral, or original, condition of no contact or proximity with urban development.

Carnivores exhibiting all of the patterns portrayed in figure 17.1 have been described in this book. In some cases, whole species or populations may exhibit a specific pattern, while in others, differences between sexes, ages, and individuals may occur. In bobcats in Northern California (chapter 10), adult females did not use the natural areas close to development (an *interior* pattern), while adult males used the area out to, but not over, the urban boundary (*edge*). In Southern California, where habitat fragmentation is more extreme, some female bobcats inhabited large but isolated patches of habitat (chapter 10; figure 10.3b; *large patch*), some females inhabited large patches but also visited much smaller patches (*island visiting*), and at the extremes of fragmentation, some female bobcats moved between small habitat fragments (chapter 10, figure 10.3a; *island hopping*), a pattern also seen in coyotes in Southern California (chapter 7, figure 7.6a). Raccoons in Wash-

ington, D.C. (Chapter 4, figure 4.4), exhibited three patterns within the same population: (1) some animals stayed within the park (*edge*); (2) others lived entirely within residential areas (*urban*); and (3) most used both park and residential habitats (*overlap*). Finally, red foxes in Britain, likely the most urban-associated carnivore in the book (if not the world), often lived entirely within residential areas (chapter 6, figure 6.3; *urban* pattern), and some coyotes in Chicago were essentially using entirely developed parts of the landscape (chapter 7, figure 7.4; although on a very fine scale they used vegetated, but highly altered, habitats).

These classifications are not intended to include every pattern of landscape use by urban carnivores—clearly, there is a continuum of different types of landscape use—but rather they may be a useful tool to describe some of the common broad patterns. Where species, populations, or individuals fall in this spectrum will have implications for management and conservation. The more flexible, or "urban compatible" (farther to the right in figure 17.1), carnivores are in their landscape use, the greater the chances of interaction with humans, their homes, and their pets, and the greater the potential for conflict. However, species that are

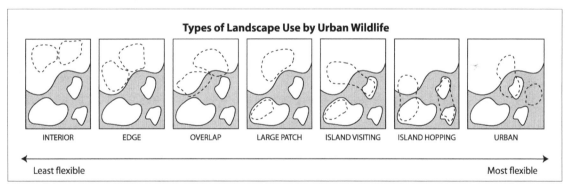

Types of Landscape Use by Urban Wildlife

INTERIOR   EDGE   OVERLAP   LARGE PATCH   ISLAND VISITING   ISLAND HOPPING   URBAN

Least flexible   Most flexible

Figure 17.1. Schematic representation of types of landscape use exhibited by carnivores in urban areas, potentially reflecting different degrees of flexibility, or "urban compatibility," in utilization of the landscape. Gray areas represent developed areas, white areas are undeveloped (i.e., generally vegetated and relatively natural), and the dashed lines represent hypothetical home ranges for carnivores. Obviously, this is a simplified scheme (as are all models), in particular in two important ways: (1) wildlife use of landscapes is a continuum, so the discrete states depicted here are only a subset of many possible patterns; and (2) land use is classified in the most simple way, i.e., with just two categories—developed and undeveloped. *Interior* animals remain in open, undeveloped spaces and do not approach the urban-wildland edge. *Edge* individuals also stay within large intact areas of natural habitat, but their home ranges may border the urban edge. Animals with home ranges that span the urban-wildland boundary and thereby include both natural and developed areas can be considered *overlap* individuals. This is the first category in which developed areas are used. *Large patch* animals may tolerate more fragmentation than interior or edge animals because they will use islands of natural habitat surrounded by urbanization, but they require patches that are larger than an average home range. Large patch animals may in fact be using less-developed area than overlap animals, but they are willing to use habitat fragments (albeit large ones), so at some point, traversing of development was required. *Island visiting* refers to animals that are willing to use both large and small (i.e., smaller than an average home range) habitat patches, as well as the intervening matrix, but they must be able to return to large patches or core natural areas. *Island hopping* animals are no longer tied to larger habitat patches but are able to move constantly between small patches of habitat. Finally, *urban* animals are able to use all parts of the landscape and may even exist in home ranges that are entirely within developed areas.

least flexible (farther to the left in figure 17.1) in their use of urban landscapes are more likely to be detrimentally affected by urbanization and fragmentation and may require conservation action.

FUTURE DIRECTIONS

In this book, we have attempted to provide detailed descriptions of the ecology of most of the better-known urban carnivores, species that have received detailed, long-term study in urban settings. We have not been able to address a number of subjects, however, and these subjects could profitably be addressed by future work.

One group of animals that we would like to have addressed more effectively in this book (and that should certainly be the subject of a full chapter in any future edition) is bears. Black bears have become a significant issue in urban areas, but unfortunately, there has been very little work on the basic ecology of bears in urban areas. Studies are needed that address the kinds of basic information about behavior and ecology available in the species chapters in this book: survival rates, mortality causes, reproduction, home range size, and use of developed areas. Given their size and therefore potential to cause property damage and to threaten human safety, bear-human conflicts in urban areas have received the most attention in the media, and some studies are beginning to address these conflicts. However, a better understanding about how individual bears behave and how bear populations function in urban areas is critical for minimizing conflicts and the impacts on both humans and bear populations.

There have been a few bear studies in urban areas. Beckmann and Berger (2003a, 2003b) studied bears in urban areas around Lake Tahoe, on the border of California and Nevada, and compared them with bears from nonurban, or "wildland," areas in other parts of Nevada. On the basis of a sample of 41 urban bears and 21 wildland bears, they found intriguing and dramatic differences in home range size, body weight, survival rates, and mortality causes between urban and nonurban bears. Specifically, the urban bears, which spent >90% of their time within city limits, had much smaller home ranges, were significantly heavier, spent 42 fewer days in hibernation, had a higher percentage of reproductive adult females, and were living in a population almost 40 times as dense (120 urban bears/100 km² vs. 3.2 wildland bears/100km²; Beckmann and Berger 2003a). Interestingly, many of these

differences, such as higher densities and smaller home ranges, are similar to those we have seen in other urban carnivores. Behaviorally, the urban bears also were more nocturnal than wildland animals (Beckmann and Berger 2003b), which is common in other urban species. In Alaska, brown bears have been radio-tracked in Anchorage and regularly make use of the salmon in city streams, alongside anglers (Alaska Department of Fish and Game, unpublished data).

We focus on species most commonly seen and studied in urban landscapes. However, as is clear from the species list in chapter 13, many more carnivores can and do exist in urban areas, about which we know very little. For many carnivores, there is not much evidence that they spend significant time in or near cities. Is this because they have not been studied there, or are they fundamentally less able to adapt to the stresses of living around people? What will this mean in the long run, as the remaining open space in the world is encroached on, or fragmented by, urban development? There is also a bias for carnivores of a certain size in our chapters. Raccoons, foxes, coyotes, badgers, bobcats, domestic cats and dogs, and skunks are in the medium-sized range of 2 kg (kit foxes) to 15 kg (coyotes). What about smaller carnivores, such as weasels, mink, mongooses, and ringtails? We know that they exist in urban areas because we may see them on occasion or we find them dead on a road, but we know practically nothing about their status or ecology. Knowledge about urban stone martens is accumulating in England and Europe (see "Stone Martens [*Martes foina*] in Urban Environments" after chapter 13), but this is rare. Perhaps, because these smaller species move over smaller distances and may exist at higher densities, they are less affected by urbanization and fragmentation. However, they may experience greater difficulty in moving through developed landscapes between areas of more suitable habitat.

Even for well-known species, many questions still need to be addressed. Some species have been studied intensively in only one or two urban areas or maybe only in one geographic region, so more remains to be learned about how their ecology varies in urban areas. For example, the bobcat chapter reviews studies in California and one in the southeastern United States, but different types of urban areas, cities, and regions are needed to get a broader picture of how bobcats respond to urbanization. Fortunately, studies are under way or planned in other areas such as Texas and Arizona. Similarly, chapter 11 on mountain lions focuses on two studies from Southern California, but results are appearing, and studies are under way or beginning in other states, including Washington, Colorado, Arizona, and Florida. Across a large geographic region such as the United States, where there is a significant range in climatic conditions, we know little about how urbanization and climate interact to affect carnivore ecology.

Even for species such as raccoons, coyotes, and red foxes, which have been studied in multiple urban regions, important gaps remain. Raccoons are widespread across North America, but our knowledge of their ecology comes almost entirely from the middle and eastern parts of the continent. Almost nothing is known about raccoons, urban or otherwise, in the West, despite their presence in many Western cities. For example, it would be particularly interesting to know where raccoons den during the day, in urban areas in the West, as these areas often lack the deciduous trees that are regular den sites in the East and Midwest. Coyotes have been studied in urban areas in the Southwest (Arizona, California), Midwest (Illinois), and Northwest (Washington). However, coyotes are increasingly moving into Eastern cities as well. How might urban coyote ecology be different in the Northeast, the Mid-Atlantic, and the Southeast? How will this important, and relatively large, carnivore affect humans, pets, and natural systems in the East? Fortunately, results about urban coyotes in a number of cities in the Eastern United States are beginning to appear from recent and ongoing studies. Finally, the study of red foxes in England (chapter 6) is an excellent example of the kinds of insight that can be gained from long-term work on urban carnivores, insights that now include fascinating results about long-term population dynamics, the effects of disease, and variation in social organization. However, red foxes are the most widespread carnivore on the planet, and we still know relatively little about their ecology in urban areas in other parts of the world, the United States and Canada in particular.

Fundamental questions about urban coyotes remain to be answered, even where coyotes have long inhabited urban areas. We still do not have a solid understanding of the significance of the use of anthropogenic foods by coyotes, and in particular the amount that they predate on pets. It is commonly believed, and stated in the media and elsewhere, that urban coyotes are relying on pets and other human-related food items. However, as demonstrated in chapter 7, diet studies to date have not shown that these items represent a significant portion of coyote diets. There may be a disjunct between people's perceptions about coyote behavior in urban areas, particularly relative to pets, and the actual biology.

Finally, along with questions related to specific taxa, conceptual issues still need to be addressed related both to basic ecology and to management and conservation. We still know little about the ecological role of carnivores in urban systems. Chapter 14 on community ecology includes case studies about carnivores interacting with other carnivores in urban landscapes. However, the interactions of carnivores and their prey species remain a relatively unstudied subject, despite the widespread attention paid to concepts such as "mesopredator release," which depends entirely on strong top-down effects from carnivores. How often and to what extent do carnivores impact populations of their prey species or the behavior of prey animals in urban areas? Conversely, how do changes in prey populations impact carnivores? These questions can have profound implications for management and conservation, particularly if prey species are of conservation or management concern (e.g., coyotes preying on Canada geese [chapter 7] or carnivores preying on rodents in landscaped areas such as golf courses or parks). Understanding predation effects on prey and predator populations is difficult, particularly in mammals, but furthering this understanding will prove valuable. A complete understanding of these community-level effects will be best achieved by including other carnivorous taxa, such as raptors or snakes.

Many issues related to management and conservation of carnivores in urban areas remain to be resolved as well. Human-carnivore coexistence and carnivore conservation in urban areas will depend on minimizing the frequency and severity of conflicts (chapter 16). Education of people already living in, or moving into, urban areas is a critical part of minimizing conflicts, but we are far from understanding how best to accomplish this education and whether current education efforts work (chapter 15). Our understanding of the biological underpinnings of these conflicts and of what management tactics may effectively reduce them is also in the nascent stages. Which specific animals are causing conflicts? Is it particular age or sex classes or particular individuals? The consensus is that too much habituation to humans, particularly resulting from activities such as feeding, can lead to conflict with carnivores. At what point is an animal habituated? Is it possible to stop or even reverse this habituation? At what point is an animal so habituated that removal is the only option? Finally, the conservation of carnivores in urban

landscapes is a new and relatively untried practice
(chapter 16), which will demand increasing attention
as the world continues to urbanize. Which carnivores
do we want to maintain in urban landscapes, and in
what parts of the landscape do we want to preserve
them? Of course, once we have made these decisions,
how can we best accomplish this?

Carnivores, fortunately, are here to stay, and so are
cities. More than 50% of the world's human population
now resides in urban areas (see chapter 1, figure 1.1).
As biologists, resource managers, planners, and urban residents, we need to do our best to understand
carnivores, cities, and how they interact, so that carnivores and urban dwellers can coexist as peacefully as
possible. A bobcat in Tucson, Arizona, a coyote in Los
Angeles, California, a raccoon in Chicago, Illinois, and
a red fox in London, England, can all add significantly
to people's experience of nature, benefiting the people,
the animals, and we hope, in the long run, the wider
world.

APPENDIX
List of Scientific Names

Common Name	Scientific Name	Common Name	Scientific Name
aardwolf	*Proteles cristatus*	dingo	*Canis familiaris dingo*
African wild dog	*Lycaon pictus*	dog, domestic	*Canis familiaris / Canis lupus familiaris*
badger, American	*Taxidea taxus*		
badger, Eurasian	*Meles meles*	elephant	*Loxodonta africana*
badger, stink	*Mydaus* sp.	elk, North American	*Cervus elaphus*
bear, Asiatic black	*Ursus thibetanus*	Ethiopian wolf	*Canis simiensis*
bear, brown	*Ursus arctos*	fisher	*Martes pennanti*
bear, European brown	*Ursus arctos arctos*	Florida panther	*Puma concolor coryi*
bear, grizzly	*Ursus arctos horribilus*	fossa	*Cryptoprocta ferox*
bear, North American black	*Ursus americanus*	fox, arctic	*Alopex lagopus*
bear, polar	*Ursus maritimus*	fox, bat-eared	*Otocyon megalotis*
bear, sloth	*Melursus ursinus*	fox, crab-eating	*Cerodocyon thous*
beaver	*Castor canadensis*	fox, gray	*Urocyon cinereoargenteus*
binturong	*Arctictis binturong*	fox, Island	*Urocyon littoralis*
bison	*Bison bison*	fox, kit	*Vulpes macrotis*
black-footed ferret	*Mustela nigripes*	fox, red	*Vulpes vulpes*
bobcat	*Lynx rufus*	fox, Ruppell's	*Vulpes rueppellii*
bonobo	*Pan paniscus*	fox, San Joaquin kit	*Vulpes macrotis mutica*
burrowing owl	*Athene cunicularia*	fox, swift	*Vulpes velox*
Canada goose	*Branta canadensis*	gaur	*Bos gaurus*
cat, domestic	*Felis catus*	genet, common	*Genetta genetta*
cat, fishing	*Prionailurus viverrinus*	genet, large-spotted	*Genetta maculata*
cat, Iriomote	*Prionailurus iriomotensis*	giant panda	*Ailuropoda melanoleuca*
cheetah	*Acinonyx jubatus*	golden eagle	*Aquila chrysaetos*
civet, Indian	*Viverricula indica*	ground squirrel	*Spermophilus* spp.
civet, Lesser Oriental	*Viverricula indica*	ground squirrel, California	*Spermophilus beechyi*
civet, Masked Palm	*Paguma larvata*	hare	*Lepus* spp.
coati, brown-nosed	*Nasua nasua*	harpy eagle	*Harpia harpyja*
coati, white-nosed	*Nasua narica*	hedgehog	*Erinaceus europaeus*
coyote	*Canis latrans*	house mouse	*Mus musculus*
deer, Key	*Odocoileus virginianus calvium*	hyena, brown	*Hyaena brunnea*
		hyena, spotted	*Crocuta crocuta*
deer, mule	*Odocoileus hemionus*	hyena, striped	*Hyaena brunnea*
deer, Roe	*Capreolus capreolis*	jackal, golden	*Canis aureus*
deer, white-tailed	*Odocoileus virginianus*	jackal, side-striped	*Canis adustus*
desert cottontail	*Sylvilagus auduboni*	jackrabbit	*Lepus* spp.
dhole	*Cuon alpinus*	jaguar	*Panthera onca*

Common Name	Scientific Name	Common Name	Scientific Name
kangaroo rat	*Dipodomys* spp.	rabbit, European	*Oryctolagus cunniculus*
killer whale	*Orcinus orca*	rabbit, North American	*Sylvilagus* spp.
kiwi	*Apteryx mantelli*	raccoon	*Procyon lotor*
least weasel	*Mustela nivalis*	raccoon dog	*Nyctereutes procyonoides*
leopard	*Panthera pardus*	red or lesser panda	*Ailurus fulgens*
leopard, snow	*Panthera uncia*	Reeves' muntjac	*Muntiacus reevesi*
lion	*Panthera leo*	ringtail	*Bassariscus astutus*
lynx	*Lynx lynx*	skunk, hooded	*Mephitis macroura*
maned wolf	*Chrysocyon brachyurus*	skunk, striped	*Mephitis mephitis*
meadow vole	*Microtus californicus*	skunk, striped hog-nosed	*Conepatus semistriatus*
meerkat or suricate	*Suricata suricatta*	skunk, western spotted	*Spilogale gracilis*
mink, American	*Mustela vison*	skunk, white-backed hog-nosed	*Conepatus leuconotus*
mongoose, brown	*Herpestes fuscus*		
mongoose, dwarf	*Elogale parvula*	squirrel (gray)	*Sciurus carolinensis*
mongoose, Indian	*Herpestes javanicus*	Stephens Island wren	*Traversia lyalli*
mongoose, ruddy	*Herpestes smithii*	stoat	*Mustela ermina*
moose	*Alces alces*	stone marten	*Martes foina*
mountain lion or puma	*Puma concolor/Felis concolor*	suricate or meerkat	*Suricata suricatta*
Norway rat	*Rattus norvegicus*	tiger	*Panthera tigris*
opossum, Virginia	*Didelphis virginiana*	tiger, Amur	*Panthera tigris altaica*
otter, Cape clawless	*Aonyx capensis*	tiger quoll	*Dasyurus maculates*
otter, Eurasion	*Lutra lutra*	turkey	*Meleagris gallopavo*
otter, North American	*Lontra canadensis*	weasel	*Mustela* spp.
otter, sea	*Enhydra lutris*	wildcat	*Felis silvestris*
pocket gopher	*Thomomys bottae*	wild horse	*Equus caballus*
porcupine	*Erethizon dorsatum*	wolf	*Canis lupus*
prairie dog	*Cynomys* spp.	wolverine	*Gulo gulo*
puma or mountain lion	*Puma concolor/Felis concolor*	woodrat	*Neotoma* spp.

BIBLIOGRAPHY

Aaris-Sørensen, J. 1977. Graevlingens forekomst i København før og nu. *Danske Vildtundersøgelser* 28:44–56.

Aaris-Sørensen, J. 1987. Past and present distribution of badgers *Meles meles* in the Copenhagen area. *Biological Conservation* 41:159–165.

Abi-Said, M. R., and D. M. Abi-Said. 2007. Distribution of the striped hyaena (*Hyaena hyaena syriaca* Matius, 1882) (Carnivora: Hyaenidae) in urban and rural areas of Lebanon. *Zoology in the Middle East* 42:3–14.

Ackerman, B. B., F. G. Lindzey, and T. P. Hemker. 1986. Predictive energetics model for cougars. Pages 333–352 *in* S. D. Miller and D. Everett, eds. *Cats of the World: Biology, Conservation, and Management*. National Wildlife Federation, Washington, D.C.

Adamec, R. E. 1976. The interaction of hunger and preying in the domestic cat (*Felis catus*): An adaptive hierarchy? *Behavioral Biology* 18:263–272.

Adams, C. E., K. J. Lindsey, and S. J. Ash. 2006. *Urban Wildlife Management*. CRC Press, Boca Raton, FL.

Adams, J. R., J. A. Leonard, and L. P. Waits. 2003. Widespread occurrence of a domestic dog mitochondrial DNA haplotype in southeastern U.S. coyotes. *Molecular Ecology* 12:541–546.

Adams, L. W., J. Hadidian, and V. Flyger. 2004. Movement and mortality of translocated urban-suburban grey squirrels. *Animal Welfare* 13:45–50.

Adams, L. W., L. W. Van Druff, and M. Luniak. 2005. Managing urban habitats and wildlife. Pages 714–739 *in* C. E. Braun, ed. *Techniques for Wildlife Investigations and Management*. Allen Press, Lawrence, KS.

Adkins, C. A., and P. Stott. 1998. Home ranges, movements and habitat associations of red foxes *Vulpes vulpes* in suburban Toronto, Ontario, Canada. *Journal of Zoology* 244:335–346.

Afonso, E., P. Thulliez, and E. Gilot-Fromont. 2006. Transmission of *Toxoplasma gondii* in an urban population of domestic cats (*Felis catus*). *International Journal for Parasitology* 36:1373–1382.

Albassam, M., R. Bhatnagar, L. Lillie, and L. Roy. 1986. Adiaspiromycosis in striped skunks in Alberta, Canada. *Journal of Wildlife Diseases* 22:13–18.

Alberti, M., and J. M. Marzluff. 2004. Ecological resilience in urban ecosystems: Linking urban patterns to human and ecological functions. *Urban Ecosystems* 7:241–265.

Allen, D. L., and W. W. Shapton. 1942. An ecological study of winter dens, with special reference to the eastern skunk. *Ecology* 23:59–68.

Alves, M. C. G. P., M. R. de Matos, M. de Lourdes Reichmann, and M. H. Dominguez. 2005. Estimation of the dog and cat population in the State of São Paulo. *Revista de Saúde Pública* 39:891–897.

Alves-Costa, C. P., G. A. B. Da Fonseca, and C. Christófaro. 2004. Variation in the diet of the brown-nosed coati (*Nasua nasua*) in southeastern Brazil. *Journal of Mammalogy* 85:478–482

Andelt, W. F. 1985. Behavioral ecology of coyotes in south Texas. *Wildlife Monographs* 94:1–45.

Andelt, W. F., and B. R. Mahan. 1980. Behavior of an urban coyote. *American Midland Naturalist* 103:399–400.

Andelt, W. F., R. L. Phillips, R. H. Schmidt, and R. B. Gill. 1999. Trapping furbearers: An overview of the biological and social issues surrounding a public policy controversy. *Wildlife Society Bulletin* 27:53–64.

Andersen, J. 2002. *The Latchkey Dog*. HarperCollins, New York.

Andersen, M. C., B. J. Martin, and G. W. Roemer. 2004. Use of matrix population models to estimate the efficacy of euthanasia versus trap-neuter-return for management of free-roaming cats. *Journal of the American Veterinary Medical Association* 12:1871–1876.

Anderson, A. E. 1983. *A Critical Review of Literature on Puma* (Felis concolor). Special Report No. 54. Colorado Division of Wildlife, Fort Collins.

Anderson, A. E., D. C. Bowden, and D. M. Kattner. 1992. *The Puma on Uncompahgre Plateau, Colorado*. Colorado Division of Wildlife Technical Publication No. 40, Fort Collins.

Anderson, E. M., and M. J. Lovallo. 2003. Bobcat and lynx. Pages 758–786 *in* G. A. Feldhamer, B. C. Thompson, and J. A. Chapman, eds. *Wild Mammals of North America: Biology, Management, and Conservation*. Johns Hopkins University Press, Baltimore.

Anderson, S. 1981. The raccoon (*Procyon lotor*) on St. Catherine's Island, Georgia: Nesting sea turtles and foraging raccoons. *American Museum Novitates* 2713:1–9.

Anon. 1995. *Badgers—Guidelines for Developers*. English Nature, Peterborough, UK.

Anon. 1996. *Menace of Moggies*. Christchurch Press, Christchurch, Australia.

Anon. 2003. *Impacts of Feral and Free-Ranging Domestic Cats on Wildlife in Florida*. Report by Feral Cat Issue Team. Florida Fish and Wildlife Conservation Commission.

Anon. 2008a. www.parliament.the-stationery-office.co.uk/pa/cm200708/cmhansrd/cm080519/test/80519w0028.htm. Accessed August 1, 2008.

Anon. 2008b. Key deer *Odocoileus virgianus clavium*. www.fws.gov/verbeach/images/pdflibrary/kede.pdf. Accessed July 18, 2008.

Ansell, R. J. 2004. The spatial organisation of a red fox (*Vulpes vulpes*) population in relation to food resources. Ph.D. diss., University of Bristol.

Ansell, R. J., P. Baker, and S. Harris. 2001. The value of gardens for wildlife—lessons from mammals and herpetofauna. *British Wildlife* 13:77–84.

Antos, M. J., G. C. Ehmke, C. L. Tzaros, and M. A. Weston. 2007. Unauthorised human use of an urban coastal wetland sanctuary: Current and future patterns. *Landscape and Urban Planning* 80:173–183.

Anvik, J. O., A. E. Hague, and A. Rahaman. 1974. A method of estimating urban dog populations and its application to the assessment of canine fecal pollution and endoparasitism in Saskatchewan. *Canadian Veterinary Journal* 15:219–223.

Apáthy, M. T. 1998. Data to the diet of the urban stone marten (*Martes foina* Erxleben) in Budapest. *Opuscula Zoologica Budapest* 31:113–118.

Aranda, M., and L. López-de Buen. 1999. Rabies in skunks from Mexico. *Journal of Wildlife Diseases* 35:574–577.

Arthur, L. M. 1981. Coyote control: The public response. *Journal of Range Management* 34:14–15.

Arthur, S. M., W. B. Krohn, and J. R. Gilbert. 1989. Home range characteristics of adult fishers. *Journal of Wildlife Management* 53:674–679.

Artsob, H., L. Spence, C. Th'ng, V. Lampotang, D. Johnston, C. MacInnes, F. Matejka, D. Voigt, and I. Watt. 1986. Arbovirus infections in several Ontario mammals, 1975–1980. *Canadian Journal of Veterinary Research* 50:42–46.

Arundel, T., D. Mattson, and J. Hart. 2006. Movements and habitat selection by mountain lions in the Flagstaff Uplands. Pages 17–30 *in* D. Mattson, ed. *Mountain Lions of the Flagstaff Uplands, 2003–2006*. Progress Report. U.S. Geological Survey, Flagstaff, AZ.

Asa, C. S., and C. Valdespino. 1998. Canid reproductive biology: An integration of proximate mechanisms and ultimate causes. *American Zoologist* 38:251–259.

Asferg, T., J. L. Jeppesen, and J. Aaris-Sørensen. 1977. Grævlingen (*Meles meles*) og grævlingejagten i Danmark 1972/73. *Danske Vildtundersøgelser* 28:1–56.

Ashman, D., G. C. Christensen, M. L. Hess, G. K. Tsukamoto, and M. S. Wickersham. 1983. *The Mountain Lion in Nevada*. Pittman-Robertson Project W-48-15. Nevada Game and Fish Department, Carson City.

Atkinson, K. T., and D. M. Shackleton. 1991. Coyote, *Canis latrans*, ecology in a rural-urban environment. *Canadian Field Naturalist* 105:49–54.

Atkinson, R. P. D., and A. J. Loveridge. 2004. Side-striped jackal *Canis adustus* Sundevall, 1847. Pages 152–156 *in* C. Sillero-Zubiri, M. Hoffman, and D. W. Macdonald, eds. *Canids: Foxes, Wolves, Jackals, and Dogs*. Status Survey and Conservation Action Plan. IUCN, Gland, Switzerland.

Atwood, T. C. 2006. The influence of habitat patch attributes on coyote group size and interaction in a fragmented landscape. *Canadian Journal of Zoology* 84:80–87.

Atwood, T. C., H. P. Weeks, and T. M. Gehring. 2004. Spatial ecology of coyotes along a suburban-to-rural gradient. *Journal of Wildlife Management* 68:1000–1009.

Bailey, T. N. 1974. Social organization in a bobcat population. *Journal of Wildlife Management* 38:435–446.

Bailey, T. N. 1993. *The African Leopard: Ecology and Behavior of a Solitary Felid*. Columbia University Press, New York.

Baird, R. W., D. J. McSweeney, C. Bane, J. Barlow, D. R. Salden, L. K. Antoine, R. G. LeDuc, and D. L. Webster. 2006. Killer whales in Hawaiian waters: Information on population identity and feeding habits. *Pacific Science* 604:523–530.

Baker, P. J., A. J. Bentley, R. J. Ansell, and S. Harris. 2005. Impact of predation by domestic cats *Felis catus* in an urban area. *Mammal Review* 35:302–312.

Baker, P. J., C. V. Dowding, S. E. Molony, P. C. L. White, and S. Harris. 2007. Activity patterns of urban red foxes (*Vulpes vulpes*) reduce the risk of traffic-induced mortality. *Behavioral Ecology* 18:716–724.

Baker, P. J., S. M. Funk, M. W. Bruford, and S. Harris. 2004a. Polygynandry in a red fox population: Implications for the evolution of group living in canids? *Behavioral Ecology* 15:766–778.

Baker, P. J., S. Funk, S. Harris, T. Newman, G. Saunders, and P. White. 2004b. The impact of human attitudes on the social and spatial organization of urban foxes (*Vulpes vulpes*) before and after an outbreak of sarcoptic mange. Pages 153–163 *in* W. W. Shaw, L. K. Harris, and L. VanDruff, eds. *Proceedings of the 4th International Symposium on Urban Wildlife Conservation*. School of Natural Resources, College of Agriculture and Life Sciences, University of Arizona, Tucson.

Baker, P. J., S. M. Funk, S. Harris, and P. C. L. White. 2000. Flexible spatial organization of urban foxes, *Vulpes vulpes*, before and during an outbreak of sarcoptic mange. *Animal Behavior* 59:127–146.

Baker, P. J., M. J. Furlong, S. Southern, and S. Harris. 2006. The potential impact of red fox *Vulpes vulpes* predation in agricultural landscapes in lowland Britain. *Wildlife Biology* 12:39–50.

Baker, P. J., and S. Harris. 2000. Interaction rates between members of a group of red foxes (*Vulpes vulpes*). *Mammal Review* 30:239–242.

Baker, P. J., and S. Harris. 2004. Red foxes: The behavioural ecology of red foxes in urban Bristol. Pages 207–216 *in* D. W. Macdonald and C. Sillero-Zubiri, eds. Biology and Conservation of Wild Canids. Oxford University Press, Oxford.

Baker, P. J., and S. Harris. 2007. Urban mammals: What does the future hold? An analysis of the factors affecting patterns of use of residential gardens in Great Britain. *Mammal Review* 37:297–315.

Baker, P. J., S. Harris, C. P. J. Robertson, G. Saunders, and P. C. L. White. 2001a. Differences in the capture rate of cage-trapped red foxes *Vulpes vulpes* and an evaluation of rabies control measures in Britain. *Journal Applied Ecology* 38:823–835.

Baker, P. J., S. E. Molony, E. Stone, I. C. Cuthill, and S. Harris. 2008. Cats about town: Is predation by free-ranging pet cats (*Felis catus*) likely to affect urban bird populations? *Ibis* 150:86–99.

Baker, P. J., T. Newman, and S. Harris. 2001b. Bristol's foxes—40 years of change. *British Wildlife* 12:411–417.

Baker, P. J., C. P. J. Robertson, S. M. Funk, and S. Harris. 1998. Potential fitness benefits of group living in the red fox, *Vulpes vulpes*. *Animal Behaviour* 56:1411–1424.

Baker, R. O., and R. M. Timm. 1998. Management of conflicts between urban coyotes and humans in Southern California. Pages 299–312 *in* R. O. Baker and A. C. Crabb, eds. *Proceedings of the 18th Vertebrate Pest Conference*. University of California, Davis.

Bakersfield Californian. 2007. With bat season upon us, official issues rabies warning. Community section. May 4.

Baldwin, R. A., A. Houston, M. Kennedy, and P. Liu. 2004. An assessment of microhabitat variables and capture success of striped skunks (*Mephitis mephitis*). *Journal of Mammalogy* 85:1068–1076.

Banks, P. B., and J. V. Bryant. 2007. Four-legged friend or foe? Dog walking displaces native birds from natural areas. *Biology Letters* 3:611–613.

Barash, D. P. 1974. Neighbor recognition in two "solitary" carnivores: The raccoon (*Procyon lotor*) and the red fox (*Vulpes fulva*). *Science* 185:794–796.

Barja, I., and R. List. 2006. Faecal marking behaviour in ringtails (*Bassariscus astutus*) during the non-breeding period: Spatial characteristics of latrines and single faeces. *Chemoecology* 16:219–222.

Barnes, T. G. 1997. State agency oversight of the nuisance wildlife control industry. *Wildlife Society Bulletin* 25:185–188.

Baron, D. 2005. *The Beast in the Garden*. W. W. Norton, New York.

Barratt, D. G. 1997a. Predation by house cats, *Felis catus* (L.), in Canberra, Australia. I. Prey composition and preference. *Wildlife Research* 24:263–277.

Barratt, D. G. 1997b. Home range size, habitat utilisation and movement patterns of suburban and farm cats *Felis catus*. *Ecography* 20:271–280.

Barratt, D. G. 1998. Predation by house cats, *Felis catus* (L.), in Canberra, Australia. II. Factors affecting the amount of prey caught and estimates of their impact on wildlife. *Wildlife Research* 25:475–487.

Bartnik, H.-C. 2001. Untersuchungen zum serologischen Nachweis der *Sarcoptes*-Räude beim Rotfuchs (*Vulpes vulpes*) in Berlin. Ph.D. diss., Freien Universität Berlin, Germany.

Bartoszewicz, M., H. Okarma, A. Zalewski, and J. Szczesna. 2008. Ecology of the raccoon (*Procyon lotor*) from western Poland. *Annual Zoological Fennici* 45:291–298.

Barutzki, D., and R. Schaper. 2003. Endoparasites in dogs and cats in Germany 1999–2002. *Parasitology Research* 90:S148–S150.

Bauer, J. W., K. A. Logan, L. L. Sweanor, and W. M. Boyce. 2005. Scavenging behavior in puma. *Southwestern Naturalist* 50:466–471.

Beasom, S. L., and R. A. Moore. 1977. Bobcat food habit response to a change in prey abundance. *Southwestern Naturalist* 21:451–457.

Beaumont, M., E. M. Barratt, D. Gottelli, A. C. Kitchener, M. J. Daniels, J. K. Pritchard, and M. W. Bruford. 2001. Genetic diversity and introgression in the Scottish wildcat. *Molecular Ecology* 10:319–336.

Beck, A. M. 1973. *The Ecology of Stray Dogs: A Study of Free-Ranging Urban Animals*. York Press, Baltimore.

Beck, A. M. 2000. The human-dog relationship: A tale of two species. Pages 1–16 *in* C. N. L. Macpherson, F. X. Meslin, and A. I. Wandeler, eds. *Dogs, Zoononses, and Public Health*. CABI, Wallingford, Oxfordshire.

Beck, T., and 10 others (Cougar Management Guidelines Working Group). 2005. *Cougar Management Guidelines*. 1st ed. Wild Futures, Bainbridge Island, WA.

Beckerman, A. P., M. Boots, and K. J. Gaston. 2007. Urban bird declines and the fear of cats. *Animal Conservation* 10:320–325.

Beckmann, J. P., and J. Berger. 2003a. Rapid ecological and behavioural changes in carnivores: The responses of black bears (*Ursus americanus*) to altered food. *Journal of Zoology* 261:207–212.

Beckmann, J. P., and J. Berger. 2003b. Using black bears to test ideal-free distribution models experimentally. *Journal of Mammalogy* 84:594–606.

Beckmann, J. P., and C. W. Lackey. 2008. Carnivores, urban landscapes, and longitudinal studies: A case history of black bears. *Human-Wildlife Conflicts* 2:168–174.

Beckmann, J. P., C. W. Lackey, and J. Berger. 2004. Evaluation of deterrent techniques and dogs to alter behavior of "nuisance" black bears. *Wildlife Society Bulletin* 32:1141–1146.

Beier, P. 1991. Cougar attacks on humans in the United States and Canada. *Wildlife Society Bulletin* 19:403–412.

Beier, P. 1993. Determining minimum habitat areas and habitat corridors for cougars. *Conservation Biology* 7:94–108.

Beier, P. 1995. Dispersal of juvenile cougars in fragmented habitat. *Journal of Wildlife Management* 59:228–237.

Beier, P. 1996. Metapopulation models, tenacious tracking, and cougar conservation. Pages 293–322 in D. R. McCullough, ed. *Metapopulations and Wildlife Conservation.* Island Press, Covelo, CA.

Beier, P. 2009. A focal species for conservation planning. Pages 177–189 in M. Hornocker and S. Negri, eds. *The Cougar: Ecology and Conservation.* University of Chicago Press, Chicago.

Beier, P., and R. H. Barrett. 1993. *The Cougar in the Santa Ana Mountain Range California.* California Department of Fish and Game, Sacramento.

Beier, P., D. Choate, and R. H. Barrett. 1995. Movement patterns of mountain lions during different behaviors. *Journal of Mammalogy* 76:1056–1070.

Beier, P., K. Penrod, C. Luke, W. Spencer, and C. Cabañero. 2006. South Coast missing linkages: Restoring connectivity to wildlands in the largest metropolitan area in the United States. Pages 555–586 in K. R. Crooks and M. A. Sanjayan, eds. *Connectivity Conservation.* Cambridge University Press, New York.

Beier, P., M. R. Vaughan, M. J. Conroy, and H. Quigley. 2003. An Analysis of Scientific Literature Related to the Florida Panther. Florida Fish and Wildlife Conservation Commission, Tallahassee. http://oak.ucc.nau.edu/pb1/publications.htm.

Bekoff, M., T. J. Daniels, and J. L. Gittleman. 1984. Life-history patterns and the comparative social ecology of carnivores. *Annual Review of Ecology and Systematics* 15:191–232.

Bekoff, M., and E. M. Gese. 2003. Coyote (*Canis latrans*). Pages 467–481 in G. A. Feldhamer, B. C. Thompson, and J. A. Chapman, eds. *Wild Mammals of North America: Biology, Management, and Conservation.* 2nd ed. Johns Hopkins University Press, Baltimore.

Bekoff, M., and M. C. Wells. 1980. Social ecology and behavior of coyotes. *Scientific American* 242:130–148.

Belcher, C.A., J. L. Nelson, and J. P. Darrant. 2007. Diet of the tiger quoll (*Dasyurus maculatus*) in south-eastern Australia. *Australian Journal of Zoology* 55:117–122.

Belden, R. C., and B. W. Hagedorn. 1993. Feasibility of translocating panthers into northern Florida. *Journal of Wildlife Management* 57:388–397.

Bender, L. C., J. C. Lewis, and D. P. Anderson. 2004. Population ecology of Columbian black-tailed deer in urban Vancouver, Washington. *Northwestern Naturalist* 85:53–59.

Benson, J. F., M. J. Chamberlain, and B. D. Leopold. 2006. Regulation of space use in a solitary felid: Population density or prey availability? *Animal Behaviour* 71:685–693.

Bentler, K. T., J. S. Hall, J. J. Root, K. Klenk, B. Schmit, B. F. Blackwell, P. C. Ramey, and L. Clark. 2007. Serologic evidence of West Nile virus exposure in North American predators. *American Journal of Tropical Medicine and Hygiene* 76:173–179.

Berdoy, M., J. P. Webster, and D. W. Macdonald. 2000. Fatal attraction in rats infected with *Toxoplasma gondii.* *Proceedings of the Royal Society of London B* 267:1591–1594.

Berg, W. E., and R. A. Chesness. 1978. Ecology of coyotes in northern Minnesota. Pages 229–247 in M. Bekoff, ed. *Coyotes: Biology, Behavior, and Management.* Academic Press, New York.

Berman, M., and I. Dunbar. 1983. The social behaviour of free-ranging suburban dogs. *Applied Animal Ethology* 10:5–17.

Berner, A., and L. W. Gysel. 1967. Raccoon use of large tree cavities and ground burrows. *Journal of Wildlife Management* 31:706–714.

Bernstein, D. A., L. A. Penner, A. Clarke-Stewart, and E. J. Roy. 2006. *Psychology.* 7th ed. Houghton Mifflin, Boston.

Berry, B. J. L. 2008. Urbanization. Pages 25–48 in J. M. Marzluff et al., eds. *Urban Ecology: An International Perspective on the Interaction between Humans and Nature.* Springer, New York.

Bester, M. N., J. P. Bloomer, P. A. Bartlett, D. D. Muller, M. van Rooyen, and H. Büchner. 2000. Final eradication of feral cats from sub-Antarctic Marion Island, southern Indian Ocean. *South African Journal of Wildlife Research* 30:53–57.

Bester, M. N., J. P. Bloomer, R. J. van Aarde, B. H. Erasmus, P. J. J. van Rensburg, J. D. Skinner, P. G. Howell, and T. W. Naude. 2002. A review of the successful eradication of feral cats from sub-Antarctic Marion Island, Southern Indian Ocean. *South African Journal of Wildlife Research* 32:65–73.

Bevanger, K., H. Brøseth, B. S. Johansen, B. Knutsen, K. V. Olsen, and T. Aarvak. 1996. Økologi og populasjonsbiologi hos europeisk grevling *Meles meles* L. i en urban-rural gradient i Sør-Trøndelag. *NINA Fagrapport* 23:1–48.

Biek, R., T. K. Ruth, K. M. Murphy, C. R. Anderson, M. Johnson, R. DeSimone, R. Gray, M. G. Hornocker, C. M. Gillin, and M. Poss. 2006. Factors associated with pathogen seroprevalence and infection in Rocky Mountain cougars. *Journal of Wildlife Diseases* 42:606–615.

Biek, R., R. L. Zarnke, C. Gillin, M. Wild, J. R. Squires, and M. Poss. 2002. Serologic survey for viral and bacterial infections in western populations of Canada lynx (*Lynx canadensis*). *Journal of Wildlife Diseases* 38:840–845.

Birch, L. C. 1957. The meanings of competition. *American Naturalist* 91:5–18.

Bissonette, J. A., and S. Broekhuizen. 1995. *Martes* populations as indicators of habitat spatial patterns: The need for a multiscale approach. Pages 95–121 in W. Z. Lidicker Jr., ed. *Landscape Approaches in Mammalian Ecology and Conservation.* University of Minnesota Press, Minneapolis.

Bixler, A. 2000. Genetic variability in striped skunks (*Mephitis mephitis*). *American Midland Naturalist* 143:370–376.

Bixler, A., and J. L. Gittleman. 2000. Variation in home

range and use of habitat in the striped skunk (*Mephitis mephitis*). *Journal of Zoology (London)* 251:525–533.

Bjerke, T., and T. Østdahl. 2004. Animal-related attitudes and activities in an urban population. *Anthrozoos* 17:109–129.

Bjerke, T., O. Reitan, and S. R. Kellert. 1998. Attitudes toward wolves in southeastern Norway. *Society and Natural Resources* 11:169–78.

Bjorge, R. R., J. R. Gunson, and W. M. Samuel. 1981. Population characteristics and movements of striped skunks (*Mephitis mephitis*) in central Alberta. *Canadian Field-Naturalist* 95:149–155.

Björnerfeldt, S., M. T. Webster, and C. Vilà. 2006. Relaxation of selective constraint on dog mitochondrial DNA following domestication. *Genome Research* 16:990–994.

Bjurlin, C. D., and B. L. Cypher. 2005. Encounter frequency with the urbanized San Joaquin kit fox correlates with public beliefs and attitudes toward the species. *Endangered Species Update* 22:107–115.

Bjurlin, C. D., B. L. Cypher, C. M. Wingert, and C. L. Van Horn Job. 2005. *Urban Roads and the Endangered San Joaquin Kit Fox*. California State University-Stanislaus. Endangered Species Recovery Program, Fresno, CA. 47 pp.

Black Bear Conservation Committee. 2006. About the Black Bear Conservation Committee. www.bbcc.org/web/index. Accessed October 16, 2007.

Blair, R. 2004. The effects of urban sprawl on birds at multiple levels of biological organization. Ecology and Society 9:2.[online] URL: http://www.ecologyandsociety.org/vol9/iss5/art2.

Blanton, J. D., J. W. Krebs, C. A. Hanlon, and C. E. Rupprecht. 2006. Rabies surveillance in the United States during 2005. *Journal of the American Veterinary Medical Association* 229:1897–1911.

Blejwas, K. M., B. N. Sacks, M. M. Jaeger, and D. R. McCullough. 2002. The effectiveness of selective removal of breeding coyotes in reducing sheep predation. *Journal of Wildlife Management* 66:451–462.

Bluett, R. D., Jr., G. F. Hubert, and C. A. Miller. 2003. Regulatory oversight and activities of wildlife control operators in Illinois. *Wildlife Society Bulletin* 31:104–116.

Blumstein, D. T., E. Fernández-Juricic, P. A. Zollner, and S. C. Garity. 2005. Inter-specific variation in avian responses to human disturbance. *Journal of Applied Ecology* 42:943–953.

Bogan, D. A. 2004. Eastern coyote home range, habitat selection and survival in the Albany Pine bush landscape. Master's thesis, State University of New York, University at Albany, Albany.

Bogan, D. A., P. D. Curtis, and G. Batcheller. 2008. Suburban coyote ecology and behavior: Implications for management and human interactions. *Northeast Natural History Conference* 10:7–8 (abstract).

Boitani, L. 2001. Carnivore introduction and invasions: Their success and management options. Pages 123–144 in J. L. Gittleman, S. M. Funk, D. Macdonald, and R. K. Wayne, eds. *Carnivore Conservation*. Cambridge University Press, Cambridge.

Boitani, L., F. Francisci, P. Ciucci, and G. Andreoli. 1995. Population biology and ecology of feral dogs in central Italy. Pages 217–244 in J. Serpell, ed. *The Domestic Dog: Its Evolution, Behaviour and Interactions with People*. Cambridge University Press, Cambridge.

Bolger, D. T., A. C. Alberts, R. M. Sauvajot, P. Potenza, C. McCalvin, D. Tran, S. Mazzoni, and M. E. Soulé. 1997. Response of rodents to habitat fragmentation in coastal southern California. *Ecological Applications* 7:552–563.

Bolger, D. T., A. Alberts, and M. E. Soulé. 1991. Occurrence patterns of bird species in habitat fragments: Sampling, extinction, and nested species subsets. *American Naturalist* 137:155–166.

Bolger, D. T., A. V. Suarez, K. R. Crooks, S. A. Morrison, and T. J. Case. 2000. Arthropod diversity in coastal sage scrub fragments: Area, age, and edge. *Ecological Applications* 10:1230–1248.

Bollin-Booth, H. A. 2007. Diet analysis of the coyote (*Canis latrans*) in metropolitan park systems of northeast Ohio. Master's thesis. Cleveland State University, Ohio.

Bolt, G., J. Monrad, P. Henriksen, H. H. Dietz, J. Koch, E. Bindseil, and A. L. Jensen. 1992. The fox (*Vulpes vulpes*) as a reservoir for canine angiostrongylosis in Denmark. *Acta Veterinaria Scandinavica* 33:357–362.

Bonanni, R., S. Cafazzo, C. Fantini, D. Pontier, and E. Natoli. 2007. Feeding-order in an urban feral domestic cat colony: Relationship to dominance rank, sex and age. *Animal Behaviour* 74:1369–1379.

Bonato, V., K. Gomes Facure, and W. Uieda. 2004. Food habits of bats of Subfamily Vampyrinae in Brazil. *Journal of Mammalogy* 85:708–713.

Bontadina, F., P. Contesse, and S. Gloor. 2001. Wie beeinflusst die persönliche Betroffenheit die Einstellung gegenüber Füchsen in der Stadt? *Forest, Snow, and Landscape Research* 76:255–266.

Booth, D., and C. Jackson. 1997. Urbanization of aquatic systems: Degradation thresholds, stormwater detention, and the limits of mitigation. *Journal of the American Water Resources Association* 33:1077–1090.

Borkenhagen, P. 1978. Prey gathered in by domestic cats (*Felis sylvestris f. catus* L., 1758). *Zeitschrift für Jagdwissenschaft* 24:27–33.

Boulanger, J. R., L. L. Bigler, P. D. Curtis, D. H. Lein, and A. J. Lembo, Jr. 2008. Evaluation of an oral vaccination program to control raccoon rabies in a suburbanized landscape. *Human-Wildlife Conflicts* 2:212–224.

Boulouis, H. J., C. Chang, J. B. Henn, R. W. Kasten, and B. B. Chomel. 2005. Factors associated with the rapid emergence of zoonotic *Bartonella* infections. *Veterinary Research* 36:383–410.

Bounds, D. L., and W. W. Shaw. 1994. Managing coyotes in U.S. National Parks: Human coyote interactions. *Natural Areas Journal* 14:280–284.

Bowen, W. D. 1982. Home range and spatial organization

of coyotes in Jasper National Park, Alberta. *Journal of Wildlife Management* 46:201–215.

Boyd, D., and B. O'Gara. 1985. Cougar predation on coyotes. *The Murrelet* 66:17.

Boydston, E. 2005. *Behavior, Ecology, and Detection Survey of Mammalian Carnivores in the Presidio*. U.S. Geological Survey Final Report. Sacramento, CA. 80 pp.

Bozek, C. K., S. Prange, and S. D. Gehrt. 2007. The influence of anthropogenic resources on multi-scale habitat selection of raccoons. *Urban Ecosystems* 10:413–425.

Braband, L. A., and K. D. Clark. 1992. Perspectives on wildlife nuisance control: Results of a wildlife damage control firm's customer survey. *Proceedings of the Eastern Wildlife Damage Control Conference* 5:34–37.

Bradley, C. A., and S. Altizer. 2006. Urbanization and the ecology of wildlife diseases. *Trends in Ecology and Evolution* 22:95–102.

Bradley, J. 2006. *Dog Bites: Problems and Solutions*. Animals and Society Institute, Ann Arbor, MI.

Bradshaw, J., and C. Cameron-Beaumont. 2000. The signalling repertoire of the domestic cat and its undomesticated relatives. Pages 67–93 *in* D. C. Turner and P. Bateson, eds. *The Domestic Cat: The Biology of Its Behaviour*. 2nd ed. Cambridge University Press, Cambridge.

Bradshaw, J. W. S. 2006. The evolutionary basis for the feeding behavior of domestic dogs (*Canis familiaris*) and cats (*Felis catus*). *Journal of Nutrition* 136:1927S–1931S.

Bradshaw, J. W. S., D. Goodwin, V. Legrand-Defrétin, and H. M. R. Nott. 1996. Food selection by the domestic cat, an obligate carnivore. *Comparative Biochemistry and Physiology* 114A:205–209.

Bradshaw, J. W. S., G. F. Horsfield, J. A. Allen, and I. H. Robinson. 1999. Feral cats: their role in the population dynamics of *Felis catus*. *Applied Animal Behaviour Science* 65:273–283.

Bradshaw, J. W. S., and H. M. R. Nott. 1995. Social and communication behaviour of companion dogs. Pages 115–130 *in* J. Serpell, ed. *The Domestic Dog: Its Evolution, Behaviour and Interactions with People*. Cambridge University Press, Cambridge.

Breheny, M. 1997. Urban compaction: Feasible and acceptable? *Sustainable Urban Development* 14:209–217.

Brickner, I. 2003. The impact of domestic cat (*Felis catus*) on wildlife welfare and conservation: A literature review, with a situation summary from Israel. www.tau.ac.il/lifesci/zoology/members/yom-tov/inbal/cats.pdf. Accessed July 27, 2007.

Bright, A. D., and M. J. Manfredo. 1996. A conceptual model of attitudes toward natural resources issues: A case study of wolf reintroduction. *Human Dimensions of Wildlife* 1:1–21.

British Broadcasting Company. 2007. Homeless badgers sett for luxury. http://news.bbc.co.uk/1/hi/scotland/north_east/6274821.stm.

Broadfoot, J. D., R. C. Rosatte, and D. T. O'Leary. 2001. Raccoon and skunk population models for urban disease control planning in Ontario, Canada. *Ecological Applications* 11:295–303.

Brochier, B., H. De Blander, R. Hanosset, D. Berkvens, B. Losson, and C. Saegerman. 2007. *Echinococcus multilocularis* and *Toxocara canis* in urban red foxes (*Vulpes vulpes*) in Brussels, Belgium. *Preventive Veterinary Medicine* 80:65–73.

Brochier, B., M. P. Kieny, F. Costy, P. Coppens, B. Bauduin, J. P. Lecocq, B. Languet, et al. 1991. Large-scale eradication of rabies using recombinant vaccinia-rabies vaccine. *Nature* 354:520–522.

Brooks, J. J., R. J. Warren, M. G. Nelms, and M. A. Tarrant. 1999. Visitor attitudes toward and knowledge of restored bobcats on Cumberland Island National Seashore, Georgia. *Wildlife Society Bulletin* 27:1089–1097.

Brooks, W. 2000. Notoedric mange. Mar Vista Animal Medical Center website. http://www.marvistavet.com/html/notoedric_mange.html. Accessed March 24, 2008.

Brown, H. D., A. R. Matzuk, I. R. Ilves, L. H. Peterson, S. A. Harris, L. H. Sarett, J. R. Egerton, et al. 1961. Antiparasitic drugs. IV.2–(4'-thiazolyl)-benzimidazole, a new anthelmintic. *Journal of American Chemistry Society* 83:1764–1765.

Brown, J. L. 2007. The influence of coyotes on an urban Canada goose population in the Chicago metropolitan area. Master's thesis, Ohio State University, Columbus.

Brudin, C. O., III. 2003. Wildlife use of existing culverts and bridges in north central Pennsylvania. International Conference on Ecology and Transportation (online, not paginated).

Buffon, Count de G. L. Le Clerc. 1828. *A Natural History*. Vol. I. Thomas Kelly, London.

Bugg, R. J., I. D. Robertson, A. D. Elliot, and R. C. A. Thompson. 1999. Gastrointestinal parasites of urban dogs in Perth, Western Australia. *Veterinary Journal* 157:295–301.

Buie, D. E., T. T. Fendley, and B. K. McNab. 1979. Fall and winter home ranges of adult bobcats on the Savannah River Plant, South Carolina. Bobcat Research Conference Proceedings. *National Wildlife Federation Scientific Technical Series* 6:42–46.

Burgess, T. 1942. *The Adventures of Bobby Coon*. Grosett and Dunlap, New York.

Burns, R. J. 1980. Evaluation of conditioned predation aversion for controlling coyote predation. *Journal of Wildlife Management* 44:938–942.

Burns, R. J. 1983. Coyote predation aversion with lithium chloride: Management implications and comments. *Wildlife Society Bulletin* 11:128–133.

Buskirk, S. W. 1999. Mesocarnivores of Yellowstone. Pages 165–187 *in* T. M. Clark, A. P. Curlee, S. C. Minta, and P. M. Kareiva, eds. *Carnivores in Ecosystems: The Yellowstone Experience*. Yale University Press, New Haven, CT.

Butcher, R. 1999. Stray dogs—a worldwide problem. *Journal of Small Animal Practice* 40:458–459.

Butler, J. R. A., and J. Bingham. 2000. Demography and dog-human relationships of the dog population in Zimbabwean communal lands. *Veterinary Record* 147:442–446.

Butler, J. R. A., and J. T. du Toit. 2002. Diet of free-ranging domestic dogs (*Canis familiaris*) in rural Zimbabwe: Implications for wild scavengers on the periphery of wildlife reserves. *Animal Conservation* 5:29–37.

Butler, J. R. A., J. T. du Toit, and J. Bingham. 2004. Free-ranging domestic dogs (*Canis familiaris*) as predators and prey in rural Zimbabwe: Threats of competition and disease to large wild carnivores. *Biological Conservation* 115:369–378.

Butler, J. S., J. Shanahan, and D. J. Decker. 2003. Public attitudes toward wildlife are changing: A trend analysis of New York residents. *Wildlife Society Bulletin* 31:1027–1036.

Calhoon, R. E., and C. Haspel. 1989. Urban cat populations compared by season, subhabitat, and supplemental feeding. *Journal of Animal Ecology* 58:321–328.

Calver, M., S. Thomas, S. Bradley, and H. McCutcheon. 2007. Reducing the rate of predation on wildlife by pet cats: The efficacy and practicability of collar-mounted pounce protectors. *Biological Conservation* 137:341–348.

Campbell, S., and G. Hartley. 2007. *Urban Fox Populations in Scottish Towns and Cities 2006.* Pesticide Usage and Wildlife Management, Scottish Agricultural Science Agency, Edinburgh, UK.

Campos, C. B., C. F. Esteves, K. M. P. M. B. Ferraz, P. G. Crawshaw, and L. M. Verdade. 2007. Diet of free-ranging cats and dogs in a suburban and rural environment, south-eastern Brazil. *Journal of Zoology* 273:14–20.

Campos-Bueno, A., G. López-Abente, and A. M. Andrés-Cercadillo. 2000. Risk factors for *Echinococcus granulosus* infection: A case-control study. *American Journal of Tropical Medicine and Hygiene* 62:329–334.

Cannon, A. R., D. E. Chamberlain, M. P. Toms, B. J. Hatchwell, and K. J. Gaston. 2005. Trends in the use of private gardens by wild birds in Great Britain 1995–2002. *Journal of Applied Ecology* 42:659–671.

Carbone, C., and J. L. Gittleman. 2002. A common rule for the scaling of carnivore density *Science* 295:2273–2276.

Carbyn, L. N. 1982. Coyote population fluctuations and spatial distribution in relation to gray wolf territories in Riding Mountain National Park, Manitoba. *Canadian Field-Naturalist* 96:176–183.

Carbyn, L. N. 1989. Coyote attacks on children in Western North America. *Wildlife Society Bulletin* 17:444–446.

Carlson, M. E. 1996. *Yersinia pestis* infection in cats. *Feline Practice* 24:22–24.

Caro, T. M. 1994. *Cheetahs of the Serengeti Plains.* University of Chicago Press, Chicago.

Carpenter, L. H., D. J. Decker, and J. F. Lipscomb. 2000. Stakeholder acceptance capacity in wildlife management. *Human Dimensions of Wildlife* 5:5–19.

Carpenter, P. J., L. C. Pope, C. Greig, D. A. Dawson, L. M. Rogers, K. Erven, G. J. Wilson, R. J. Delahay, C. L. Cheeseman, and T. Burke. 2005. Mating system

of the Eurasian badger, *Meles meles*, in a high density population. *Molecular Ecology* 14:273–284.

Carss, D. N. 1995. Prey brought home by two domestic cats (*Felis catus*) in northern Scotland. *Journal of Zoology* 237:678–686.

Carter, J., and J. Finn. 1999. MOAB: A spatially explicit individual-based expert system for creating animal foraging models. *Ecological Modelling* 119:29–42.

Casey, A. L., P. R. Krausman, W. W. Shaw, and H. G. Shaw. 2005. Knowledge of and attitudes toward mountain lions: A public survey of residents adjacent to Saguaro National Park, Arizona. *Human Dimensions of Wildlife* 10:29–38.

Cauley, D. L. 1970. The effects of urbanization on raccoon (*Procyon lotor*) populations. Master's thesis, University of Cincinnati, Cincinnati, Ohio.

Cauley, D. L., and J. R. Schinner. 1973. The Cincinnati raccoons. *Natural History* 82:158–160.

Cavallini, P. 1995. Variation in the body size of the red fox. *Annales Zoologici Fennici* 32:421–427.

Cavallini, P. 1996. Variation in the social system of the red fox. *Ethology Ecology & Evolution* 8:323–342.

Cavallini, P., and S. Santini. 1995. Timing of reproduction in the red fox, *Vulpes vulpes*. *Zeitschrift für Säugetierkunde* 60:337–342.

Cave, T. A., H. Thompson, S. W. J. Reid, D. R. Hodgson, and D. D. Addie. 2002. Kitten mortality in the United Kingdom: A retrospective analysis of 274 histopathological examinations (1986 to 2000). *Veterinary Record* 151:497–501.

Ceballos, G., and A. Miranda. 1986. *Los mamíferos de Chamela, Jalisco.* Universidad Nacional Autónoma de México, México, D.F. 436 pp.

Centers for Disease Control and Prevention. 2003. Nonfatal dog bite—related injuries treated in hospital emergency departments—United States, 2001. *Morbidity and Mortality Weekly Report* 52:605–610.

Centonze, L. A., and J. K. Levy. 2002. Characteristics of free-roaming cats and their caretakers. *Journal of the American Veterinary Medical Association* 220:1627–1633.

Cepek, J. D. 2004. Diet composition of coyotes in the Cuyahoga Valley National Park, Ohio. *Ohio Journal of Science* 104:60–64.

Chace, J. F., and J. J. Walsh. 2006. Urban effects on native avifauna: A review. *Landscape and Urban Planning* 74:46–69.

Chamberlain, M. J., B. D. Leopold, L. W. Burger, B. W. Plowman, and L. M. Conner. 1999. Survival and cause-specific mortality of adult bobcats in central Mississippi. *Journal of Wildlife Management* 63:613–620.

Cheeseman, C. L., W. J. Cresswell, S. Harris, and P. J. Mallinson. 1988. Comparison of dispersal and other movements in two badger (*Meles meles*) populations. *Mammal Review* 18:51–59.

Cheeseman, C. L., J. W. Wilesmith, J. Ryan, and P. J. Mallinson. 1987. Badger population dynamics in a high-density area. *Symposia of the Zoological Society of London* 58:279–294.

Childs, J. E., L. E. Robinson, R. Sadek, A. Madden, M. E. Miranda, and N. L. Miranda. 1998. Density estimates of rural dog populations and an assessment of marking methods during a rabies vaccination campaign in the Philippines. *Preventive Veterinary Medicine* 33:207–218.

Childs, J. E., and L. Ross. 1986. Urban cats: Characteristics and estimation of mortality due to motor vehicles. *American Journal of Veterinary Research* 47:1643–1648.

Chomel, B. B., H. J. Boulouis, H. Petersen, R. W. Kasten, K. Yamamoto, C. Chang, C. Gandoin, C. Bouillin, and C. M. Hew. 2002. Prevalence of *Bartonella* infection in domestic cats in Denmark. *Veterinary Research* 33:205–213.

Chomel, B. B., and J. Trotignon. 1992. Epidemiologic surveys of dog and cat bites in the Lyon area, France. *European Journal of Epidemiology* 8:619–624.

Christensen, H. 1985. Urban fox population in Oslo. *Revue d'Ecologie (Terre et la Vie)* 40:185–186.

Churcher, J. B., and J. H. Lawton. 1987. Predation by domestic cats in an English village. *Journal of Zoology* 212:439–456.

Cialdini, R. B., R. R. Reno, and C. A. Kallgren. 1990. A focus theory of normative conduct: Recycling the concept of norms to reduce littering in public places. *Journal of Personality and Social Psychology* 58:1015–1026.

Cignini, B., and F. Riga. 1997. Red fox sightings in Rome. *Hystrix* 9:71–74.

Clark, H. O., Jr. 2005. *Otocyon megalotis. Mammalian Species* 766:1–5.

Clark, H. O., Jr., G. D. Warrick, B. L. Cypher, P. A. Kelly, D. F. Williams, and D. E. Grubbs. 2005. Competitive interactions between endangered kit foxes and nonnative red foxes. *Western North American Naturalist* 65:153–163.

Clark, W. R., J. J. Hasbrouck, J. M. Kienzler, and T. F. Glueck. 1989. Vital statistics and harvest of an Iowa raccoon population. *Journal of Wildlife Management* 53:982–990.

Clarke, A. L., and T. Pacin. 2002. Domestic cat "colonies" in natural areas: A growing exotic species threat. *Natural Areas Journal* 22:154–159.

Clarke, G. P., P. C. L. White, and S. Harris. 1998. Effects of roads on badger *Meles meles* populations in south-west England. *Biological Conservation* 86:117–124.

Cleaveland, S., M. Kaare, D. Knobel, and M. K. Laurenson. 2006. Canine vaccination: Providing broader benefits for disease control. *Veterinary Microbiology* 117:43–50.

Clements, D. K. 1992. Brock's defence—a brief examination of the law relating to badgers. *British Wildlife* 3:193–199.

Clevenger, A. P., B. Chruszcz, and K. E. Gunson. 2001. Drainage culverts as habitat linkages and factors affecting passage by mammals. *Journal of Applied Ecology* 38:1340–1349.

Clevenger, A. P., and N. Waltho. 2005. Performance indices to identify attributes of highway crossing structures facilitating movement of large mammals. *Biological Conservation* 121:453–464.

Clevenger, A. P., and J. Wierzchowski. 2006. Maintaining and restoring connectivity in landscapes fragmented by roads. Pages 502–535 *in* K. R. Crooks and M. A. Sanjayan, eds. *Connectivity Conservation*. Cambridge University Press, New York.

Clutton-Brock, J. 1995. Origins of the dog: Domestication and early history. Pages 7–20 *in* J. Serpell, ed. *The Domestic Dog: Its Evolution, Behaviour and Interactions with People*. Cambridge University Press, Cambridge.

Clutton-Brock, J. 1996. Competitors, companions, status symbols or pests: A review of human associations with other carnivores. Pages 375–392 *in* J. L. Gittleman, ed. *Carnivore Behavior, Ecology, and Evolution*. Vol. 2. Cornell University Press, Ithaca, NY.

Clutton-Brock, J. 1999. *A Natural History of Domesticated Mammals*. 2nd ed. Cambridge University Press, Cambridge.

Clutton-Brock, T. H., D. Gaynor, G. M. McIlrath, A. D. C. Maccoll, R. Kansky, P. Chadwick, M. Manser, J. D. Skinner, and P. N. M. Brotherton. 2001. Predation, group size and mortality in a cooperative mongoose, *Suricata suricatta. Journal of Animal Ecology* 68:672–683.

Cohn, J. 1998. A dog-eat-dog world. *BioScience* 48:430–434.

Coleman, J. S., and S. A. Temple. 1993. Rural residents' free-ranging domestic cats: A survey. *Wildlife Society Bulletin* 21:381–390.

Coleman, J., S. Temple, and S. Craven. 1996. A conservation dilemma—the free-ranging domestic cat. *The Probe* 172:1–5.

Coman, B. J., and J. L. Robinson. 1989. Some aspects of stray dog behaviour in an urban fringe area. *Australian Veterinary Journal* 66:30–32.

Coman, B. J., J. Robinson, and C. Beaumont. 1991. Home range, dispersal and density of red foxes (*Vulpes vulpes* L.) in central Victoria. *Wildlife Research* 18:215–223.

Conforti, V. A., and F. C. C. de Azevedo. 2003. Local perceptions of jaguars (*Panthera onca*) and pumas (*Puma concolor*) in the Iguacu National Park area, south Brazil. *Biological Conservation* 111:215–221.

Conover, M. R. 2002. *Resolving Human-Wildlife Conflicts: The Science of Wildlife Damage Management*. CRC Press, Boca Raton, FL.

Conrad, P. A., M. A. Miller, C. Kreuder, E. R. James, J. Mazet, H. Dabritz, D. A. Jessup, F. Gulland, and M. E. Grigg. 2005. Transmission of *Toxoplasma*: Clues from the study of sea otters as sentinels of *Toxoplasma gondii* flow into the marine environment. *International Journal of Parasitology* 35:1155–1168.

Contesse, P., D. Hegglin, S. Gloor, F. Bontadina, and P. Deplazes. 2004. The diet of urban foxes (*Vulpes vulpes*) and the availability of anthropogenic food in the city of Zurich, Switzerland. *Mammalian Biology* 69:81–95.

Coppinger, R., and L. Coppinger. 2001. *Dogs: A New*

Understanding of Canine Origin, Behavior, and Evolution. University of Chicago Press, Chicago.

Corbett, J. B. 1992. Rural and urban newspaper coverage of wildlife: Conflict, community, and bureaucracy. *Journalism Quarterly* 69:929–37.

Corbett, J. B. 1995a. When wildlife make the news: An analysis of rural and urban north-central U.S. newspapers. *Public Understanding of Science* 4:397–410.

Corbett, L. K. 1979. Feeding ecology and social organisation of wild cats and domestic cats in Scotland. Ph.D. diss., University of Aberdeen, Scotland.

Corbett, L. K. 1995b. *The Dingo in Australia and Asia.* Cornell University Press, Ithaca, NY.

Corbett, L. K. 2004. Dingo (*Canis lupus dingo*). Pages 223–230 *in* C. Sillero-Zubiri, M. Hoffmann, and D. W. Macdonald, eds. *Canids: Foxes, Wolves, Jackals and Dogs.* Status Survey and Conservation Action Plan. IUCN Canid Specialist Group, IUCN, Gland, Switzerland.

Corbett, L. K., and A. E. Newsome. 1997. The feeding ecology of the dingo. III. Dietary relationships with widely fluctuating prey populations in arid Australia: An hypothesis of alternation of predation. *Oecologia* 74:215–227.

Cornell, D., and J. E. Cornley. 1979. Aversive conditioning of campground coyotes, *Canis latrans*, in Joshua-Tree National Monument, California, USA. *Wildlife Society Bulletin* 7:129–131.

Courchamp, F., M. Langlais, and G. Sugihara. 1999. Cats protecting birds: Modelling the mesopredator release effect. *Journal of Animal Ecology* 68:282–292.

Courchamp, F., and D. Pontier. 1994. Feline immunodeficiency virus: An epidemiologic review. *Comptes Rendus de l'Académie des Sciences Série III* 317:1123–1134.

Courchamp, F., C. Suppo, E. Fromont, and C. Bouloux. 1997. Dynamics of two feline retroviruses (FIV and FeLV) within one population of cats. *Proceedings of the Royal Society of London B* 264:785–794.

Courtenay, O., and L. Maffei. 2004. Crab-eating fox *Cerdocyon thous* (Linnaeus, 1766). Pages 32–38 *in* C. Sillero-Zubiri, M. Hoffman, and D. W. Macdonald, eds. *Canids: Foxes, Wolves, Jackals, and Dogs.* Status Survey and Conservation Action Plan. IUCN, Gland, Switzerland.

Cox, J. J., D. S. Maehr, and J. L. Larkin. 2006. Florida panther habitat use: New approach to an old problem. *Journal of Wildlife Management* 70:1778–1785.

Cramer, P. C., and J. A. Bissonette. 2007. Integrating wildlife crossings into transportation plans and projects in North America. Pages 328–334 in *Proceedings of the 2007 International Conference on Ecology and Transportation.* Center for Transportation and the Environment, North Carolina State University, Raleigh.

Crandell, R. A. 1975. Arctic fox rabies. Pages 23–40 *in* G. M. Baer, ed. *The Natural History of Rabies.* Academic Press, New York.

Cranfield, M. R., I. K. Barker, K. G. Mehren, and W. A. Rapley. 1984. Canine distemper in wild raccoons (*Procyon lotor*) at the metropolitan Toronto zoo. *Canadian Veterinary Journal* 25:63–66.

Creel, S. 1996. Behavioral endocrinology and social organization in dwarf mongooses. Pages 46–77 *in* J. L. Gittleman, ed. *Carnivore Behavior, Ecology and Evolution.* Vol. 2. Cornell University Press, Ithaca, NY.

Cresswell, P., S. Harris, R. G. H. Bunce, and D. J. Jefferies. 1989. The badger (*Meles meles*) in Britain: Present status and future population changes. *Biological Journal of the Linnean Society* 38:91–101.

Cresswell, P., S. Harris, and D. Jefferies. 1990. *The History, Distribution, Status and Habitat Requirements of the Badger in Britain.* Nature Conservancy Council, Peterborough, UK.

Cresswell, W. J., and S. Harris. 1988a. Foraging behaviour and home-range utilization in a suburban badger (*Meles meles*) population. *Mammal Review* 18:37–49.

Cresswell, W. J., and S. Harris. 1988b. The effects of weather conditions on the movements and activity of badgers (*Meles meles*) in a suburban environment. *Journal of Zoology* 216:187–194.

Cresswell, W. J., S. Harris, C. L. Cheeseman, and P. J. Mallinson. 1992. To breed or not to breed: An analysis of the social and density-dependent constraints on the fecundity of female badgers (*Meles meles*). *Philosophical Transactions of the Royal Society London B* 338:393–407.

Crooks, K. R. 2002. Relative sensitivities of mammalian carnivores to habitat fragmentation. *Conservation Biology* 16:488–502.

Crooks, K. R., and M. E. Soulé. 1999. Mesopredator release and avifaunal extinctions in a fragmented system. *Nature* 400:563–566.

Crooks, K. R., A. V. Suarez, and D. T. Bolger. 2004. Avian communities along a gradient of urbanization impact in a highly fragmented landscape. *Biological Conservation* 115:451–462.

Crooks, K. R., A. V. Suarez, D. T. Bolger, and M. E. Soulé. 2001. Extinction and colonization of birds on habitat islands. *Conservation Biology* 15:159–172.

Crowell-Davis, S. L., T. M. Curtis, and R. J. Knowles. 2004. Social organization in the cat: a modern understanding. *Journal of Feline Medicine and Surgery* 6:19–28.

Cunningham, S. C., L. A. Haynes, C. Gustavson, and D. D. Haywood. 1995. *Evaluation of the Interaction between Mountain Lions and Cattle in the Aravaipa-Klondyke Area of Southeast Arizona.* Arizona Game and Fish Department Technical Report 17. Phoenix.

Curt, M. 1999. Restoring the gray wolf in Idaho. *Endangered Species Bulletin* 24:12–13.

Cushing, B. S. 1985. Estrous mice and vulnerability to weasel predation. *Ecology* 66:1976–1978.

Cuyler, W. K. 1924. Observations on the habits of the striped skunk (*Mephitis mesomelas vaarians*). *Journal of Mammalogy* 5:180–189.

Cuzin, F., and D. M. Lenain. 2004. Ruppell's fox. Pages 201–205 *in* C. Sillero-Zubiri, M. Hoffman, and D. W. Macdonald, eds. *Canids: Foxes, Wolves, Jackals, and Dogs.*

Status Survey and Conservation Action Plan. IUCN/ SSC Canid Specialist Group. Gland, Switzerland.

Cypher, B. L. 2003. Foxes. Pages 511–546 in G. A. Feldhamer, B. C. Thompson, and J. A. Chapman, eds. *Wild Mammals of North America: Biology, Management, and Conservation*. 2nd ed. Johns Hopkins University Press, Baltimore.

Cypher, B. L., C. D. Bjurlin, and J. L. Nelson. 2005. *Effects of Two-Lane Roads on Endangered San Joaquin Kit Foxes*. California State University–Stanislaus, Endangered Species Recovery Program, Fresno, CA. 21 pp.

Cypher, B. L., H. O. Clark, Jr., P. A. Kelly, C. Van Horn Job, G. W. Warrick, and D. F. Williams. 2001. Interspecific interactions among mammalian predators: Implications for the conservation of endangered San Joaquin kit foxes. *Endangered Species UPDATE* 18:171–174.

Cypher, B. L., and N. Frost. 1999. Condition of San Joaquin kit foxes in urban and exurban habitats. *Journal of Wildlife Management* 63:930–938.

Cypher, B. L., P. A. Kelly, and D. F. Williams. 2003. Factors influencing populations of endangered San Joaquin kit foxes: Implications for conservation and recovery. Pages 125–137 in M. A. Sovada and L. N. Carbyn, eds. *The Swift Fox: Ecology and Conservation in a Changing World*. Canadian Plains Research Center, Regina, Saskatchewan.

Cypher, B. L., and G. D. Warrick. 1993. Use of human-derived food items by urban kit foxes. *Transactions of the Western Section of the Wildlife Society* 29:34–37.

Cypher, B. L., G. D. Warrick, M. R. M. Otten, T. P. O'Farrell, W. H. Berry, C. E. Harris, T. T. Kato, P. M. McCue, J. H. Scrivner, and B. W. Zoellick. 2000. Population dynamics of San Joaquin kit foxes at the Naval Petroleum Reserves in California. *Wildlife Monographs* 145:1–43.

Czech, B., P. R. Krausman, and P. K. Devers. 2000. Economic associations among causes of species endangerment in the United States. *BioScience* 50:593–601.

Daag, A. 1970. Wildlife in an urban area. *Le Naturaliste Canadien* 97:201–212.

Dabritz, H. A., E. R. Atwill, I. A. Gardner, M. A. Miller, and P. A. Conrad. 2006. Outdoor fecal deposition by free-roaming cats and attitudes of cat owners and nonowners toward stray pets, wildlife, and water pollution. *Journal of the American Veterinary Medical Association* 229:74–81.

Dahle, B., and J. E. Swenson. 2003. Factors influencing length of maternal care in brown bears (*Ursus arctos*) and its effect on offspring. *Behavioral Ecology and Sociobiology* 54:352–358.

Daigle, J. J., R. M. Muth, R. R. Zwick, and R. R. Glass. 1998. Sociocultural dimensions of trapping: A factor analytic study of trappers in six northeastern states. *Wildlife Society Bulletin* 26:614–625.

Dalgish, J., and S. Anderson. 1979. A field experiment on learning by raccoons. *Journal of Mammalogy* 60:620–622.

Daniels, T. J. 1983a. The social organization of free-ranging urban dogs. II. Estrous groups and the mating system. *Applied Animal Ethology* 10:365–373.

Daniels, T. J. 1983b. The social organization of free-ranging urban dogs. I. Non-estrous social behaviour. *Applied Animal Ethology* 10:341–363.

Daniels, T. J., and M. Bekoff. 1989. Population and social biology of free-ranging dogs, *Canis familiaris. Journal of Mammalogy* 70:754–762.

Dards, J. L. 1978. Home ranges of feral cats in Portsmouth dockyard. *Carnivore Genetics Newsletter* 3:242–255.

Dards, J. L. 1983. The behaviour of dockyard cats: Interaction of adult males. *Applied Animal Ethology* 10:133–153.

Davidson, W., and V. Nettles. 1997. *Field Manual of Wildlife Diseases in the Southeastern United States*. 2nd ed. Southeastern Cooperative Wildlife Disease Study. University of Georgia, Athens.

Davison, J., M. Huck, R. J. Delahay, and T. J. Roper. 2008a. Restricted ranging behaviour in a high-density population of urban badgers. *Journal of Zoology* 277:45–53.

Davison, J., M. Huck, R. J. Delahay, and T. J. Roper. 2008b. Urban badger setts: Characteristics, patterns of use, and management implications. *Journal of Zoology* 275:190–200.

De Almeida, M. H. 1987. Nuisance furbearer damage control in urban and suburban areas. Pages 996–1006 in M. Novak, J. A. Baker, M. E. Obbard, and B. Malloch, eds. *Wild Furbearer Management and Conservation in North America*. Ministry of Natural Resources, Toronto, Ontario.

De Balogh, K. K., A. I. Wandeler, and F. X. Meslin. 1993. A dog ecology study in an urban and a semi-rural area of Zambia. *Onderstepoort Journal of Veterinary Research* 60:437–443.

De Blander, H., T. Kervyn, B. Gaubicher, and B. Brochier. 2004. *Le renard roux Vulpes vulpes en Région Bruxelles-Capitale*. Institut Bruxellois pour la Gestion de l'Environment, Brussels, Belgium.

Decker, D. J., T. L. Brown, and W. F. Siemer, eds. 2001. *Human Dimensions of Wildlife Management in North America*. The Wildlife Society, Bethesda, MD.

Decker, D. J., T. L. Brown, J. J. Vaske, and M. J. Manfredo. 2004a. Human dimensions of wildlife management. Pages 187–198 in M. Manfredo, J. J. Vaske, B. L. Bruyere, D. R. Field, and P. J. Brown, eds. *Society and Natural Resources: A Summary of Knowledge*. Modern Litho, Jefferson, MO.

Decker, D. J., C. A. Jacobson, and T. L. Brown. 2006. Situation-specific "impact dependency" as a determinant of management acceptability: Insights from wolf and grizzly bear management in Alaska. *Wildlife Society Bulletin* 34:426–432.

Decker, D. J., T. B. Lauber, and W. F. Siemer. 2002. *Human-Wildlife Conflict Management: A Practitioners' Guide*. Northeast Wildlife Damage Management Research and Outreach Cooperative, Ithaca, NY.

Decker, D. J., and J. O'Pezio. 1989. Consideration of bear-people conflicts in black bear management for the

Catskill region of New York: Application of a comprehensive management model. Pages 181–187 *in* M. Bromley, ed. *Bear-People Conflicts: Proceedings of a Symposium On Management Strategies*. Department of Renewable Resources, Yellowknife, Northwest Territories, Canada.

Decker, D. J., and K. G. Purdy. 1988. Toward a concept of wildlife acceptance capacity in wildlife management. *Wildlife Society Bulletin* 16:53–57.

Decker, D. J., D. B. Raik, and W. F. Siemer. 2004b. *Community-Based Suburban Deer Management: A Practitioner's Guide*. Northeast Wildlife Damage Management Research and Outreach Cooperative, Ithaca, NY.

Dekker, D. 1983. Denning and foraging habits of red foxes, *Vulpes vulpes*, and their interaction with coyotes, *Canis latrans*, in central Alberta, 1972–1981. *Canadian Field-Naturalist* 97:303–306.

Delahay, R. J., J. Davison, D. W. Poole, A. J. Matthews, C. J. Wilson, M. J. Heydon, and T. J. Roper. 2009. Managing conflict between humans and wildlife: Trends in licensed operations to resolve problems with badgers *Meles meles* in England. *Mammal Review* 39:53–66.

Dennis, C. 2002. Baiting plan to remove fox threat to Tasmanian wildlife. *Nature* 416:357.

Denny, E., P. Yakovlevich, M. D. B. Eldridge, and C. Dickman. 2002. Social and genetic analysis of a population of free-living cats (*Felis catus* L.) exploiting a resource-rich habitat. *Wildlife Research* 29:405–413.

Deplazes, P., D. Hegglin, S. Gloor, and T. Romig. 2004. Wilderness in the city: The urbanization of *Echinococcus multilocularis*. *Trends in Parasitology* 20:77–84.

DeStefano, S., and R. M. DeGraaf. 2003. Exploring the ecology of suburban wildlife. *Frontiers in Ecology and the Environment* 1:95–101.

DeVault, T. L., O. E. Rhodes, Jr., and J. A. Shivik. 2003. Scavenging by vertebrates: Behavioral, ecological, and evolutionary perspectives on an important energy transfer pathway in terrestrial ecosystems. *Oikos* 102:225–234.

Devillard, S., L. Say, and D. Pontier. 2003. Dispersal pattern of domestic cats (*Felis catus*) in a promiscuous urban population: Do females disperse or die? *Journal of Animal Ecology* 72:203–211.

Devillard, S., L. Say, and D. Pontier. 2004. Molecular and behavioural analyses reveal male-biased dispersal between social groups of domestic cats. *Ecoscience* 11:175–180.

Dibello, F. J., S. M. Arthur, and W. B. Krohn. 1990. Food habits of sympatric coyotes, *Canis latrans*, red foxes, *Vulpes vulpes*, and bobcats, *Lynx rufus*, in Maine. *Canadian Field-Naturalist* 104:403–408.

Dickson, B. G., and P. Beier. 2002. Home range and habitat selection by adult cougars in southern California. *Journal of Wildlife Management* 66:1235–1245.

Dickson, B. G., and P. Beier. 2007. Quantifying the influence of topographic position on cougar (*Puma concolor*) movement in Southern California, USA. *Journal of Zoology* 271:270–277.

Dickson, B. G., J. S. Jenness, and P. Beier. 2005. Influence of vegetation, topography, and roads on cougar movement in Southern California. *Journal of Wildlife Management* 69:264–276.

Dierschke, V. 2003. Predation hazard during migratory stopover: Are light or heavy birds under risk? *Journal of Avian Biology* 34:24–29.

Dijak, W. D., and F. R. Thompson. 2000. Landscape and edge effects on the distribution of mammalian predators in Missouri. *Journal of Wildlife Management* 64:209–216.

Disney, M., and L. K. Spiegel. 1992. Sources and rates of San Joaquin kit fox mortality in western Kern County, California. *Transactions of the Western Section of the Wildlife Society* 28:73–82.

Ditchkoff, S. S., S. T. Saalfeld, and C. J. Gibson. 2006. Animal behavior in urban ecosystems: Modifications due to human-induced stress. *Urban Ecosystems* 9:5–12.

Dobson, A. P., J. P. Rodriguez, and W. M. Roberts. 2001. Synoptic tinkering: Integrating strategies for large-scale conservation. *Ecological Applications* 11:1019–1026.

Dobson, A. P., J. P. Rodriguez, W. M. Roberts, and D. S. Wilcove. 1997. Geographic distribution of endangered species in the United States. *Science* 275:550–553.

Doncaster, C. P., C. R. Dickman, and D. W. Macdonald. 1990. Feeding ecology of red foxes (*Vulpes vulpes*) in the city of Oxford, England. *Journal of Mammalogy* 71:188–194.

Doncaster, C. P., and D. W. Macdonald. 1991. Drifting territoriality in the red fox *Vulpes vulpes*. *Journal of Animal Ecology* 60:423–439.

Doncaster, C. P., and D. W. Macdonald. 1997. Activity patterns and interactions of red foxes (*Vulpes vulpes*) in Oxford city. *Journal of Zoology* 241:73–87.

Doty, J. B., and R. C. Dowler. 2006. Denning ecology in sympatric populations of skunks (*Spilogale gracilis and Mephitis mephitis*) in west-central Texas. *Journal of Mammalogy* 87:131–138.

Dragoo, J. W., D. B. Fagre, D. J. Schmidly, and L. B. Penry. 1989. First specimen of a hog-nosed skunk (*Conepatus mesolucus*) from Bexar County, Texas. *Texas Journal of Science* 41:331–333.

Dragoo, J. W., and R. L. Honeycutt. 1997. Systematics of mustelid-like carnivores. *Journal of Mammalogy* 78:426–443.

Dragoo, J. W., R. Honeycutt, and D. Schmidly. 2003. Taxonomic status of white-backed hog-nosed skunks, genus *Conepatus* (Carnivora: Mephitidae). *Journal of Mammalogy* 84:159–176.

Dragoo, J. W., D. K. Matthes, A. Aragon, C. C. Hass, and T. L. Yates. 2004. Identification of skunk species submitted for rabies testing in the Desert Southwest. *Journal of Wildlife Diseases* 40:371–376.

Driscoll, C. A., M. Menotti-Raymond, A. L. Roca, K. Hupe, W. E. Johnson, E. Geffen, E. H. Harley, et al. 2007. The Near Eastern origin of cat domestication. *Science* 317:519–523.

Dubey, J. P., R. J. Cawthorn, C. A. Speer, and G. A. Wobeser. 2003. Redescription of the Sarcocysts of *Sarcocystis rileyi* (Apicomplexa: Sarcocystidae). *Journal of Eukaryotic Microbiology* 50:476–482.

Dubey, J. P., P. Parnell, C. Sreekumar, M. Vianna, R. Young, E. Dahl, and T. Lehmann. 2004. Molecular charasteristics of *Toxoplasma gondii* isolates from striped skunk (*Mephitis mephitis*), Canada goose (*Branta canadensis*), black-winged lorry (*Eos cyanogenia*), and cats (*Felis catus*). *Journal of Parasitology* 90:1171–1173.

Dugdale, H. L., D. W. Macdonald, L. C. Pope, and T. Burke. 2007. Polygynandry, extra-group paternity and multiple-paternity litters in European badger (*Meles meles*) social groups. *Molecular Ecology* 16:5294–5306.

Dunn, E. H., and D. L. Tessaglia. 1994. Predation of birds at feeders in winter. *Journal of Field Ornithology* 65:8–16.

Duscher, G., T. Steineck, P. Günter, H. Prosl, and A. Joachim. 2005. *Echinococcus multilocularis* in foxes in Vienna and surrounding territories. *Wiener Tierärztliche Monatsschrift* 92:16–20.

Eason, C. T., E. C. Murphy, G. R. G. Wright, and E. B. Spurr. 2002. Assessment of risks of brodifacoum to non-target birds and mammals in New Zealand. *Ecotoxicology* 11:35–48.

Egenvall, A., B. N. Bonnett, Å. Hedhammar, and P. Olson. 2005. Mortality in over 350,000 insured Swedish dogs from 1995–2000. II. Breed-specific age and survival patterns and relative risk for causes of death. *Acta Veterinaria Scandinavica* 46:121–136.

Egoscue, H. J. 1956. Preliminary studies of the kit fox in Utah. *Journal of Mammalogy* 37:351–357.

Egoscue, H. J. 1962. Ecology and life history of the kit fox in Tooele County, Utah. *Ecology* 43:481–497.

Egoscue, H. J. 1975. Population dynamics of the kit fox in western Utah. *Bulletin of the Southern California Academy of Science* 74:122–127.

Eisenberg, J. F. 1989. An introduction to the Carnivora. Pages 1–9 *in* J. L. Gittleman, ed. 1989. *Carnivore Behavior, Ecology, and Evolution.* Cornell University Press, Ithaca, NY.

Ellis, R. J. 1964. Tracking raccoons by radio. *Journal of Wildlife Management* 28:363–368.

Elmhagen, B., and S. P. Rushton. 2007. Trophic control of mesopredators in terrestrial ecosystems: Top-down or bottom-up? *Ecology Letters* 10:197–206.

El-Yuguda, A. D., A. A. Babba, and S. S. A. Babba. 2007. Dog population structure and cases of rabies among dog bite victims in urban and rural areas of Borno State, Nigeria. *Tropical Veterinarian* 25:34–40.

Enck, J. W., and T. L. Brown. 2002. New Yorkers' attitudes toward restoring wolves to the Adirondack Park. *Wildlife Society Bulletin* 30:16–28.

Enck, J. W., and T. L. Brown. 2004. *Landowner and Hunter Response to Implementation of a Quality Deer Management (QDM) Cooperative Near King Ferry, New York.* Human Dimensions Research Unit. Series No. 03–7. Department of Natural Resources, Cornell University, Ithaca, NY.

Endres, K. M., and W. P. Smith. 1993. Influence of age, sex, season, and availability on den selection by raccoons within the Central Basin of Tennessee. *American Midland Naturalist* 129:116–131.

Engeman, R. M., K. Christensen, M. Pipas, and D. L. Bergman. 2003. Population monitoring in support of a rabies vaccination program for skunks in Arizona. *Journal of Wildlife Diseases* 39:746–750.

Epe, C., M. Meuwissen, M. Stoye, and T. Schnieder. 1999. Transmission trials, ITS2-PCR and RAPD-PCR show identity of *Toxocara canis* isolates from red fox and dog. *Veterinary Parasitology* 84:101–112.

Ericsson, G., and T. A. Heberlein. 2003. Attitudes of hunters, locals, and the general public in Sweden now that the wolves are back. *Biological Conservation* 111:149–59.

Ernest, H. B., W. M. Boyce, V. C. Bleich, B. May, S. J. Stiver, and S. G. Torres. 2003. Genetic structure of mountain lion (*Puma concolor*) populations in California. *Conservation Genetics* 4:353–366.

Estes, J., K. R. Crooks, and R. Holt. 2001. Ecological role of predators. Pages 857–878 *in* S. Levin, ed. *Encyclopedia of Biodiversity.* Academic Press, San Diego, CA.

Evans, P. G. H., D. W. Macdonald, and C. L. Cheeseman. 1989. Social structure of the Eurasian badger (*Meles meles*): Genetic evidence. *Journal of Zoology* 218:587–595.

Ewer, R. F. 1973. *The Carnivores.* Cornell University Press, Ithaca, NY.

Facure, K. G., and E. L. A. Monteiro-Filho. 1996. Feeding habits of the crab-eating fox, *Cerdocyon thous* (Carnivora, Canidae), in a suburban area of southeastern Brazil. *Mammalia* 60:147–149.

Faeth, S. H., P. S. Warren, E. Shochat, and W. A. Marussich. 2005. Trophic dynamics in urban communities. *BioScience* 55:399–407.

Farias, V. 2000. Gray fox distribution in Southern California: Detecting the effects of intraguild predation. Master's thesis, University of Massachusetts, Amherst.

Farias, V., T. K. Fuller, R. K. Wayne, and R. M. Sauvajot. 2005. Survival and cause-specific mortality of grey foxes (*Urocyon cinereoargenteus*) in Southern California. *Journal of Zoology* 266:249–254.

Fecske, D. M. 2006. Mountain Lions in North Dakota. *Wild Cats News* (online).

Fedriani, J. M., T. K. Fuller, and R. M. Sauvajot. 2001. Does the availability of anthropogenic food enhance densities of omnivorous animals? An example with coyotes in Southern California. *Ecography* 24:325–331.

Fedriani, J. M., T. K. Fuller, R. M. Sauvajot, and E. C. York. 2000. Competition and intraguild predation among three sympatric carnivores. *Oecologia* 125:258–270.

Ferguson, S. H. 2000. Influence of sea ice dynamics on habitat selection by polar bears. *Ecology* 81:761–772.

Fernández-Juricic, E., A. Sallent, R. Sanz, and I. Rodríguez-Prieto. 2003. Testing the risk-disturbance hypothesis in a fragmented landscape: Nonlinear responses of house sparrows to humans. *Condor* 105:316–326.

Fèvre, E. M., R. W. Kaboyo, V. Persson, M. Edelsten, P. G. Coleman, and S. Cleaveland. 2005. The epidemiology of animal bite injuries in Uganda and projections of the burden of rabies. *Tropical Medicine and International Health* 10:790–798.

Fielding, W. J., and S. J. Plumridge. 2005. Characteristics of owned dogs on the island of New Providence, the Bahamas. *Journal of Applied Animal Welfare Science* 8:245–260.

Filan, S. L., and R. H. Llewellyn-Jones. 2006. Animal-assisted therapy for dementia: A review of the literature. *International Pyschogeriatrics* 18:597–611.

Filho, F. A., P. P. Chieffi, C. R. S. Correa, E. D. Camargo, E. P. R. da Silveira, and J. J. B. Aranha. 2003. Human toxocariasis: Incidence among residents in the outskirts of Campinas, State of São Paulo, Brazil. *Revista do Instituto de Medicina Tropical de São Paulo* 45:293–294.

Fiore, C. A., and K. B. Sullivan. n.d. Domestic cat (*Felis catus*) predation of birds in an urban environment. www .geocities.com/the_srco/Article.html?200713. Accessed July 12, 2007.

Fischer, C., L. A. Reperant, J. M. Weber, D. Hegglin, and P. Deplazes. 2005. *Echinococcus multilocularis* infections of rural, residential, and urban foxes (*Vulpes vulpes*) in the canton of Geneva, Switzerland. *Parasite* 12:339–346.

Fishbein, M., and I. Ajzen. 1975. *Belief, Attitude, Intention, and Behavior*. Addison-Wesley, Reading, MA.

Fisher, P. M., and C. A. Marks, eds. 1996. *Humaneness and Vertebrate Pest Control*. Report Series No. 2. Victorian Institute of Animal Science, Victoria, Australia.

Fitter, R. S. R. 1945. *London's Natural History*. Collins, London.

Fitzgerald, B. M. 1990. Is cat control needed to protect urban wildlife? *Environmental Conservation* 17:168–169.

Fitzgerald, B. M., and D. C. Turner. 2000. Hunting behaviour of domestic cats and their impact on prey populations. Pages 151–175 *in* D. C. Turner and P. Bateson, eds. *The Domestic Cat: The Biology of Its Behaviour*. 2nd ed. Cambridge University Press, Cambridge.

Fitzhugh, E. L., M. W. Kenyon, and K. Etling. 2003. Lessening the impact of a cougar attack on a human. Pages 89–103 *in* S. A. Becker, ed. *Proceedings of the Seventh Mountain Lion Workshop*. Wyoming Game and Fish Department, Lander.

Flamand, A., P. Coulon, F. Lafay, A. Kappeler, M. Artois, M. Aubert, J. Blancou, and A. I. Wandeler. 1992. Eradication of rabies in Europe. *Nature* 360:115–116.

Flux, J. E. C. 2007. Seventeen years of predation by one suburban cat in New Zealand. *New Zealand Journal of Ecology* 34:289–296.

Flynn, J. J. 1996. Carnivoran phylogeny and rates of evolution: Morphological, taxic, and molecular. Pages 542–581 *in* J. L. Gittleman, ed. *Carnivore Behavior, Ecology, and Evolution*. Vol. 2. Cornell University Press, Ithaca, NY.

Fogle, B. 2000. *The New Encyclopedia of the Dog*. Dorling Kindersley, London.

Fogle, B. 2001. *The New Encyclopedia of the Cat*. Dorling Kindersley, London.

Foley, P., J. E. Foley, J. K. Levy, and T. Paik. 2005. Analysis of the impact of trap-neuter-return programs on populations of feral cats. *Journal of the American Veterinary Medical Association* 227:1775–1781.

Font, E. 1987. Spacing and social organization: Urban stray dogs revisited. *Applied Animal Behaviour Science* 17:319–328.

Forman, R. T. T. 2008. The urban region: Natural systems in our place, our nourishment, our home range, our future. *Landscape Ecology* 23:251–253.

Forman, R. T. T., D. Sperling, J. A. Bissonette, A. P. Clevenger, C. D. Cutshall, V. H. Dale, L. Fahrig, R. France, C. R. Goldman, K. Heanue, J. A. Jones, F. J. Swanson, T. Turrentine, and T. C. Winter. 2003. *Road Ecology*. Island Press, Washington, D.C.

Foster, M. L., and S. R. Humphrey. 1995. Use of highway underpasses by Florida panthers and other wildlife. *Wildlife Society Bulletin* 23:95–100.

Fournier, A. K., and E. S. Geller. 2004. Behavior analysis of companion-animal overpopulation: A conceptualization of the problem and suggestions for intervention. *Behavior and Social Issues* 13:51–68.

Fowler, P. A., and P. A. Racey. 1988. Overwintering strategies of the badger, *Meles meles*, at 57°N. *Journal of Zoology* 214:635–651.

Fox, C. H., and C. M. Papouchis. 2005. *Coyotes in Our Midst: Coexisting with an Adaptable and Resilient Carnivore*. Animal Protection Institute, Sacramento, CA.

Fox, J. S. 1983. Relationships of diseases and parasites to the distribution and abundance of bobcats in New York. Ph.D. diss., State University of New York, Syracuse.

Fox, M. W., A. M. Beck, and E. Blackman. 1975. Behavior and ecology of a small group of urban dogs (*Canis familiaris*). *Applied Animal Ethology* 1:119–137.

Fretwell, S. D. 1987. Food chain dynamics: The central theory of ecology? *Oikos* 50:291–301.

Fritts, S. H., and J. A. Sealander. 1978. Reproductive biology and population characteristics of bobcats (*Lynx rufus*) in Arkansas. *Journal of Mammalogy* 59:347–353.

Fritzell, E. K. 1978. Aspects of raccoon (*Procyon lotor*) social organization. *Canadian Journal of Zoology* 56:260–271.

Fromont, E., M. Artois, and D. Pontier. 1996. Cat population structure and circulation of feline viruses. *Acta Oecologica* 17:609–620.

Fromont, E., F. Courchamp, M. Artois, and D. Pontier. 1997. Infection strategies of retroviruses and social grouping of domestic cats. *Canadian Journal of Zoology* 75:1994–2002.

Fromont, E., D. Pontier, and M. Langlais. 1998. Dynamics of a feline retrovirus (FeLV) in host populations with variable spatial structure. *Proceedings of the Royal Society of London B* 265:1097–1104.

Frost, N. 2005. San Joaquin kit fox home range, habitat use, and movements in urban Bakersfield. Master's thesis, Humboldt State University, Arcata, CA.

Fuller, R. A., K. N. Irvine, P. Devine-Wright, P. H. Warren, and K. J. Gaston. 2007. Psychological benefits of greenspace increase with biodiversity. *Biology Letters* 3:390–394.

Fuller, R. A., P. H. Warren, P. R. Armsworth, O. Barbosa, and K. J. Gaston. 2008. Garden bird feeding predicts the structure of urban avian assemblages. *Diversity and Distributions* 14:131–137.

Fuller, T. K. 1989a. Denning behavior of wolves in north-central Minnesota. *American Midland Naturalist* 121:184–188.

Fuller, T. K. 1989b. Population dynamics of wolves in north-central Minnesota. *Wildlife Monographs* 105:1–41.

Fuller, T. K. 2004. *Wolves of the World*. Voyageur Press, Stillwater, MN.

Fuller, T. K., W. E. Berg, and D. W. Kuehn. 1985. Bobcat home range size and daytime cover-type use in north-central Minnesota. *Journal of Mammalogy* 66:568–571.

Fuller, T. K., A. R. Bikenvicius, and P. W. Kat. 1990. Movement and behavior of large-spotted genets (*Genetta maculata*) near Elmenteita, Kenya (Mammalia Viverridae). *Tropical Zoology* 3:13–19.

Fuller, T. K., A. R. Biknevicius, P. W. Kat, B. van Valkenburgh, and R. K. Wayne. 1989. The ecology of three sympatric jackal species in the Rift Valley of Kenya. *African Journal of Ecology* 27:313–323.

Fuller, T. K., and L. B. Keith. 1981. Non-overlapping ranges of coyotes and wolves in northeastern Alberta. *Journal of Mammalogy* 62:403–405.

Fuller, T. K., and P. R. Sievert. 2001. Carnivore demography and the consequences of changes in prey availability. Pages 163–178 *in* J. L. Gittleman, S. M. Funk, D. Macdonald, and R. K. Wayne, eds. *Carnivore Conservation*. Cambridge University Press, Cambridge.

Fulton, D. C., M. J. Manfredo, and J. Lipscomb. 1996. Wildlife value orientations: A conceptual and measurement approach. *Human Dimensions of Wildlife* 1:24–47.

Funk, S. M., C. V. Fiorello, S. Cleaveland, and M. E. Gompper. 2001. The role of disease in carnivore ecology and conservation. Pages 443–466 *in* J. L. Gittleman, S. M. Funk, D. Macdonald, and R. K. Wayne, eds. *Carnivore Conservation*. Cambridge University Press, Cambridge.

Gade, D. W. 2006. Hyenas and humans in the horn of Africa. *Geographical Review* 96:609–632.

Gage, K. L., D. T. Dennis, K. A. Orloski, P. Ettestad, T. L. Brown, P. J. Reynolds, W. J. Pape, C. L. Fritz, L. G. Carter, and J. D. Stein. 2000. Cases of cat-associated human plague in the Western U.S., 1977–1998. *Clinical Infectious Diseases* 30:893–900.

Galbreath, R., and D. Brown. 2004. The tale of the lighthouse keeper's cat: Discovery and extinction of the Stephens Island wren (*Traversia lyalli*). *Notornis* 51:193–200.

Gander, F. F. 1986. Raccoons: Natural history in a back yard. *Animal Kingdom* June:84–89.

Garrott, R. A., and L. E. Eberhardt. 1987. Arctic fox. Pages 395–406 *in* M. Nowak, J. A. Baker, M. E. Obbard, and B. Malloch, eds. *Wild Furbearer Management and Conservation in North America*. Ontario Trappers Association and Ontario Ministry of Natural Resources, Toronto, Ontario.

Garshelis, D. L., M. L. Gibeau, and S. Herrero. 2005. Grizzly bear demographics in and around Banff National Park and Kananaskis Country, Alberta. *Journal of Wildlife Management* 69:277–297.

Gaston, K. J., P. H. Warren, K. Thompson, and R. M. Smith. 2005. Urban domestic gardens (IV): The extent of the resource and its associated features. *Biodiversity and Conservation* 14:3327–3349.

Gates, B., J. Hadidian, and L. Simon. 2006. "Nuisance" wildlife control trapping: Another perspective. *Proceedings of the Vertebrate Pest Conference* 22:505–509.

Gaubert, P., F. Jiguet, P. Bayle, and F. M. Angelici. 2008. Has the common genet (*Genetta genetta*) spread into south-eastern France and Italy? *Italian Journal of Zoology* 75:43–57.

Geffen, E., A. Mercure, D. J. Girman, D. W. Macdonald, and R. K. Wayne. 1992. Phylogenetic relationships of the fox-like canids: Mitochondrial DNA restriction fragment site and cytochrome b sequence analyses. *Journal of Zoology (London)* 228:27–39.

Gehrt, S. D. 2003. Raccoon and allies. Pages 611–634 *in* G. A. Feldhamer, B. C. Thompson, and J. A. Chapman, eds. *Wild Mammals of North America*. 2nd ed. Johns Hopkins University Press, Baltimore.

Gehrt, S. D. 2004a. *Winter Denning Ecology, Juvenile Dispersal, and Disease of Striped Skunks*. Max McGraw Wildlife Foundation Project Report No. 01-R02F. Columbus, OH. 45 pp.

Gehrt, S. D. 2004b. Ecology and management of striped skunks, raccoons, and coyotes in urban landscapes. Pages 81–104 *in* N. Fascione, A. Delach, and M. Smith, eds. *Predators and People: From Conflict to Conservation*. Island Press, Washington, D.C.

Gehrt, S. D. 2005. Seasonal survival and cause-specific mortality of urban and rural striped skunks in the absence of rabies. *Journal of Mammalogy* 86:1164–1170.

Gehrt, S. D. 2006. *Urban Coyote Ecology and Management: The Cook County, Illinois Coyote Project*. Bulletin 929. Ohio State University Extension, Columbus.

Gehrt, S. D. 2007. Biology of coyotes in urban landscapes. Pages 303–311 *in* D. L. Nolte, W. M. Arjo, and D. H. Stalman, eds. *Proceedings of the 12th Wildlife Damage Management Conference*. Corpus Christi, TX.

Gehrt, S. D., C. Anchor, and L. A. White. 2009. Home range and landscape use of coyotes in a major metropolitan landscape: Coexistence or conflict? *Journal of Mammalogy* 90:1045–1057.

Gehrt, S. D., and J. E. Chelsvig. 2003. Bat activity and distribution patterns in an urban landscape. *Ecological Applications* 13:939–950.

Gehrt, S. D., and J. E. Chelsvig. 2004. Bat activity in an urban landscape: Species-specific patterns along an urban gradient. *Ecological Applications* 14:625–635.

Gehrt, S. D., and W. R. Clark. 2003. Raccoons, coyotes, and reflections on the mesopredator release hypothesis. *Wildlife Society Bulletin* 31:836–842.

Gehrt, S. D., and E. K. Fritzell. 1998a. Duration of familial bonds and dispersal patterns for raccoons in south Texas. *Journal of Mammalogy* 79:859–872.

Gehrt, S. D., and E. K. Fritzell. 1998b. Resource distribution, female home range dispersion and male spatial interactions: Group structure in a solitary carnivore. *Animal Behaviour* 55:1211–1227.

Gehrt, S. D., and E. K. Fritzell. 1999. Survivorship of a non-harvested raccoon population in south Texas. *Journal of Wildlife Management* 63:889–894.

Gehrt, S. D., G. F. Hubert, Jr., and J. A. Ellis. 2002. Long-term population trends of raccoons in Illinois. *Wildlife Society Bulletin* 30:457–463.

Gehrt, S. D., G. F. Hubert, Jr., and J. A. Ellis. 2006. Extrinsic effects on long-term population trends of Virginia opossums and striped skunks at a large spatial scale. *American Midland Naturalist* 155:168–180.

Gehrt, S. D., and S. Prange. 2007. Interference competition between coyotes and raccoons: A test of the mesopredator release hypothesis. *Behavioral Ecology* 18:204–214.

Geist, V. 2008. The danger of wolves. *The Wildlife Professional* 2:34–35.

Genchi, C., L. Rinaldi, C. Cascone, M. Mortarino, and G. Cringoli. 2005. Is heartworm disease really spreading in Europe? *Veterinary Parasitology* 133:137–148.

Gentry, C. 1983. *When Dogs Run Wild: The Sociology of Feral Dogs and Wildlife*. McFarland, Jefferson, NC.

George, S. L., and K. R. Crooks. 2006. Recreation and large mammal activity in an urban nature reserve. *Biological Conservation* 133:107–117.

George, W. G. 1974. Domestic cats as predators and factors in winter shortages of raptor prey. *Wilson Bulletin* 86:384–396.

Gese, E. M. 2001. Territorial defense by coyotes (*Canis latrans*) in Yellowstone National Park, Wyoming: Who, how, where, when, and why. *Canadian Journal of Zoology* 79:980–987.

Gese, E. M., and S. Grother. 1995. Analysis of coyote predation on deer and elk during winter in Yellowstone National Park, Wyoming. *American Midland Naturalist* 133:36–43.

Gese, E. M., and R. L. Ruff. 1997. Scent-marking by coyotes, *Canis latrans*: The influence of social and ecological factors. *Animal Behaviour* 54:1155–1166.

Gese, E. M., and R. L. Ruff. 1998. Howling by coyotes (*Canis latrans*): Variation among social classes, seasons, and pack sizes. *Canadian Journal of Zoology* 76:1037–1043.

Gibeau, M. L. 1998. Use of urban habitats by coyotes in the vicinity of Banff Alberta. *Urban Ecosystems* 2:129–139.

Gier, H. T. 1968. *Coyotes in Kansas*. Kansas Agricultural Experiment Station, Kansas State University, Manhattan.

Gilbert, B. S., and S. Boutin. 1991. Effect of moonlight on winter activity of snowshoe hares. *Arctic Alpine Research* 23:61–65.

Gilbert, F. F. 1982. Public attitudes toward urban wildlife: A pilot study in Guelph, Ontario. *Wildlife Society Bulletin* 10:245–253.

Gilbert, O. L. 1989. *The Ecology of Urban Habitats*. Chapman and Hall, New York.

Gill, D. 1970. The coyote and the sequential occupants of the Los Angeles Basin. *American Anthropologist* 72:821–826.

Gillies, C., and M. Clout. 2003. The prey of domestic cats (*Felis catus*) in two suburbs of Auckland City, New Zealand. *Journal of Zoology* 259:309–315.

Ginger, S. M., E. C. Hellgren, M. A. Kasparian, L. P. Levesque, D. M. Engle, and D. M. Leslie, Jr. 2003. Niche shift by Virginia opossum following reduction of a putative competitor, the raccoon. *Journal of Mammalogy* 84:1279–1291.

Gittleman, J. L. 1985. Carnivore body size: Ecological and taxonomic correlates. *Oecologia* 67:540–554.

Gittleman, J. L. 1986. Carnivore life history patterns: Allometric, phylogenetic, and ecological associations. *American Naturalist* 127:744–771.

Gittleman, J. L., ed. 1989a. *Carnivore Behavior, Ecology, and Evolution*. Cornell University Press, Ithaca, NY.

Gittleman, J. L. 1989b. Carnivore group living: Comparative trends. Pages 183–207 *in* J. L. Gittleman, ed. 1989. *Carnivore Behavior, Ecology, and Evolution*. Cornell University Press, Ithaca, NY.

Gittleman, J. L., S. M. Funk, D. Macdonald, and R. K. Wayne. 2001. *Carnivore Conservation*. Conservation Biology Series 5. Cambridge University Press, Cambridge.

Gittleman, J. L., and P. H. Harvey. 1982. Carnivore home range size, metabolic needs and ecology. *Behavioral Ecology and Sociobiology* 10:57–63.

Glaeser, E. I. 1998. Are cities dying? *Journal of Economic Perspectives* 12:139–161.

Glista, D. J., T. L. DeVault, and J. A. DeWoody. 2008. Vertebrate road mortality predominantly impacts amphibians. *Herpetological Conservation and Biology* 3:77–87.

Gloor, S. 2002. The rise of urban foxes (*Vulpes vulpes*) in Switzerland and ecological and parasitological aspects of a fox population in the recently colonised city of Zurich. Ph.D. diss., University of Zürich, Switzerland.

Gloor, S., F. Bontadina, D. Hegglin, P. Deplazes, and U. Breitenmoser. 2001. The rise of urban fox populations in Switzerland. *Mammalian Biology* 66:155–164.

Godin, A. J. 1982. Striped and hooded skunks (*Mephitis mephitis* and allies). Pages 674–687 *in* J. A. Chapman and G. A. Feldhamer, eds. *Wild Mammals of North America: Biology, Management, and Economics*. Johns Hopkins University Press, Baltimore.

Godman, J. D. 1826. Raccoon. Pages 161–176 *in* *American Natural History*. H. C. Carey and I. Lea, Philadelphia.

Goedeke, T. L. 2005. Devils, angels, or animals: The social

construction of otters in conflict over management. Pages 25–50 *in* A. Herda-Rapp and T. L. Goedeke, eds. *Mad about Wildlife: Looking at Social Conflict over Wildlife.* Brill, Boston.

Gompper, M. E. 1997. Population ecology of the white-nosed coati (*Nasua narica*) on Barro Colorado Island, Panama. *Journal of Zoology* 241:441–455.

Gompper, M. E. 2002. Top carnivores in the suburbs? Ecological and conservation issues raised by colonization of north-eastern North America by coyotes. *BioScience* 52:185–190.

Gore, M. L., B. A. Knuth, P. D. Curtis, and J. E. Shanahan. 2006. Stakeholder perceptions of risk associated with human-black bear conflicts in New York's Adirondack Park campgrounds: Implications for theory and practice. *Wildlife Society Bulletin* 34:36–43.

Gore, M. L., B. A. Knuth, P. D. Curtis, and J. E. Shanahan. 2007. Campground manager and user perceptions of risk associated with negative human-black bear interactions. *Human Dimensions of Wildlife* 12:31–43.

Gore, M. L., B. A. Knuth, C. W. Scherer, and P. D. Curtis. 2008. Evaluating a conservation investment designed to reduce human-wildlife conflict. *Conservation Letters* 1:136–145.

Gore, M. L., W. F. Siemer, J. E. Shanahan, D. Schuefele, and D. J. Decker. 2005. Effects on risk perception of media coverage of a black bear-related human fatality. *Wildlife Society Bulletin* 33:507–516.

Gorenzel, W. P., and T. P. Salmon. 1995. Characteristics of American crow urban roosts in California. *Journal of Wildlife Management* 59:638–645.

Gorman, M. L. 1975. The diet of feral *Herpestes auropunctatus* (Carnivora: Viverridae) in the Fijian Islands. *Journal of Zoology* 175:273–278.

Gorman, M. L., and B. J. Trowbridge. 1989. The role of odor in the social lives of carnivores. Pages 57–88 *in* J. L. Gittleman, ed. *Carnivore Behavior, Ecology, and Evolution.* Cornell University Press, Ithaca, NY.

Gosselink, T. E. 1999. Seasonal variations in habitat use and home range of sympatric coyotes and red foxes in agricultural and urban areas of east-central Illinois. Master's thesis, University of Illinois at Urbana–Champaign.

Gosselink, T. E. 2002. Social organization, natal dispersal, survival, and cause-specific mortality of red foxes in agricultural and urban areas of east-central Illinois. Ph.D. diss., University of Illinois at Urbana–Champaign.

Gosselink, T. E., T. R. van Deelen, R. E. Warner, and M. G. Joselyn. 2003. Temporal habitat partitioning and spatial use of coyotes and red foxes in east-central Illinois. *Journal of Wildlife Management* 67:90–103.

Gosselink, T. E., T. R. van Deelen, R. E. Warner, and P. C. Mankin. 2007. Survival and cause-specific mortality of red foxes in agricultural and urban areas of Illinois. *Journal of Wildlife Management* 71:1862–1873.

Goszczyński, J. 2002. Home ranges in red fox: Territoriality diminishes with increasing area. *Acta Theriologica* 47, Supplement 1:103–114.

Grassman, L. I., Jr., M. E. Tewes, and N. J. Silvy. 2005. Ranging, habitat use and activity patterns of binturong *Arctictis binturong* and yellow-throated marten *Martes flavigula* in north-central Thailand. *Wildlife Biology* 11:49–57.

Grayson, J., M. Calver, and A. Lymbery. 2007. Species richness and community composition of passerine birds in suburban Perth: Is predation by pet cats the most important factor? Pages 195–207 *in* D. Lunney, P. Eby, P. Hutchings, and S. Burgin, eds. *Pest or Guest: The Zoology of Overabundance.* Royal Zoological Society of New South Wales, Mosman, New South Wales, Australia.

Greenberg, J. 2002. *A Natural History of the Chicago Region.* University of Chicago Press, Chicago.

Greenwood, R. J. 1979. Relating residue in raccoon feces to food consumed. *American Midland Naturalist* 102:191–193.

Greenwood, R. J., W. E. Newton, G. Pearson, and G. J. Schamber. 1997. Population and movement characteristics of radio collared striped skunks in North Dakota during an epizootic of rabies. *Journal of Wildlife Diseases* 33:226–241.

Greenwood, R. J., and A. B. Sargeant. 1998. Age related reproduction in striped skunks (*Mephitis mephitis*) in the upper Midwest. *Journal of Mammalogy* 75:657–662.

Greenwood, R. J., A. B. Sargeant, J. L. Piehl, D. Abuhl, and B. A. Hansen. 1999. Foods and foraging of prairie striped skunks during the avian nesting season. *Wildlife Society Bulletin* 27:823–832.

Gremillion-Smith C., and A. Woolf. 1988. Epizootiology of skunk rabies in North America. *Journal of Wildlife Diseases* 24:620–626.

Griffin, J. C. 2001. Bobcat ecology on developed and less-developed portions of Kiawah Island, South Carolina. Master's thesis, University of Georgia, Athens.

Griffiths, H. I., and D. H. Thomas. 1993. The status of the badger *Meles meles* (L., 1758) (Carnivora, Mustelidae) in Europe. *Mammal Review* 23:17–58.

Grimm, N. B., J. M. Grove, S. T. A. Pickett, and C. L. Redman. 2000. Integrated approaches to long-term studies of urban ecological systems. *BioScience* 50:571–584.

Grinder, M. I., and P. R. Krausman. 2001a. Morbidity-mortality factors and survival of an urban coyote population in Arizona. *Journal of Wildlife Diseases* 37:312–317.

Grinder, M. I., and P. R. Krausman. 2001b. Home range, habitat use, and nocturnal activity of coyotes in an urban environment. *Journal of Wildlife Management* 65:887–898.

Grinnell, J., D. S. Dixon, and J. M. Linsdale. 1937. *Fur-Bearing Mammals of California.* Vol. 2. University of California Press, Berkeley.

Grubbs, S. E., and P. R. Krausman. 2009. Use of urban landscape by coyotes. *Southwestern Naturalist* 54:1–12.

Gunther, I., and J. Terkel. 2002. Regulation of free-roaming cat (*Felis silvestris catus*) populations: A survey of the literature and its application to Israel. *Animal Welfare* 11:171–188.

Gustavson, C. R., J. R. Jowsey, and D. N. Milligan. 1982. A 3-year evaluation of taste aversion coyote control in Saskatchewan. *Journal of Range Management* 35:57–59.

Gustavson, C. R., D. J. Kelly, M. Sweeny, and J. Garcia. 1976. Prey lithium aversions. I. Coyotes and wolves. *Behavioral Biology* 17:61–72.

Haas, C. D. 2001. Responses of mammals to roadway underpasses across an urban wildlife corridor, the Puente-Chino Hills, California. Page 580 *in* C. Irwin, P. Garrett, and K. McDermott, eds. *Proceedings of the 2001 International Conference on Ecology and Transportation.* North Carolina State University, Raleigh.

Haas, S. K., V. Hayssen, and P. R. Krausman. 2005. *Panthera leo. Mammalian Species* 762:1–11.

Hadidian, J., M. R. Childs, R. H. Schmidt, L. J. Simon, and A. Church. 2001. Nuisance wildlife control practices, policies, and procedures in the United States. Pages 165–168 *in* R. Field, R. J. Warren, H. Okarma, and P. R. Sievert, eds. *Wildlife, Land, and People: Priorities for the 21st Century.* The Wildlife Society, Valko, Hungary.

Hadidian, J., C. H. Fox, and W. S. Lynn. 2006. The ethics of wildlife control in humanized landscapes. *Proceedings of the 22nd Vertebrate Pest Conference* 22:500–504.

Hadidian, J., D. A. Manski, and S. P. D. Riley. 1991. Daytime resting site selection in an urban raccoon population. Pages 39–45 *in* L.W. Adams and D. L. Leedy, eds. *Wildlife Conservation in Metropolitan Environments.* National Institute for Urban Wildlife, Columbia, MD.

Hadidian, J., and S. P. D. Riley. 1996. Home range and activity patterns of a three-legged adult female raccoon. Pages 138–146 *in* M. D. Reynolds, ed. *Uncommon Care for Common Animals.* International Wildlife Rehabilitation Council, Suisun, CA.

Hairston, N. G., F. E. Smith, and L. B. Slobodkin. 1960. Community structure, population control, and competition. *American Naturalist* 44:421–425.

Hall, E. R. 1981. *The Mammals of North America.* 2nd ed. John Wiley and Sons, New York.

Hall, E. R., and K. R. Kelson. 1959. *The Mammals of North America.* 1st ed. Ronald Press, New York.

Hall, H. T., and J. D. Newsom. 1976. Summer home ranges and movements of bobcats in bottomland hardwoods of southern Louisiana. *Proceedings of the Annual Conference of the Southeast Association of Fish and Wildlife Agencies* 30:427–436.

Hamilton, D., D. J. Witter, and T. L. Goedeke. 2000. Balancing public opinion in managing river otters in Missouri. *Transactions of the North American Wildlife and Natural Resources Conference* 65:292–299.

Hamilton, W. J., Jr. 1936. Seasonal food of skunks in New York. *Journal of Mammalogy* 17:240–246.

Hamilton, W. J., Jr. 1937. Winter activity of the skunk. *Ecology* 18:326–327.

Hancock, K., L. Thiele, A. Zajac, F. Evlinger, and D. Lindsay. 2005. Antibodies to *Toxoplasma gondii* in raccoons (*Procyon lotor*) from an urban area of Virginia. *Journal of Parasitology* 90:694.

Hanlon, C., M. Niezgoda, P. Morrill, and C. Rupprecht. 2002. Oral efficacy of an attenuated rabies vaccine in skunks and raccoons. *Journal of Wildlife Diseases* 38:420–427.

Hansen, L., N. Mathews, R. Hansen, B. Vander Lee, and R. Lutz. 2003. Genetic structure in striped skunks (*Mephitis mephitis*) on the Southern High Plains of Texas. *Western North American Naturalist* 63:80–87.

Hansen, L., N. Mathews, B. Vander Lee, and R. Lutz. 2004. Population characteristics, survival rates, and causes of mortality in striped skunks (*Mephitis mephitis*) on the southern High Plains, Texas. *Southwestern Naturalist* 49:54–60.

Harris, S. 1977. Distribution, habitat utilization, and age structure of a suburban fox (*Vulpes vulpes*) population. *Mammal Review* 7:25–39.

Harris, S. 1979. Age-related fertility and productivity in red foxes, *Vulpes vulpes*, in suburban London. *Journal of Zoology* 187:195–199.

Harris, S. 1981a. The food of suburban foxes (*Vulpes vulpes*), with special reference to London. *Mammal Review* 11:151–168.

Harris, S. 1981b. An estimation of the number of foxes (*Vulpes vulpes*) in the city of Bristol and some possible factors affecting their distribution. *Journal of Applied Ecology* 18:455–465.

Harris, S. 1982. Activity patterns and habitat utilization of badgers (*Meles meles*) in suburban Bristol: A radio-tracking study. *Symposia of the Zoological Society of London* 49:301–323.

Harris, S. 1984. Ecology of urban badgers *Meles meles*: Distribution in Britain and habitat selection, persecution, food and damage in the city of Bristol. *Biological Conservation* 28:349–375.

Harris, S. 1985. Humane control of foxes. Pages 63–74 *in* D. P. Britt, ed. *Humane Control of Land Mammals and Birds.* Universities Federation for Animal Welfare, Potters Bar, UK.

Harris, S., and P. Baker. 2001. *Urban Foxes.* Whittet Books, Stowmarket, Suffolk, UK.

Harris, S., and W. J. Cresswell. 1987. Dynamics of a suburban badger (*Meles meles*) population. *Symposia of the Zoological Society of London* 58:295–311.

Harris, S., and W. J. Cresswell. 1988. Bristol's badgers. *Proceedings of the Bristol Naturalists' Society* 48:17–30.

Harris, S., W. J. Cresswell, P. Reason, and P. Cresswell. 1992. An integrated approach to monitoring badger (*Meles meles*) population changes in Britain. Pages 945–953 *in* D. R. McCullough and R. H. Barrett, eds.

Wildlife 2001: Populations. Elsevier Applied Science, London.

Harris, S., D. Jefferies, C. Cheeseman, and C. Booty. 1994. *Problems with Badgers?* 3rd ed. Royal Society for the Prevention of Cruelty to Animals, Horsham, UK.

Harris, S., and H. G. Lloyd. 1991. Fox *Vulpes vulpes*. Pages 351–367 in G. C. Corbett and S. Harris, eds. *Handbook of British Mammals*. 3rd ed. Blackwell Scientific, Oxford.

Harris, S., P. Morris, S. Wray, and D. Yalden. 1995. *A Review of British Mammals: Population Estimates and Conservation Status of British Mammals Other Than Cetaceans*. Joint Nature Conservation Committee, Peterborough, UK.

Harris, S., and J. M. V. Rayner. 1986a. Urban fox (*Vulpes vulpes*) population estimates and habitat requirements in several British cities. *Journal of Animal Ecology* 55:575–591.

Harris, S., and J. M. V. Rayner. 1986b. Models for predicting urban fox (*Vulpes vulpes*) numbers in British cities and their application for rabies control. *Journal of Animal Ecology* 55:593–603.

Harris, S., and J. M. V. Rayner. 1986c. A discriminant analysis of the current distribution of urban foxes (*Vulpes vulpes*) in Britain. *Journal of Animal Ecology* 55:605–611.

Harris, S., and P. Skinner. 2002. The badger sett at Lustrell's Crescent/Winton Avenue, Saltdean, Sussex. Unpublished report to Elliot Morley MP.

Harris, S., and G. C. Smith. 1987a. Demography of two urban fox (*Vulpes vulpes*) populations. *Journal of Applied Ecology* 24:75–86.

Harris, S., and G. C. Smith. 1987b. The use of sociological data to explain the distribution and numbers of urban foxes (*Vulpes vulpes*) in England and Wales. *Symposia of the Zoological Society of London* 58:313–328.

Harris, S., G. C. Smith, and W. J. Trewhella. 1988. Rabies in urban foxes (*Vulpes vulpes*): Developing a Control Strategy. *State Veterinary Journal* 42:149–161.

Harris, S., and W. J. Trewhella. 1988. An analysis of some of the factors affecting dispersal in an urban fox (*Vulpes vulpes*) population. *Journal of Applied Ecology* 25:409–422.

Harris, S., and P. C. L. White. 1992. Is reduced affiliative rather than increased agonistic behaviour associated with dispersal in red foxes? *Animal Behaviour* 44:1085–1089.

Harris, S., and T. Woollard. 1988. Bristol's foxes. *Proceedings of the Bristol Naturalists' Society* 48:3–15.

Harrison, D. J., J. A. Bissonette, and J. A. Sherburne. 1989. Spatial relationships between coyotes and red foxes in Eastern Maine. *Journal of Wildlife Management* 53:181–185.

Harrison, R. L. 1997. A comparison of gray fox ecology between residential and undeveloped rural landscapes. *Journal of Wildlife Management* 61:112–122.

Harrison, R. L. 1998. Bobcats in residential areas: Distribution and homeowner attitudes. *Southwestern Naturalist* 43:469–75.

Harrison, S. 1990. Cougar predation on bighorn sheep in the Junction Wildlife Management Area, British Columbia. Master's thesis, University of British Columbia, Vancouver.

Harrison, S., and A. D. Taylor. 1997. Empirical evidence for metapopulations. Pages 27–42 in I. A. Hanski and M. E. Gilpin, eds. *Metapopulation Biology*. Academic Press, San Diego, CA.

Hartley, F. G. L., B. K. Follett, S. Harris, D. Hirst, and A. S. McNeilly. 1994. The endocrinology of gestation failure in foxes (*Vulpes vulpes*). *Journal of Reproduction and Fertility* 100:341–346.

Hartman, L. H., A. J. Gaston, and D. S. Eastman. 1997. Raccoon predation on ancient murrelets on East Limestone Island, British Columbia. *Journal of Wildlife Management* 61:377–388.

Haspel, C., and R. E. Calhoon. 1989. Home ranges of free-ranging cats (*Felis catus*) in Brooklyn, New York. *Canadian Journal of Zoology* 67:178–181.

Hass, C. C. 2002. *Reduction of Nuisance Skunks in an Urbanized Area*. Final Report to Arizona Game and Fish Department, Phoenix. 33 pp.

Hass, C. C. 2003. *Ecology of Hooded and Striped Skunks in Southeastern Arizona*. Final Report to Arizona Game and Fish Department, Phoenix. 52 pp.

Hass, C. C., and J. W. Dragoo. 2006. Rabies in hooded and striped skunks in Arizona. *Journal of Wildlife Diseases* 42:825–829.

Hatten, S. 2000. The effects of urbanization on raccoon population demographics, home range, and spatial distribution patterns. Ph.D. diss., University of Missouri, Columbia.

Haugen, O. L. 1954. Longevity of the raccoon in the wild. *Journal of Mammalogy* 35:439.

Hawkins, C. C., W. E. Grant, and M. T. Longnecker. 2004. Effect of house cats, being fed in parks, on California birds and rodents. Pages 164–170 in W. W. Shaw, L. K. Harris, and L. Vandruff, eds. *Proceedings of the 4th International Symposium on Urban Wildlife Conservation*. School of Natural Resources, College of Agriculture and Life Sciences, University of Arizona, Tucson.

Hawkins, C. E., and P. A. Racey. 2008. Food habits of an endangered carnivore, *Cryptoprocta ferox*, in the dry deciduous forests of western Madagascar. *Journal of Mammalogy* 89:64–74.

Haydon, D. T., D. A. Randall, L. Matthews, D. L. Knobel, L. A. Tallents, M. B. Gravenor, S. D. Williams, et al. 2006. Low-coverage vaccination strategies for the conservation of endangered species. *Nature* 443:692–695.

Hayes, C. J., E. S. Rubin, M. C. Jorgensen, R. A. Botta, and W. M. Boyce. 2000. Mountain lion predation of bighorn sheep in the peninsular ranges, California. *Journal of Wildlife Management* 64:954–959.

Hegglin, D., F. Bontadina, S. Contesse, S. Gloor, and P. Deplazes. 2007. Plasticity of predation behaviour as a putative driving force for parasite life-cycle dynamics: The case of urban foxes and *Echinococcus multilocularis* tapeworm. *Functional Ecology* 21:552–560.

Hegglin, D., P. I. Ward, and P. Deplazes. 2003. Antihelmintic baiting of foxes against urban contamination with *Echinococcus multilocularis*. *Emerging Infectious Diseases* 9:1266–1272.

Hemker, T. P., F. G. Lindzey, and B. B. Ackerman. 1984. Population characteristics and movement patterns of cougars in southern Utah. *Journal of Wildlife Management* 48:1275–1284.

Hemker, T. P., F. G. Lindzey, B. B. Ackerman, and A. J. Button. 1982. Survival of cougar cubs in a non-hunted population. Pages 327–332 *in* S. D. Miller, and D. D. Everett, eds. *Cats of the World: Biology, Conservation, and Management*. National Wildlife Federation, Washington, D.C.

Hendry, A. P., T. J. Farrugia, and M. T. Kinnison. 2008. Human influences on rates of phenotypic change in wild animal populations. *Molecular Ecology* 17:20–29.

Henke, S. E., and F. C. Bryant. 1999. Effects of coyote removal on the faunal community in western Texas. *Journal of Wildlife Management* 63:1066–1081.

Hennessy, C.A. 2007. Mating strategies and pack structure of coyotes in an urban landscape: A genetic investigation. Master's thesis, Ohio State University, Columbus.

Henry, J. D. 1980. The urine marking behavior and movement patterns of red foxes (*Vulpes vulpes*) during a breeding and post-breeding period. Pages 11–27 *in* D. Müller-Schwarze and R. M. Silverstein, eds. *Chemical Signals: Vertebrates and Aquatic Invertebrates*. Plenum Press, New York.

Henry, J. D. 1986. *Red Fox: The Catlike Canine*. Smithsonian Institution Press, Washington D.C.

Herr, J. 2008. Ecology and behaviour of urban stone martens (*Martes foina*) in Luxembourg. Ph.D. thesis, University of Sussex, UK.

Herr, J., and T. Roper. 2007. Behavioural ecology of urban stone martens (*Martes foina*) in Luxembourg. 2007 Easter Conference and AGM of the Mammal Society, April 13–15, 2007, Cirencester, UK.

Herrero, S. 1985. *Bear Attacks: Their Causes and Avoidance*. Nick Lyons Books / Winchester Press, Piscataway, NJ.

Herrero, S. 2005. During 2005 more people killed by bears in North America than in any previous year. *International Bear News* 14:34–35.

Herrero, S., and S. Fleck. 1990. Injury to people inflicted by black, grizzly, and polar bears: Recent trends and new insights. *International Conference on Bear Resources and Management* 8:25–32.

Herrero, S., T. Smith, T. D. DeBruyn, K. Gunther, and C. A. Matt. 2005. From the field: Brown bear habituation to people—safety, risks, and benefits. *Wildlife Society Bulletin* 33:362–73.

Heydon, M. J., and J. C. Reynolds. 2000. Demography of rural foxes (*Vulpes vulpes*) in relation to cull intensity in three contrasting regions of Britain. *Journal of Zoology* 251:265–276.

Hilton, H. 1978. Systematics and ecology of the eastern coyote. Pages 209–228 *in* M. Bekoff, ed. *Coyotes: Biology, Behavior, and Management*. Academic Press, New York.

Hoagland, D. B., and C. W. Kilpatrick. 1999. Genetic variation and differentiation among insular populations of the small Indian mongoose (*Herpestes javanicus*). *Journal of Mammalogy* 80:169–179.

Hofer, H., and M. L. East. 1993. The commuting system of Serengeti spotted hyaenas: How a predator copes with migratory prey. I. Social organization. *Animal Behaviour* 46:547–557.

Hofer, S., S. Gloor, U. Müller, A. Mathis, D. Hegglin, and P. Deplazes. 2000. High prevalence of *Echinococcus multilocularis* in urban red foxes (*Vulpes vulpes*) and voles (*Arvicola terrestris*) in the city of Zürich, Switzerland. *Parasitology* 120:135–142.

Hoffmann, C. O. 1979. Weights of suburban raccoons in southwestern Ohio. *Ohio Journal of Science* 79:139–142.

Hoffmann, C. O., and J. L. Gottschang. 1977. Numbers, distribution, and movements of a raccoon population in a suburban residential community. *Journal of Mammalogy* 58:623–636.

Hohmann, G., and B. Fruth. 2008. New records on prey capture and meat eating by bonobos at Lui Kotale, Salonga National Park, Democratic Republic of Congo. *Folia Primatologica* 79:103–110.

Hohmann, U., S. Voigt, and U. Andreas. 2001. Quo vadis raccoon? New visitors in our backyards—on the urbanization of an allochthone carnivore in Germany. *Naturschutz und Verhalten* 2:143–148.

Hohmann, U., S. Voigt, and U. Andreas. 2002. Raccoons take the offensive: A current assessment. *Neobiota* 1:191–192.

Holišová, V., and R. Obrtel. 1982. Scat analytical data on the diet of urban stone martens, *Martes foina* (Mustelidae, Mammalia). *Folia Zoologica* 31:21–30.

Holland, C. V., P. O'Lorcain, M. R. H. Taylor, and A. Kelly. 1995. Sero-epidemiology of toxocariasis in school children. *Parasitology* 110:535–545.

Holt, R. D. 1977. Predation, apparent competition, and the structure of prey communities. *Theoretical Population Biology* 12:197–229.

Hopkins, R. A., M. J. Kutilek, and G. L. Shreve. 1986. The density and home range characteristics of cougars in the Diablo Range of California. Pages 223–226 *in* S. D. Miller and D. D. Everett, eds. *Cats of the World: Biology, Conservation, and Management*. National Wildlife Federation, Washington, D.C.

Horne, J. S., E. O. Garton, S. M. Krone, and J. S. Lewis. 2007. Analyzing animal movements using Brownian bridges. *Ecology* 88: 2354–2363.

Hornocker, M. G. 1970. An analysis of mountain lion predation upon mule deer and elk in the Idaho primitive area. *Wildlife Monographs* 21:1–39.

Howell, R. G. 1982. The urban coyote problem in Los Angeles County. Pages 21–23 *in* R. E. Marsh, ed. *Proceedings of the Tenth Vertebrate Pest Conference*. University of California, Davis.

Howes, C. A. 2002. Red in tooth and claw. 2. Studies on the natural history of the domestic cat *Felis catus* Lin. in Yorkshire. *Naturalist* 127:101–130.

Hsu, Y., L. L. Severinghaus, and J. A. Serpell. 2003. Dog keeping in Taiwan: Its contribution to the problem of free-roaming dogs. *Journal of Applied Animal Welfare Science* 6:1–23.

Huck, M., J. Davison, and T. J. Roper. 2008a. Predicting European badger *Meles meles* sett distribution in urban environments. *Wildlife Biology* 14:188–198.

Huck, M., A. C. Frantz, D. A. Dawson, T. Burke, and T. J. Roper. 2008b. Low genetic variability, female-biased dispersal, and high movement rates in an urban population of Eurasian badgers *Meles meles*. *Journal of Animal Ecology* 77:905–915.

Huffaker, C. B. 1958. Experimental studies on predation: Dispersion factors and predator-prey oscillations. *Hilgardia* 27:343–83.

Hunt, R. M., Jr. 1996. Biogeography of the order Carnivora. Pages 485–541 *in* J. L. Gittleman, ed. *Carnivore Behavior, Ecology, and Evolution*. Vol. 2. Cornell University Press, Ithaca, NY.

Hunziker, M., C. W. Hoffmann, and S. Wild-Eck. 2001. Die Akzeptanz von Wolf, Luchs und "Stadtfuchs"—Ergebnisse einer gesamtschweizerisch-repräsentativen Umfrage. *Forest, Snow, and Landscape Research* 76:301–326.

Husseman, J. S., D. L. Murray, G. Power, C. Mack, C. R. Wenger, and H. Quigley. 2003. Assessing differential prey selection patterns between two sympatric large carnivores. *Oikos* 101:591–601.

Hwang, Y., and S. Larivière. 2001. *Mephitis macroura*. *Mammalian Species* 686:1–14.

Hwang, Y. T., S. Larivière, and F. Messier. 2007. Local and landscape level den selection of striped skunks on the Canadian prairies. *Canadian Journal of Zoology* 85:33–39.

Hwang, Y., G. Wobeser, S. Larivière, and F. Messier. 2002. *Streptococcus equisimilis* infection in striped skunks (*Mephitis mephitis*) in Saskatchewan. *Journal of Wildlife Diseases* 38:641–643.

Ikeda, T., M. Asano, Y. Matoba, and G. Abe. 2004. Present status of invasive alien raccoon and its impact in Japan. *Global Environmental Research* 8:125–131.

Iossa, G., C. D. Soulsbury, P. J. Baker, and S. Harris. 2008. Body mass, territory size, and life-history tactics in a socially monogamous canid, the red fox *Vulpes vulpes*. *Journal of Mammalogy* 89:1481–1490.

Iriarte, J. A., W. L. Franklin, W. E. Johnson, and K. H. Redford. 1990. Biogeographic variation of food habits and body size of the America puma. *Oecologia* 85:185–190.

Ishida, Y., T. Yahara, E. Kasuya, and A. Yamane. 2001. Female control of paternity during copulation: Inbreeding avoidance in feral cats. *Behaviour* 138:235–250.

Iuell, B., ed. 2003. *Wildlife and Traffic: A European Handbook for Identifying Conflicts and Designing Solutions*. European Cooperation in the Field of Scientific and Technical Research (COST) 341. Office for Official Publications of the European Communities, Luxembourg.

Ivanter, E. V., and N. A. Sedova. 2008. Ecological monitoring of urban groups of stray dogs: An example of the city of Petrozavodsk. *Russian Journal of Ecology* 39:105–110.

Izawa, M. 1984. Daily activities of the feral cat *Felis catus*. *Linnaean Journal of the Mammalogical Society of Japan* 9:219–228.

Izawa, M., T. Doi, and Y. Ono. 1982. Grouping patterns of feral cats living on a small island in Japan. *Japanese Journal of Ecology* 32:373–382.

Jackson, J. M. 1965. Structural characteristics of norms. Pages 301–309 *in* I. D. Steiner and M. F. Fishbein, eds. *Current Studies in Social Psychology*. Holt, Reinhart, and Winston, New York.

Jackson, P., and K. Nowell, eds. 1996. *Wild Cats: Status Survey and Conservation Action Plan*. IUCN/SSC Felid Specialist Group. Gland, Switzerland.

Jacobsen, J. E., K. R. Kazacos, and F. H. Montague. 1982. Prevalence of eggs of *Baylisascaris procyonis* (Nemotoda: Ascarodiea) in raccoon scats from an urban and rural community. *Journal of Wildlife Diseases* 18:461–464.

Jarvis, P. J. 1990. Urban cats as pests and pets. *Environmental Conservation* 17:169–171.

Jenkins, D. J., and N. A. Craig. 1992. The role of foxes *Vulpes vulpes* in the epidemiology of *Echinococcus granulosus* in urban environments. *Medical Journal of Australia* 157:754–756.

Jenkins, S. R., B. D. Perry, and W. G. Winkler. 1988. Ecology and epidemiology of raccoon rabies. *Reviews of Infectious Diseases* 10:620–625.

Jenkinson, S., and C. P. Wheater. 1998. The influence of public access and sett visibility on badger (*Meles meles*) sett disturbance and persistence. *Journal of Zoology* 246:478–482.

Jessup, D. A. 2004a. Believes feral cat welfare has dark side. *Journal of the American Veterinary Medical Association* 224:1070.

Jessup, D. A. 2004b. The welfare of feral cats and wildlife. *Journal of the American Veterinary Medical Association* 225:1377–1383.

Jessup, D. A. 2004c. Some common ground on feral cats may be emerging. . . . Speaking out against painful, unnecessary surgeries. *Journal of the American Veterinary Medical Association* 225:1662.

Jessup, D. A. 2004d. [Untitled]. *Journal of the American Veterinary Medical Association* 224:1752–1754.

Jessup, D. A. 2004e. [Untitled]. *Journal of the American Veterinary Medical Association* 225:199–200.

Jessup, D. A. 2005. [Untitled]. *Journal of the American Veterinary Medical Association* 226:30–31.

Jhala, Y. V., and P. D. Moehlman. 2004. Golden jackal *Canis aureus* Linnaeus, 1758. Pages 156–161 *in* C. Sillero-Zubiri, M. Hoffmann and D. W. Macdonald, eds.

Canids: Foxes, Wolves, Jackals, and Dogs. Status Survey and Conservation Action Plan. IUCN, Gland, Switzerland.

Jobin, B., and J. Picman. 1997. Factors affecting predation on artificial nests in marshes. *Journal of Wildlife Management* 61:792–800.

Johansen, B. S. 1993. Home-range size, movement patterns and biotope preference of European badgers (*Meles meles* L.) along an urban-rural gradient—effects of biotope fragmentation. Master's thesis, Department of Zoology, University of Trondheim, Norway.

Johnson, A. S. 1970. Biology of the raccoon (*Procyon lotor varius,* Nelson and Goldman) in Alabama. *Auburn University Agriculture Experiment Station Bulletin* 402:1–148.

Johnson, C. N., J. L. Isaac, and D. O. Fisher. 2007. Rarity of a top predator triggers continent-wide collapse of mammal prey: Dingoes and marsupials in Australia. *Proceedings of the Royal Society B* 274:341–346.

Johnson, W. E., E. Eizirik, J. Pecon-Slattery, W. J. Murphy, A. Antunes, E. Teeling, and S. J. O'Brien. 2006. The late Miocene radiation of modern Felidae: A genetic assessment. *Science* 311:73–77.

Johnson, W. E., T. K. Fuller, and W. L. Franklin. 1996. Sympatry in canids: A review and assessment. Pages 189–218 in J. L. Gittleman, ed. *Carnivore Behavior, Ecology, and Evolution.* Vol. 2. Cornell University Press, Ithaca, NY.

Jokimäki, J., M. L. Kaisanlahti-Jokimäki, A. Sorace, E. Fernández-Juricic, I. Rodriguez-Prieto, and M. D. Jimenez. 2005. Evaluation of the "safe nesting zone" hypothesis across an urban gradient: A multi-scale study. *Ecography* 28:59–70.

Jones, J. L., D. Kruszon-Moran, and M. Wilson. 2003. *Toxoplasma gondii* infection in the United States, 1999–2000. *Emerging Infectious Diseases* 9:1371–1374.

Jonker, S. A., J. F. Organ, R. M. Muth, R. R. Zwick, and W. F. Siemer. 2009. Stakeholder norms toward beaver management in Massachusetts. *Journal of Wildlife Management* 73, in press.

Joshi, A. R., D. L. Garshelis, and J. L. D. Smith. 1997. Seasonal and habitat-related diets of sloth bears in Nepal. *Journal of Mammalogy* 78:584–597.

Kaeuffer, R., D. Pontier, S. Devillard, and N. Perrin. 2004. Effective size of two feral domestic cat populations (*Felis catus* L.): Effect of the mating system. *Molecular Ecology* 13:483–490.

Kalz, B., K. M. Scheibe, I. Wegner, and J. Priemer. 2000. Health status and causes of mortality in feral cats in a delimited area of the inner-city of Berlin. [In German.] *Berliner und Münchener Tierärztliche Wochenschrift* 113:417–422.

Kamler, J. F., W. B. Ballard, R. L. Gilliland, P. R. Lemons, and K. Mote. 2003a. Impacts of coyotes on swift foxes in northwestern Texas. *Journal of Wildlife Management* 67:317–323.

Kamler, J. F., W. B. Ballard, R. L. Gilliland, and K. Mote.

2003b. Spatial relationships between swift foxes and coyotes in northwestern Texas. *Canadian Journal of Zoology* 81:168–172.

Kamler, J. F., R. M. Lee, J. C. deVos, Jr., W. B. Ballard, and H. A. Whitlaw. 2002. Survival and cougar predation of translocated bighorn sheep in Arizona. *Journal of Wildlife Management* 66:1267–1272.

Kanda, L. L., T. K. Fuller, P. R. Sievert, and K. D. Friedland. 2005. Variation in winter microclimate and its potential influence on Virginia opossum (*Didelphis virginiana*) survival in Amherst, Massachusetts. *Urban Ecosystems* 8:215–225.

Kaneko, Y., and N. Maruyama. 2005. Changes in Japanese badger (*Meles meles anakuma*) body weight and condition caused by the provision of food by local people in a Tokyo suburb. *Mammalian Science* 45:157–164.

Kaneko, Y., N. Maruyama, and N. Kanzaki. 1996. Growth and seasonal changes in body weight and size of Japanese badger in Hinodecho, suburb of Tokyo. *Journal of Wildlife Research* 1:42–46.

Kaneko, Y., N. Maruyama, and D. W. Macdonald. 2006. Food habits and habitat selection of suburban badgers (*Meles meles*) in Japan. *Journal of Zoology* 270:78–89.

Karanth, K. U., and M. E. Sunquist. 1995. Prey selection by tiger, leopard, and dhole in tropical forests. *Journal of Animal Ecology* 64:439–450.

Kasworm, W. F., and T. J. Thier. 1994. Adult black bear reproduction, survival, and mortality sources in Northwest Montana. *Bears: Their Biology and Management* 9:223–230.

Kato, M., H. Yamamoto, Y. Inukai, and S. Kira. 2003. Survey of the stray dog population and the health education program on the prevention of dog bites and dog-acquired infections: A comparative study in Nepal and Okayama Prefecture, Japan. *Acta Medica Okayama* 57:261–266.

Katti, M., and P. S. Warren. 2004. Tits, noise, and urban bioacoustics. *Trends in Ecology and Evolution* 19:109–110.

Kaufmann, J. H. 1982. Raccoons and allies. Pages 567–585 in J. A. Chapman and G. A. Feldhammer, eds. *Wild Mammals of North America,* eds. Johns Hopkins University Press, Baltimore.

Kauhala, K., and M. Saeki. 2004. Raccoon dog. Pages 136–142 in C. Sillero-Zubiri, M. Hoffman, and D. W. Macdonald, eds. *Canids: Foxes, Wolves, Jackals, and Dogs.* Status Survey and Conservation Action Plan. IUCN/SSC Canid Specialist Group. Gland, Switzerland and Cambridge.

Kautz, R., R. Kawula, T. Hoctor, J. Comiskey, D. Jansen, D. Jennings, J. Kasbohm, et al. 2006. How much is enough? Landscape level conservation for the Florida panther. *Biological Conservation* 130:118–133.

Kaye, J. P., P. M. Groffman, N. B. Grimm, L. A. Baker, R. V. Pouyat. 2006. A distinct urban biogeochemistry? *Trends in Ecology and Evolution* 21:192–199.

Kays, R. W., and A. A. DeWan. 2004. Ecological impact of

inside/outside house cats around a suburban nature reserve. *Animal Conservation* 7:273–283.

Kazacos, K. R. 1983. Raccoon roundworms (*Baylisascaris procyonis*): A cause of human and animal disease. *Agricultural Experiment Station of Purdue University Bulletin* 422.

Kazacos, K. R. 2001. *Baylisascaris procyonis* and related species. Pages 301–341 *in* W. Samuel, M. Pybus and A. Kocan, eds. *Parasitic Diseases of Wild Mammals*. Iowa State University Press, Ames.

Kellert, S. R. 1984. Urban American perceptions of animals and the natural environment. *Urban Ecology* 8:209–228.

Kellert, S. R. 1985. Public perceptions of predators, particularly the wolf and coyote. *Biological Conservation* 31:167–189.

Kellert, S. R. 1996. The value of life: Biological diversity and human society. Island Press/Shearwater Books, Washington, D.C.

Kellert. S. R., M. Black, C. Reid Rush, and A. J. Bath. 1996. Human culture and large carnivore conservation in North America. *Conservation Biology* 10:977–990.

Kelly, B. T., and E. O. Garton. 1997. Effects of prey size, meal size, meal composition, and daily frequency of feeding on the recovery of rodent remains from carnivore scats. *Canadian Journal of Zoology* 75:1811–1817.

Kennelly, J. J. 1978. Coyote reproduction. Pages 73–93 *in* M. Bekoff, ed. Coyotes: *Biology, Behavior, and Management*. Academic Press, New York.

Kerby, G., and D. W. Macdonald. 1988. Cat society and the consequences of colony size. Pages 67–81 *in* D. C. Turner and P. Bateson, eds. *The Domestic Cat: The Biology of Its Behaviour*. Cambridge University Press, Cambridge.

Kieran, J. 1959. *A Natural History of New York*. Houghton Mifflin, Boston.

Kinlaw, A. 1995. *Spilogale putorius. Mammalian Species* 511:1–7.

Kitchen, A. M., E. M. Gese, and E. R. Schauster. 1999. Resource partitioning between coyotes and swift foxes; space, time, and diet. *Canadian Journal of Zoology* 77:1645–1656.

Kitchen, A. M., E. M. Gese, and E. R. Schauster. 2000. Changes in coyote activity patterns due to reduced exposure to human persecution. *Canadian Journal of Zoology* 78:853–857.

Kleiman, D. G. 1977. Monogamy in mammals. *Quarterly Review of Biology* 52:39–69.

Kleiven, J., T. Bjerke, and B. P. Kaltenborn. 2004. Factors influencing the social acceptability of large carnivore behaviours. *Biodiversity and Conservation* 13:1647–1658.

Knapp, D. K. 1978. *Effects of Agricultural Development in Kern County, California, on the San Joaquin Kit Fox in 1977*. California Department of Fish and Game, Non-game Wildlife Investigations Final Report, Project E-1-1, Job V-1.21. Sacramento, CA.

Knapp, S., I. Kühn, V. Mosbrugger, and S. Klotz. 2008. Do protected areas in urban and rural landscapes differ in species diversity? *Biodiversity and Conservation* 17:1595–1612.

Knick, S. T. 1990. Ecology of bobcats relative to exploitation and a prey decline in southeastern Idaho. *Wildlife Monographs* 108:1–42.

Knobel, D. L., S. Cleaveland, P. G. Coleman, E. M. Fèvre, M. I. Meltzer, M. E. G. Miranda, A. Shaw, J. Zinsstag, and F.-X. Meslin. 2005. Re-evaluating the burden of rabies in Africa and Asia. *Bulletin of the World Health Organization* 83:360–368.

Knobel, D. L., M. K. Laurenson, R. R. Kazwala, L. A. Boden, and S. Cleaveland. 2008. A cross-sectional study of factors associated with dog ownership in Tanzania. *BMC Veterinary Research* 4: article 5. doi: 10.1186/1746-6148-4-5.

Knowlton, F. F. 1972. Preliminary interpretations of coyote population mechanics with some management implications. *Journal of Wildlife Management* 36:369–382.

Knowlton, F. F., E. M. Gese, and M. M Jaeger. 1999. Coyote depredation control: An interface between biology and management. *Journal of Range Management* 52:398–412.

Knuth, B. A., R. J. Stout, W. F. Siemer, D. J. Decker, and R. C. Stedman. 1992. Risk management concepts for improving wildlife population decisions and public communication strategies. *Transactions of the North American Wildlife and Natural Resources Conference* 57:63–74.

Koehler, G. M., and M. G. Hornocker. 1991. Seasonal resource use among mountain lions, bobcats, and coyotes. *Journal of Mammalogy* 72:391–396.

Kohn, M. H., E. C. York, D. A. Kamradt, G. Haught, R. M. Sauvajot and R. K. Wayne. 1999. Estimating population size by genotyping feces. *Proceedings of the Royal Society of London, Series B* 266:657–663.

Kolata, R. J., and D. E. Johnston. 1975. Motor vehicle accidents in urban dogs: A study of 600 cases. *Journal of the American Veterinary Medical Association* 167:938–941.

Kolb, H. H. 1984. Factors affecting the movements of dog foxes in Edinburgh. *Journal of Applied Ecology* 21:161–173.

Kolb, H. H. 1985. Habitat use by foxes in Edinburgh. *Revue D' Ecologie la Terre et la Vie* 40:139–143.

Kolb, H. H. 1986. Some observations on the home ranges of vixens (*Vulpes vulpes*) in the suburbs of Edinburgh. *Journal of Zoology* 210:636–639.

Kolowski, J. M., D. Katan, K. R. Theis, and K. E. Holekamp. 2007. Daily patterns of activity in the spotted hyena. *Journal of Mammalogy* 88:1017–1028.

König, A. 2008. Fears, attitudes, and opinions of suburban residents with regards to their urban foxes: A case study in the community of Grünwald—a suburb of Munich. *European Journal of Wildlife Research* 54:101–109.

Koopman, M. E. 1995. Food habits, space use, and movements of the San Joaquin kit fox on the Elk Hills Naval Petroleum Reserves in California. Master's thesis, University of California, Berkeley.

Koopman, M. E., B. L. Cypher, and J. H. Scrivner. 2000. Dispersal patterns of San Joaquin kit foxes (*Vulpes macrotis mutica*). *Journal of Mammalogy* 81:213–222.

Koopman, M. E., J. H. Scrivner, and T. T. Kato. 1998. Patterns of den use by San Joaquin kit foxes. *Journal of Wildlife Management* 62:373–379.

Kowalczyk, R., A. N. Bunevich, and B. Jędrzejewska. 2000. Badger density and distribution of setts in Białowieża Primeval Forest (Poland and Belarus) compared to other Eurasian populations. *Acta Theriologica* 45:395–408.

Kowalczyk, R., A. Zalewski, B. Jedrzejewska, and W. Jedrze-jewski. 2003. Spatial organization and demography of badgers (*Meles meles*) in Bialowieza Primeval Forest, Poland, and the influence of earthworms on badger densities in Europe. *Canadian Journal of Zoology* 81:74–87.

Kravetz, J. D., and D. G. Federman. 2002. Cat-associated zoonoses. *Archives of Internal Medicine* 162:1945–1952.

Krebs, C. J. 2001. *Ecology*. 5th ed. Benjamin Cumming, San Francisco.

Krebs, J. W., S. M. Williams, J. S. Smith, C. E. Rupprecht, and J. E. Childs. 2003. Rabies among infrequently reported mammalian carnivores in the United States, 1960–2000. *Journal of Wildlife Disease* 39:253–261.

Kreidl, P., F. Allerberger, G. Judmaier, H. Auer, H. Aspöck, and A. J. Hall. 1998. Domestic pets as risk factors for alveolar hydatid disease in Austria. *American Journal of Epidemiology* 147:978–981.

Kruuk, H. 1978. Foraging and spatial organisation of the European badger, *Meles meles* L. *Behavioural Ecology and Sociobiology* 4:75–89.

Kruuk, H. 1989. *The Social Badger: Ecology and Behaviour of a Group-Living Carnivore (Meles meles)*. Oxford University Press, Oxford.

Kruuk, H. 2002. *Hunter and Hunted: Relationships between Carnivores and People*. Cambridge University Press, Cambridge.

Kruuk, H., and W. A. Sands. 1972. The aardwolf (*Proteles cristatus*, Sparrman 1783) as a predator of termites. *East African Wildlife Journal* 10:211–227.

Kugelschafter, K., S. Deeg, W. Kümmerle, and H. Rehm. 1984/1985. Steinmarderschäden (*Martes foina* (Erxleben, 1777)) an Kraftfahrzeugen: Schadensanalyse und verhaltensbiologische Untersuchungsmethodik. *Säugetierkundliche Mitteilungen* 32:35–48.

Kunkel, K. E., T. K. Ruth, D. H. Pletscher, and M. G. Hornocker. 1999. Winter prey selection by wolves and cougars in and near Glacier National Park, Montana. *Journal of Wildlife Management* 63:901–910.

Lafferty, K. D. 2006. Can the common brain parasite, *Toxoplasma gondii*, influence human culture? *Proceedings of the Royal Society of London B* 273:2749–2755.

Laing, S. P. 1988. Cougar habitat selection and spatial use patterns in southern Utah. Master's thesis, University of Wyoming, Laramie.

Laliberte, A. S., and W. J. Ripple. 2004. Range contractions of North American carnivores and ungulates. *BioScience* 54:123–138.

Lambert, C. M. S., R. B. Wielgus, H. S. Robinson, D. D. Katnik, H. S. Cruickshank, R. Clarke, and J. Almack. 2006. Cougar population dynamics and viability in the Pacific Northwest. *Journal of Wildlife Management* 70:246–254.

Landriault, L. J., M. E. Obbard, and W. J. Rettie. 2000. *Nuisance Black Bears and What to Do with Them*. OMNR, Northeast Science & Technology. TN-017. 20 pp.

Langston, R. H. W., D. Liley, G. Murison, E. Woodfield, and R. T. Clarke. 2007. What effects do walkers and dogs have on the distribution and productivity of breeding European nightjar *Caprimulgus europaeus*? *Ibis* 149:27–36.

Lanszki, J. 2003. Feeding habits of stone martens in a Hungarian village and its surroundings. *Folia Zoologica* 52:367–377.

Larivière, S., and S. H. Ferguson. 2003. Evolution of induced ovulation in North American carnivores. *Journal of Mammalogy* 84:937–947.

Larivière, S., and F. Messier. 1996. Aposematic behavior in the striped skunk, *Mephitis mephitis*. *Ethology* 102:986–992.

Larivière, S., and F. Messier. 1997. Seasonal and daily activity patterns of striped skunks (*Mephitis mephitis*) in the Canadian prairies. *Journal of the Zoological Society of London* 243:255–262.

Larivière, S., and F. Messier. 1998. Spatial organization of a prairie striped skunk population during waterfowl nesting season. *Journal of Wildlife Management* 62:199–204.

Larivière, S., and M. Pasitschniak-Arts. 1996. *Vulpes vulpes*. *Mammalian Species* 537:1–11.

Lavin, S. R., T. R. Van Deelen, P. W. Brown, R. E. Warner, and S. H. Ambrose. 2003. Prey use by red foxes (*Vulpes vulpes*) in urban and rural areas of Illinois. *Canadian Journal of Zoology* 81:1070–1082.

Lee, I. T., J. K. Levy, S. P. Gorman, P. C. Crawford, and M. R. Slater. 2002. Prevalence of feline leukemia virus infection and serum antibodies against feline immunodeficiency virus in unowned free-roaming cats. *Journal of the American Veterinary Medical Association* 220:620–622.

Lehner, P. N., C. McCluggage, D. R. Mitchell, and D. H. Neil. 1983. Selected parameters of the Fort Collins, Colorado, dog population, 1979–80. *Applied Animal Ethology* 10:19–25.

Le Lay, G., and T. Lóde. 2004. Stone marten's urban habitat: Do red fox activities explain their distribution? Page 35 in 4th International Martes Symposium. Martes in Carnivore Communities. University of Lisbon, July 20–24, 2004, Lisbon, Portugal.

Lembeck, M. 1978. *Bobcat Study, San Diego County, California*. State of California Department of Fish and Game. Report E-W-2, IV-1.7.

Lembeck, M., and G. I. Gould. 1979. Dynamics of harvested and unharvested bobcat populations in California. Bobcat Research Conference Proceedings. *National Wildlife Federation Scientific Technical Series* 6:53–54.

Leonard, J. A., R. K. Wayne, J. Wheeler, R. Valadez, S. Guillén, and C. Vilà. 2002. Ancient DNA evidence for old world origin of new world dogs. *Science* 298:1613–1616.

Leong, K. M., D. J. Decker, J. Forester, P. D. Curtis, and M. A. Wild. 2007. Expanding problem frames to understand human-wildlife conflicts in urban proximate parks. *Journal of Park and Recreation Administration* 25:62–78.

Lepczyk, C. A., A. G. Mertig, and J. Liu. 2003. Landowners and cat predation across rural-to-urban landscapes. *Biological Conservation* 115:191–201.

Leslie, M., S. Messenger, R. Rohde, J. Smith, R. Cheshier, C. Hanlon, and C. Rupprecht. 2006. Bat-associated rabies virus in skunks. *Emerging Infectious Diseases* 12:1274–1277.

Levins, R. 1969. Some demographic and genetic consequences of environmental heterogeneity for biological control. *Bulletin of the Entomological Society of America* 15:237–240.

Levy, J. K., and P. C. Crawford. 2004. Humane strategies for controlling feral cat populations. *Journal of the American Veterinary Medical Association* 225:1354–1360.

Levy, J. K., D. W. Gale, and L. A. Gale. 2003a. Evaluation of the effect of a long-term trap-neuter return and adoption program on a free-roaming cat population. *Journal of the American Veterinary Medical Association* 222:42–46.

Levy, J. K., J. E. Woods, S. L. Turick, and D. L. Etheridge. 2003b. Number of unowned free-roaming cats in a college community in the southern United States and characteristics of community residents who feed them. *Journal of the American Veterinary Medical Association* 223:202–205.

Lewis, J. C. 1994. Dispersal of introduced red foxes in urban Southern California. Ph.D. diss., Humboldt State University, Arcata.

Lewis, J. C., K. L. Sallee, and R. T. Golightly. 1993. *Introduced Red Fox in California*. Department of Fish and Game Wildlife Management Division, Nongame Bird and Mammal Section Report 93–10, Sacramento, CA.

Lewis, J. C., K. L. Sallee, and R. T. Golightly. 1999. Introduction and range expansion of nonnative red foxes (*Vulpes vulpes*) in California. *American Midland Naturalist* 142:372–381.

Leyhausen, P. 1979. *Cat Behaviour: The Predatory and Social Behaviour of Domestic and Wild Cats*. Garland STPM Press, London.

Liberg, O. 1980. Spacing patterns in a population of rural free roaming domestic cats. *Oikos* 35:336–349.

Liberg, O. 1981. Predation and social behaviour in a population of domestic cats: An evolutionary perspective. Ph.D. diss., University of Lund, Sweden.

Liberg, O. 1983. Courtship behaviour and sexual selection in the domestic cat. *Applied Animal Ethology* 10:117–132.

Liberg, O. 1984. Social behaviour in free ranging domestic and feral cats. Pages 175–181 *in* R. S. Anderson, ed. *Nutrition and Behaviour in Dogs and Cats*. Pergamon Press, Oxford.

Liberg, O., and M. Sandell. 1998. Spatial organisation and reproductive tactics in the domestic cat and other felids.

Pages 83–98 *in* D. C. Turner and P. Bateson, eds. *The Domestic Cat: The Biology of Its Behaviour*. Cambridge University Press, Cambridge.

Liberg, O., M. Sandell, D. Pontier, and E. Natoli. 2000. Density, spatial organisation and reproductive tactics in the domestic cat and other felids. Pages 119–147 *in* D. C. Turner and P. Bateson, eds. *The Domestic Cat: The Biology of Its Behaviour*. 2nd ed. Cambridge University Press, Cambridge.

Lidicker, W. Z., Jr. 1975. The role of dispersal in the demography of small mammals. Pages 103–128 *in* F. B. Golley, K. Petrusewicz, and L Ryszkaowski, eds. *Small Mammals: Their Productivity and Population Dynamics*. Cambridge University Press, Cambridge.

Lilith, M., M. Calver, I. Styles, and M. Garkaklis. 2006. Protecting wildlife from predation by owned domestic cats: Application of a precautionary approach to the acceptability of proposed cat regulations. *Austral Ecology* 31:176–189.

Lima, S. L. 1998. Stress and decision-making under the risk of predation: Recent developments from behavioral, reproductive, and ecological perspectives. *Advances in the Study of Behaviour* 27:215–290.

Lindblad-Toh, K., et al. 2005. Genome sequence, comparative analysis, and haplotype structure of the domestic dog. *Nature* 438:803–819.

Lindström, E. R. 1994. Large prey for small cubs—on crucial resources of a boreal red fox population. *Ecography* 17:17–22.

Lindström, E. R., H. Andrén, P. Angelstam, G. Cederlund, B. Hörnfeldt, L. Jäderberg, P.-A. Lemnell, B. Martinsson, K. Sköld, and J. E. Swenson. 1994. Disease reveals the predator: Sarcoptic mange, red fox predation, and prey populations. *Ecology* 75:1042–1049.

Lindzey, F. G. 1987. Mountain lion. Pages 656–668 *in* M. Novak, J. A. Baker, M. E. Obbard, and B. Malloch eds. *Wild Furbearer Management and Conservation in North America*. Ontario Trappers Association and Ontario Ministry of Natural Resources, Toronto, Canada.

Lindzey, F. G., B. B. Ackerman, D. Barnhurst, and T. P. Hemker. 1988. Survival rates of mountain lions in southern Utah. *Journal of Wildlife Management* 52:664–667.

Lindzey, F. G., W. D. Vansickle, B. B. Ackerman, D. Barnhurst, T. P. Hemker, and S. P. Laing. 1994. Cougar population dynamics in southern Utah. *Journal of Wildlife Management* 58:619–624.

Lischka, S. A., S. J. Riley, and B. A. Rudolph. 2008. Effects of impact perception on acceptance capacity for white-tailed deer. *Journal of Wildlife Management* 72:502–509.

Littin, K. E., D. J. Mellor, and C. T. Eason. 2004. Animal welfare and ethical issues relevant to the humane control of vertebrate pests. *New Zealand Veterinary Journal* 52:1–10.

Littin, K. E., C. E. O'Connor, N. G. Gregory, D. J. Mellor, and C. T. Eason. 2002. Behaviour, coagulopathy, and pathology of brushtail possums (*Trichosurus vulpecula*) poisoned with brodifacoum. *Wildlife Research* 29:259–267.

Litvaitis, J. A., and D. J. Harrison. 1989. Bobcat-coyote niche relationships during a period of coyote population increase. *Canadian Journal of Zoology* 67:1180–1188.

Litvaitis, J. A., J. A. Sherburne, and J. A. Bissonette. 1986. Bobcat habitat use and home range size in relation to prey density. *Journal of Wildlife Management* 50:110–117.

Lloyd, H. G. 1980. *The Red Fox*. Batsford, London.

Loe, J., and E. Roskraft. 2004. Large carnivores and human safety: A review. *Ambio* 33:283–288.

Logan, K. A. 1983. Mountain lion population and habitat characteristics in the Big Horn Mountains of Wyoming. Master's thesis, University of Wyoming, Laramie.

Logan, K. A., and L. L. Irwin. 1985. Mountain lion habitats in the Big Horn Mountains, Wyoming. *Wildlife Society Bulletin* 13:257–262.

Logan, K. A., L. L. Irwin, and R. Skinner. 1986. Characteristics of a hunted mountain lion population in Wyoming. *Journal of Wildlife Management* 50:648–654.

Logan, K. A., and L. L. Sweanor. 2001. *Desert Puma: Evolutionary Ecology and Conservation of an Enduring Carnivore*. Island Press, Washington D.C.

Lohr, C., W. B. Ballard, and A. Bath. 1996. Attitudes toward gray wolf reintroduction to New Brunswick. *Wildlife Society Bulletin* 24:414–20.

Loker, C. A., D. J. Decker, and S. J. Schwager. 1999. Social acceptability of wildlife management actions in suburban areas: 3 cases from New York. *Wildlife Society Bulletin* 27:152–159.

Long, J. L. 2003. *Introduced Mammals of the World: Their History, Distribution, and Influence*. CABI Publishing, Wallingford, Oxfordshire.

Longcore, T., and C. Rich. 2004. Ecological light pollution. *Frontiers in Ecology and the Environment* 2:191–198.

Loomis, J. B., and D. S. White. 1996. Economic benefits of rare and endangered species: Summary and meta-analysis. *Ecological Economics* 18:197–206.

Lord, L.K. 2008. Attitudes toward and perceptions of free-roaming cats among individuals living in Ohio. *Journal of the American Veterinary Medical Association* 232:1159–1167.

Lovallo, M. J., G. L. Storm, D. S. Klute, and W. M. Tzilkowski. 2001. Multivariate models of bobcat habitat selection for Pennsylvania landscapes. Pages 4–17 *in* A. Woolf and C. K. Nielsen, eds. *Proceedings of a Symposium on the Ecology and Management of Bobcats*. Illinois Department of Natural Resources, Champaign.

Lovari, S., and L. Parigi. 1995. The red fox as a gamebird killer or a considerate parent? *Mammalia* 59:455–459.

Loveridge, A. J., J. E. Hunt, F. Murindagomo, and D. W. Macdonald. 2006. Influence of drought on predation of elephant (*Loxodonta africana*) calves by lions (*Panthera leo*) in an African wooded savannah. *Journal of Zoology* 270:523–530.

Loyola, R. D., G. de Oliveira, J. A. F. Diniz-Filho, and T. M. Lewinsohn. 2008. Conservation of Neotropical carnivores under different prioritization scenarios: Mapping species traits to minimize conservation conflicts. *Diversity and Distributions* 14:949–960.

Lucherini, M., and G. Crema. 1993. Diet of urban stone martens in Italy. *Mammalia* 57:274–277.

Ludwig, D. R., and S. M. Mikolajczak. 1985. Post-release behavior of captive-reared raccoons. Pages 144–155 *in* P. A. Beaver, ed. *Wildlife Rehabilitation*. Suffolk Book Manufacturers, Hauppauge, NY.

Luoma, J. R. 1997. Catfight: Feral cats are dining on birds and other small wild animals by the millions. *Audubon* 99:84–91.

Luria, B. J., J. K. Levy, M. R. Lappin, E. B. Breitschwerdt, A. M. Legendre, J. A. Hernandez, S. P. Gorman, and I. T. Lee. 2004. Prevalence of infectious diseases in feral cats in Northern Florida. *Journal of Feline Medicine and Surgery* 6:287–296.

Lyall-Watson, M. 1963. A critical re-examination of food "washing" behaviour in the raccoon (*Procyon lotor* Linn.). *Proceedings of the Zoological Society of London* 141:371–393.

Lyons, A. J. 2005. Activity patterns of urban American black bears in the San Gabriel Mountains of Suthern California. *Ursus* 16:255–262.

Lyren, L. M. 2001. Movement patterns of coyotes and bobcats relative to roads and underpasses in the Chino Hills area of southern California. Master's thesis, California State Polytechnic University, Pomona.

MacClintock, D. 1981. *A Natural History of Raccoons*. Charles Scribner's Sons, New York.

MacCracken, J. G. 1982. Coyote foods in a Southern California suburb. *Wildlife Society Bulletin* 10:280–281.

Macdonald, D. W. 1979a. Flexible social system of the golden jackal, *Canis aureus*. *Behavioral Ecology and Sociobiology* 5:17–38.

Macdonald, D. W. 1979b. "Helpers" in fox society. *Nature* 282:69–71.

Macdonald, D. W. 1979c. Some observations and field experiments on the urine marking behaviour of the red fox, *Vulpes vulpes* L. *Zeitschrift für Tierpsychologie* 51:1–22.

Macdonald, D. W. 1980a. Social factors affecting reproduction amongst red foxes (*Vulpes vulpes* L., 1758). Pages 123–175 *in* E. Zimen, ed. *The Red Fox: Symposium on Behaviour and Ecology*. Junk, The Hague, Netherlands.

Macdonald, D. W. 1980b. The red fox, *Vulpes vulpes*, as a predator upon earthworms, *Lumbricus terrestris*. *Zeitschrift für Tierpsychologie* 52:171–200.

Macdonald, D. W. 1980c. Patterns of scent marking with urine and faeces amongst carnivore communities. *Symposia of the Zoological Society of London* 45:107–139.

Macdonald, D. W. 1992. *The Velvet Claw*. BBC Books, London.

Macdonald, D. W. 2006. *Encyclopaedia of Mammals*. 6th ed. Oxford University Press, Oxford.

Macdonald, D. W., and P. J. Apps. 1978. The social behaviour of a group of semi-dependent farm cats, *Felis catus*: A progress report. *Carnivore Genetics Newsletter* 3:256–268.

Macdonald, D. W., P. J. Apps, G. M. Carr, and G. Kerby. 1987. Social dynamics, nursing coalitions, and infanticide among farm cats (*Felis catus*). *Advances in Ethology* 28:1–64.

Macdonald, D. W., and G. M. Carr. 1981. Foxes beware: You are back in fashion. *New Scientist* 89:9–11.

Macdonald, D. W., and G. M. Carr. 1995. Variation in dog society: Between resource dispersion and social flux. Pages 199–216 in J. Serpell, ed. *The Domestic Dog: Its Evolution, Behaviour and Interactions with People*. Cambridge University Press, Cambridge.

Macdonald, D. W., and C. Newman. 2002. Population dynamics of badgers (*Meles meles*) in Oxfordshire, U.K.: Numbers, density and cohort life histories, and a possible role of climate change in population growth. *Journal of Zoology* 256:121–138.

Macdonald, D. W., and J. C. Reynolds. 2004. Red fox. Pages 129–136 in C. Sillero-Zubiri, M. Hoffman, and D. W. Macdonald, eds. *Canids: Foxes, Wolves, Jackals, and Dogs: Status Survey and Conservation Action Plan*. IUCN/SSC Canid Specialist Group. Gland, Switzerland and Cambridge, UK.

Macdonald, D. W., and D. R. Voigt. 1985. The biological basis of rabies models. Pages 71–108 in P. J. Bacon, ed. *Population Dynamics of Rabies in Wildlife*. Academic Press, London.

Macdonald, D. W., N. Yamaguchi, and G. Kerby. 2000. Group-living in the domestic cat: Its socio-biology and epidemiology. Pages 95–118 in D. C. Turner and P. Bateson, eds. *The Domestic Cat: The biology of Its Behaviour*. 2nd ed. Cambridge University Press, Cambridge.

Machlis, G. E., J. E. Force, and W. R. Burch, Jr. 1997. The human ecosystem. I. The human ecosystem as an organizing concept in ecosystem management. *Society and Natural Resources* 10:347–367.

MacInnes, C. D., and C. A. LeBer. 2000. Wildlife management agencies should participate in rabies control. *Wildlife Society Bulletin* 28:1156–1167.

Macpherson, A. H. 1969. The dynamics of Canadian arctic fox populations. *Canadian Wildlife Service Report*. Series 8. 52 pp.

Macpherson, C. N. L. 2005. Human behaviour and the epidemiology of parasitic zoonoses. *International Journal for Parasitology* 35:1319–1331.

Macpherson, C. N. L., F. X. Meslin, and A. I. Wandeler. 2000. *Dogs, Zoonoses and Public Health*. CABI, Wallingford, Oxfordshire.

Maehr, D. S., and J. A. Cox. 1995. Landscape features and panthers in Florida. *Conservation Biology* 9:1008–1019.

Maehr, D. S., E. D. Land, and J. C. Roof. 1991. Social ecology of Florida panthers. *National Geographic Research and Exploration* 7:414–431.

Maehr, D. S., E. D. Land, J. C. Roof, and J. W. McCown.

1989a. Early maternal behavior in the Florida panther *Felis concolor coryi*. *American Midland Naturalist* 22:34–43.

Maehr, D. S., E. D. Land, D. B. Shindle, O. L. Bass, and T. S. Hoctor. 2002. Florida panther dispersal and conservation. *Biological Conservation* 106:187–197.

Maehr, D. S., J. C. Roof, E. D. Land, and J. W. McCown. 1989b. First reproduction of a panther (*Felis concolor coryi*) in southwestern Florida, U.S.A. *Mammalia* 53:129–131.

Maguiña, C., and E. Gotuzzo. 2000. Emerging and re-emerging diseases in Latin America: Bartonellosis, new and old. *Infectious Disease Clinics of North America* 14:1–22.

Magurran, A. E. 1988. *Ecological Diversity and Its Measurement*. Croom Helm, London.

Mahlow, J. C., and M. R. Slater. 1996. Current issues in the control of stray and feral cats. *Journal of the American Veterinary Medical Association* 209:2016–2020.

Major, J. T., and J. A. Sherburne. 1987. Interspecific relationships of coyotes, bobcats, and red foxes in western Maine. *Journal of Wildlife Management* 51:606–616.

Manfredo, M. J., J. J. Vaske, and D. J. Decker. 1995. Human dimensions of wildlife management: Basic concepts. Pages 17–31 in R. L. Knight and K. J. Gutzwiller, eds. *Wildlife and Recreationists: Coexistence through Management and Research*. Island Press, Washington D.C.

Manfredo, M. J., H. C. Zinn, L. Sikorowski, and J. Jones. 1998. Public acceptance of mountain lion management: A case study of Denver, Colorado, and nearby foothills areas. *Wildlife Society Bulletin* 26:964–70.

Manning, A. M., and A. N. Rowan. 1992. Companion animal demographics and sterilization status: Results from a survey in four Massachusetts towns. *Anthrozoos* 5:192–201.

Manor, R., and D. Saltz. 2004. The impact of free-roaming dogs on gazelle kid/female ratio in a fragmented area. *Biological Conservation* 119:231–236.

Mansfield, T. M., and K. G. Charlton. 1998. Trends in mountain lion depredation and public safety incidents in California. Pages 118–121 in R. O. Baker and A. C. Crabb, eds. *Proceedings of the 18th Vertebrate Pest Conference*, University of California-Davis.

Manski, D., and J. Hadidian. 1985. *Rock Creek Raccoons: Movements and Resource Utilization in an Urban Environment*. National Park Service, Washington, D.C.

Marker, L. L., and A. J., Dickman. 2005. Factors affecting leopard (*Panthera pardus*) spatial ecology, with particular reference to Namibian farmlands. *South African Journal of Wildlife Research* 35:105–115.

Marks, C. 1999. Ethical issues in vertebrate pest management: Can we balance the welfare of individuals and ecosystems? Pages 79–89 in D. Mellor and V. Monamy, eds. *Proceedings of the Conference Held at the Western Plains Zoo, Dubbo, NSW*. ANZCCART, Australia.

Marks, C. A. 2001. Bait-delivered cabergoline for the reproductive control of the red fox (*Vulpes vulpes*): Estimating mammalian non-target risk in south-eastern Australia. *Reproduction, Fertility, and Development* 13:499–510.

Marks, C. A., and T. E. Bloomfield. 1998. Canine heartworm (*Dirofilaria immitis*) detected in red foxes (*Vulpes vulpes*) in urban Melbourne. *Veterinary Parasitology* 78:147–154.

Marks, C. A., and T. E. Bloomfield. 1999a. Distribution and density estimates for urban foxes (*Vulpes vulpes*) in Melbourne: Implications for rabies control. *Wildlife Research* 26:763–775.

Marks, C. A., and T. E. Bloomfield. 1999b. Bait uptake by foxes (*Vulpes vulpes*) in urban Melbourne: The potential of oral vaccination for rabies control. *Wildlife Research* 26:777–787.

Marks, C. A., and T. E. Bloomfield. 2006. Home-range size and selection of natal den and diurnal shelter sites by urban red foxes (*Vulpes vulpes*) in Melbourne. *Wildlife Research* 33:339–347.

Marshall, A. D., and J. H. Jenkins. 1966. Movements and home ranges of bobcats as determined by radiotracking in the upper coastal plain of west-central South Carolina. *Proceedings of the Annual Conference of the Southeast Association of Fish and Wildlife Agencies* 20:206–214.

Martinez-Espineira, R. 2006. Public attitudes toward lethal coyote control. *Human Dimensions of Wildlife* 11:89–100.

Martínez-Moreno, F. J., S. Hernández, E. López-Cobos, C. Becerra, I. Acosta, and A. Martínez-Moreno. 2007. Estimation of canine intestinal parasites in Córdoba (Spain) and their risk to human health. *Veterinary Parasitology* 143:7–13.

Marzluff, J. M. 2001. Worldwide urbanization and its effects on birds. Pages 19–47 *in* J. M. Marzluff, R. Bowman, and R. Donnelly, eds. *Avian Ecology and Conservation in an Urbanizing World.* Kluwer Academic, Norwell, MA.

Marzluff, J. M., E. Shulenberger, W. Endlicher, M. Alberti, G. Bradley, C. Ryan, C. ZumBrunnen, and U. Simon. 2008. *Urban Ecology: An International Perspective on the Interaction between Humans and Nature.* Springer, New York.

Maslow, A. H. 1943. A theory of human motivation. *Psychology Review* 50:370–96.

Mason, C. F., S. M. Macdonald, H. C. Bland, and J. Ratford. 1992. Organochlorine pesticide and PCB contents in otter (*Lutra lutra*) scats from western Scotland. *Water, Air, and Soil Pollution* 64:617–626.

Matter, H. C., and T. J. Daniels. 2000. Dog ecology and population biology. Pages 17–62 *in* C. N. L. Macpherson, F. X. Meslin, and A. I. Wandeler, eds. *Dogs, Zoonoses, and Public Health.* CABI Publishing, Wallingford, Oxfordshire.

Matter, H. C., C. L. Schumacher, H. Kharmachi, S. Hammami, A. Tlatli, J. Jemli, L. Mrabet, F. X. Meslin, M. F. A. Aubert, B. E. Neuenschwander, and K. E. Hicheri. 1998. Field evaluation of two bait delivery systems for the oral immunization of dogs against rabies in Tunisia. *Vaccine* 16:657–665.

May, R. M. 1988. Control of feline delinquency. *Nature* 332:392–393.

McAtee, W. L. 1918. A sketch of the natural history of the District of Columbia. *Bulletin of the Biological Society of Washington* 1:1–142.

McCarthy, T. M., and G. Chapron. 2003. *Snow Leopard Survival Strategy.* ISLT and SLN, Seattle.

McClennen, N., R. L. Wigglesworth, S. H. Anderson, and D. G. Wachob. 2001. The effect of suburban and agricultural development on the activity patterns of coyotes (*Canis latrans*). *American Midland Naturalist* 146:27–36.

McClure, M. F., N. S. Smith, and W. W. Shaw. 1995. Diets of coyotes near the boundary of Saguaro National Monument and Tucson, Arizona. *Southwestern Naturalist* 40:101–104.

McComb, W. C. 1981. Effects of land use upon food habits, productivity, and gastro-intestinal parasites of raccoons. Pages 642–651 *in* J. A. Chapman and D. Pursley, eds. *Worldwide Furbearer Conference Proceedings.* Worldwide Furbearer Conference, Frostburg, MD.

McCombie, B. 1999. Those dirty raccoons. *Field and Stream* 104:8.

McDonald, R. A., C. Webbon, and S. Harris. 2000. The diet of stoats (*Mustela erminea*) and weasels (*Mustela nivalis*) in Great Britain. *Journal of Zoology* 252:363–371.

McDonnell, M. J., S. T. A. Pickett, P. Groffman, P. Bohlen, R. V. Pouyat, W. C. Zipperer, R. W. Parmelee, M. M. Carriero, and K. Medley. 1997. Ecosystem processes along an urban-to-rural gradient. *Urban Ecosystems* 1:21–36.

McDougal, C. 1987. The man-eating tiger in geographical and historic perspective. Pages 435–448 *in* R. L. Tilson and U. S. Seal, eds. *Tigers of the World: The Biology, Biopolitics, Management, and Conservation of an Endangered Species.* Noyes Publications, Park Ridge, NJ.

McDowall, D. 2006. *Richmond Park: The Walker's Guide.* 2nd ed. Privately published, Richmond, Surrey.

McGrew, J. C. 1979. *Vulpes macrotis. Mammalian Species* 123:1–6.

McIlroy, J., G. Saunders, and L. A. Hinds. 2001. The reproductive performance of female red foxes, *Vulpes vulpes*, in central-western New South Wales during and after a drought. *Canadian Journal of Zoology* 79:545–553.

McIntyre, N. E., K. Knowled-Yanez, and D. Hope. 2000. Urban ecology as an interdisciplinary field: Differences in the use of "urban" between the social and natural sciences. *Urban Ecology* 4:5–24.

McIvor, D. E., and M. R. Conover. 1994. Perceptions of farmers and non-farmers toward management of problem wildlife. *Wildlife Society Bulletin* 22:212–19.

McKinney, M. L. 2002. Urbanization, biodiversity, and conservation. *BioScience* 52:883–890.

McKinney, M. L. 2006. Urbanization as a major cause of biotic homogenization. *Biological Conservation* 127:247–260.

McLaren, B. E., and R. O. Peterson. 1994. Wolves, moose, and tree rings on Isle Royale. *Science* 266:1555–1558.

McNay, M. E. 2002. Wolf-human interactions in Alaska and Canada: A review of the case history. *Wildlife Society Bulletin* 30:831–843.

McQuillan, G. M., D. Kruszon-Moran, B. J. Kottiri, L. R. Curtin, J. W. Lucas, and R. S. Kington. 2004. Racial and ethnic differences in the seroprevalence of 6 infectious diseases in the United States: Data from NHANES III, 1988–1994. *American Journal of Public Health* 94:1952–1958.

McRae, B. H., P. Beier, L. E. DeWald, L. H. Huynh, and P. Keim. 2005. Habitat barriers limit gene flow and illuminate historical events in a wide-ranging carnivore, the American puma. *Molecular Ecology* 14:1965–1977.

Mead, C. J. 1982. Ringed birds killed by cats. *Mammal Review* 12:183–186.

Mead, R. A. 1989. The physiology and evolution of delayed implantation in carnivores. Pages 437–464 in J. L. Gittleman, ed. *Carnivore Behavior, Ecology, and Evolution.* Cornell University Press, Ithaca, NY.

Mech, L. D. 1970. *The Wolf: The Ecology and Behavior of an Endangered Species.* University of Minnesota Press, Minneapolis.

Mech, L. D. 2003. Incidence of mink, *Mustela vison,* and river otter, *Lutra canadensis,* in a highly urbanized area. *Canadian Field-Naturalist* 117:115–116.

Mech, L. D., and M. Korb. 1978. An unusually long pursuit of a deer by a wolf. *Journal of Mammalogy* 59:860–861.

Mech, L. D., and F. J. Turkowski. 1966. Twenty-three raccoons in one winter den. *Journal of Mammalogy* 47:529–530.

Medway, D. G. 2004. The land bird fauna of Stephens Island, New Zealand in the early 1890s, and the cause of its demise. *Notornis* 51:201–211.

Meek, P. D. 1998. Food items brought home by domestic cats *Felis catus* (L) living in Booderee National Park, Jervis Bay. *Proceedings of the Linnean Society of New South Wales* 120:43–47.

Meltzer, M. I., and C. E. Rupprecht. 1998a. A review of the economics of the prevention and control of rabies. 1. Global impact and rabies in humans. *Pharmacoeconomics* 14:365–383.

Meltzer, M. I., and C. E. Rupprecht. 1998b. A review of the economics of the prevention and control of rabies. 2. Rabies in dogs, livestock, and wildlife. *Pharmacoeconomics* 14:481–498.

Mendenhall, V. M., and L. F. Pank. 1980. Secondary poisoning of owls by anticoagulant rodenticides. *Wildlife Society Bulletin* 8:311–315.

Mendes-de-Almeida, F., N. Labarthe, J. Guerrero, M. C. F. Faria, A. S. Branco, C. D. Pereira, J. D. Barreira, and M. J. S. Pereira. 2007. Follow-up of the health conditions of an urban colony of free-roaming cats (*Felis catus* Linnaeus, 1758) in the city of Rio de Janeiro, Brazil. *Veterinary Parasitology* 147:9–15.

Mengel, R.M. 1971. A study of dog-coyote hybrids and implications concerning hybridization in *Canis. Journal of Mammalogy* 52:316–336.

Mercure, A., K. Ralls, K. P. Koepeli, and R. K. Wayne. 1993. Genetic subdivisions among small canids: Mitrochondrial DNA differentiation of swift, kit, and arctic foxes. *Evolution* 47:1313–1328.

Metropolitan Bakersfield Habitat Conservation Plan. 1994. *Metropolitan Bakersfield Habitat Conservation Plan.* Metropolitan Bakersfield Habitat Conservation Plan Steering Committee, Bakersfield, CA.

Meyer, S., and J. M. Weber. 1996. Ontogeny of dominance in free-living red foxes. *Ethology* 102:1008–1019.

Meyer, W. 1999. Feeding biology of stray cats. *Kleintierpraxis* 44:15–26.

Meyer, W., A. Schnapper, and G. Eilers. 2003. Garbage-dependent nutrition of wild canids and stray dogs. 2. Stray dogs and dangers. *Kleintierpraxis* 48:419–426.

Mezquida, E. T., S. S. Slater, and C. W. Benkman. 2006. Sage-grouse and indirect interactions: Potential implications of coyote control on sage-grouse populations. *Condor* 108:747–759.

Middleton, A. L. V., and R. J. Paget. 1974. *Badgers of Yorkshire and Humberside.* William Sessions, York, UK.

Miller, B., R. P. Reading, and S. Forrest. 1996. *Prairie Night: Black-Footed Ferrets and the Recovery of Endangered Species.* Smithsonian Institution Press, Washington, D.C.

Miller, C. A., L. K. Campbell, and J. A. Yeagle. 2001a . *Attitudes of Homeowners in the Greater Chicago Metropolitan Region toward Nuisance Wildlife.* Human Dimensions Program Report SR-00-02 Project No. IL W-112-R-9/Job 103.1/Study 103. Illinois Department of Natural Resources, Springfield.

Miller, J. R., and R. J. Hobbs. 2002. Conservation where people live and work. *Conservation Biology* 16:330–337.

Miller, R., and G. V. J. Howell. 2008. Regulating consumption with bite: Building a contemporary framework for urban dog management. *Journal of Business Research* 61:525–531.

Miller, S. D., and D. W. Speake. 1979. Progress report: Demography and home range of the bobcat in south Alabama. Bobcat Research Conference Proceedings. *National Wildlife Federation Scientific Technical Series* 6:123–124.

Miller, S. G., R. L. Knight, and C. K. Miller. 2001b. Wildlife responses to pedestrians and dogs. *Wildlife Society Bulletin* 29:124–132.

Mills, D. S., S. L. Bailey, and R. E. Thurstans. 2000. Evaluation of the welfare implications and efficacy of an ultrasonic "deterrent" for cats. *Veterinary Record* 147:678–680.

Mills, M. G. L., S. Freitag, and A. S. Van Jaarsveld. 2004. Geographic priorities for carnivore conservation in Africa. Pages 467–483 in J. L. Gittleman, S. M. Funk, D. Macdonald, and R. K. Wayne, eds. *Carnivore Conservation.* Cambridge University Press, Cambridge.

Mills, M. G. L., and H. Hofer, compilers. 1998. *Hyaenas.* Status Survey and Conservation Action Plan. IUCN/SSC Hyaena Specialist Group. IUCN, Gland, Switzerland.

Mills, M. G. L., J. M. Juritz, and W. Zucchini. 2001. Estimating the size of spotted hyaena (*Crocuta crocuta*) populations through playback recordings. *Animal Conservation* 4:335–343.

Mills, M. G. L., and M. E. J. Mills. 1978. The diet of the brown hyaena *Hyaena brunnea* in the southern Kalahari. *Koedoe* 21:125–149.

Miquelle, D., I. Nikolaev, J. Goodrich, B. Litvinov, E. Smirnov, and E. Suvorov. 2005. Searching for the coexistence recipe: A case study of conflicts between people and tigers in the Russian Far East. Pages 305–322 *in* R. Woodroffe, S. Thirgood, and A. Rabinowitz, eds. *People and Wildlife: Conflict or Coexistence?* Cambridge University Press, New York.

Mirmovitch, V. 1995. Spatial organisation of urban feral cats (*Felis catus*) in Jerusalem. *Wildlife Research* 22:299–310.

Mishra, C. 1997. Livestock depredation by large carnivores in the Indian trans-Himalaya: Conflict perceptions and conservation prospects. *Environmental Conservation* 24:338–343.

Mishra, C. 2001. High altitude survival: Conflicts between pastoralism and wildlife in the trans-Himalaya. Master's thesis, Wageningen University, The Netherlands.

Mitchell, B. R., M. M. Jaeger, and R. H. Barrett. 2004. Coyote depredation management: Current methods and research needs. *Wildlife Society Bulletin* 32:1209–1218.

Mitchell, M. S., D. E. Ausband, C. A. Sime, E. E. Bangs, J. A. Gude, M. D. Jimenez, C. M. Mack, T. J. Meier, M. S. Nadeau, and D. W. Smith. 2008a. Estimation of successful breeding pairs for wolves in the Northern Rocky Mountains, USA. *Journal of Wildlife Management* 72:881–891.

Mitchell, N., R. Pratt, and L. Malone 2008b. The Narragansett Bay coyote study. *Northeast Natural History Conference* 10:36 (abstract).

Møller, A. P., and J. Erritzøe. 2000. Predation against birds with low immunocompetence. *Oecologia* 122:500–504.

Montag, J. M., M. E. Patterson, and W. A. Freimund. 2005. The wolf viewing experience in the Lamar Valley of Yellowstone National Park. *Human Dimensions of Wildlife* 10:273–84.

Moodie, E. 1995. *The Potential for Biological Control of Feral Cats in Australia.* Report to Australian Nature Conservancy Agency, Canberra, Australia.

Moore, G. C., and G. R. Parker. 1992. Colonization by the eastern coyote (*Canis latrans*). Pages 23–37 *in* A. Boer, ed. *Ecology and Management of the Eastern Coyote.* Wildlife Research Unit, University of New Brunswick, Fredericton, Canada.

Moore, L. 1963. *Little Raccoon and the Thing in the Pool.* Scholastic Press, New York.

Morey, P. S. 2004. Landscape use and diet of coyotes, *Canis latrans*, in the Chicago Metropolitan Area. Master's thesis, Utah State University, Logan.

Morey, P. S., E. M. Gese, and S. Gehrt. 2007. Spatial and temporal variation in the diet of coyotes in the Chicago Metropolitan Area. *American Midland Naturalist* 158:147–161.

Morgan, E. R., S. E. Shaw, S. F. Brennan, T. D. De Waal, B. R. Jones, and G. Mulcahy. 2005. *Angiostrongylus vasorum*: A real heartbreaker. *Trends in Parasitology* 21:49–51.

Morgan, E. R., A. Tomlinson, S. Hunter, T. Nichols, E. Roberts, M. T. Fox, and M. A. Taylor. 2008. *Angiostrongylus vasorum* and *Eucoleus aerophilus* in foxes (*Vulpes vulpes*) in Great Britain. *Veterinary Parasitology* 154:48–57.

Moriarty, J. M. 2007. Female bobcat reproductive behavior and kitten survival in an urban fragmented landscape. Master's thesis, California State University, Northridge.

Morrell, S. 1972. Life history of the San Joaquin kit fox. *California Fish and Game* 58:162–174.

Morris, D. 1986. *Dogwatching.* Jonathan Cape, London.

Morris, D. 1999. *Cat Breeds of the World.* Viking, New York.

Morris, J. A. 1959. *The Adventures of Rick Raccoon.* National Wildlife Federation, Washington, D.C.

Morris, J. G. 2002. Idiosyncratic nutrient requirements of cats appear to be diet-induced evolutionary adaptations. *Nutrition Research Reviews* 15:153–168.

Morrison, S. A., and D. T. Bolger. 2002. Variation in a sparrow's reproductive success with rainfall: Food and predator-mediated processes. *Oecologia* 133:315–324.

Mosillo, M., E. J. Heske, and J. D. Thompson. 1999. Survival and movements of translocated raccoons in northcentral Illinois. *Journal of Wildlife Management* 63:278–286.

Moyer, M. A., J. W. McCown, T. H. Eason, and M. K. Oli. 2006. Does genetic relatedness influence space use pattern? A test on Florida black bears. *Journal of Mammalogy* 87:255–261.

Mullens, A. 1996. "I think we have a problem in Victoria": MDs respond quickly to toxoplasmosis outbreak in BC. *Canadian Medical Association Journal* 154:1721–1724.

Munson, L., L. Marker, E. Dubovi, J. A. Spencer, J. F. Evermann, and S. J. O'Brien. 2004. Serosurvey of viral infections in free-ranging Namibian cheetahs (*Acinonyx jubatus*). *Journal of Wildlife Diseases* 40:23–31.

Murphy, K. 1983. *Characteristics of a Hunted Population of Mountain Lions in Western Montana.* Final Job Report. Project W-120-R-13 and 14. Montana Fish Wildlife and Parks, Missoula.

Murphy, K. M. 1998. The ecology of the cougar (*Puma concolor*) in the northern Yellowstone ecosystem: Interactions with prey, bears, and humans. Ph.D. diss., University of Idaho, Moscow.

Murray, D. L., C. A. Kapke, J. F. Evermann, and T. K. Fuller. 1999. Infectious disease and the conservation of free-ranging large carnivores. *Animal Conservation* 2:241–254.

Myers, N. 1990. The biodiversity challenge: Expanded hotspots analysis. *The Environmentalist* 10:243–256.

Nass, R. D., G. Lynch, and J. Theade. 1984. Circumstances associated with predation rates on sheep and goats. *Journal of Range Management* 37:423–426.

Nassar, R., and J. E. Mosier. 1980. Canine population dynamics: A study of the Manhattan, Kansas, canine population. *American Journal of Veterinary Research* 41:1798–1803.

Nassar, R., and J. E. Mosier. 1982. Feline population dynamics: A study of the Manhattan, Kansas, feline population. *American Journal of Veterinary Research* 43:167–170.

Nassar, R., and J. E. Mosier. 1991. Projections of pet population from census demographic data. *Journal of the American Veterinary Medical Association* 198:1157–1159.

Nassar, R., J. E. Mosier, and L. W. Williams. 1984. Study of the feline and canine populations in the greater Las Vegas area. *American Journal of Veterinary Research* 45:282–287.

Natoli, E. 1985. Spacing pattern in a colony of urban stray cats (*Felis catus* L.) in the historic centre of Rome. *Applied Animal Behaviour Science* 14:289–304.

Natoli, E. 1990. Mating strategies in cats: A comparison of the role and importance of infanticide in domestic cats, *Felis catus* L., and lions, *Panthera leo* L. *Animal Behaviour* 40:183–186.

Natoli, E., A. Baggio, and D. Pontier. 2001. Male and female agonistic and affiliative relationships in a social group of farm cats (*Felis catus* L.). *Behavioural Processes* 53:137–143.

Natoli, E., and E. De Vito. 1988. The mating system of feral cats living in a group. Pages 99–108 *in* D. C. Turner and P. Bateson, eds. *The Domestic Cat: The Biology of Its Behaviour.* Cambridge University Press, Cambridge.

Natoli, E., and E. De Vito. 1991. Agonistic behaviour, dominance rank and copulatory success in a large multi-male feral cat, *Felis catus* L., colony in central Rome. *Animal Behaviour* 42:227–241.

Natoli, E., E. De Vito, and D. Pontier. 2000. Mate choice in the domestic cat (*Felis silvestris catus* L.). *Aggressive Behavior* 26:455–465.

Natoli, E., L. Maragliano, G. Cariola, A. Faini, R. Bonanni, S. Cafazzo, and C. Fantini. 2006. Management of feral domestic cats in the urban environment of Rome (Italy). *Preventive Veterinary Medicine* 77:180–185.

Natoli, E., L. Say, S. Cafazzo, R. Bonanni, M. Schmid, and D. Pontier. 2005. Bold attitude makes male urban feral domestic cats more vulnerable to feline immunodeficiency virus. *Neuroscience and Biobehavioral Reviews* 29:151–157.

Natoli, E., M. Schmid, L. Say, and D. Pontier. 2007. Male reproductive success in a social group of urban feral cats (*Felis catus* L.). *Ethology* 113:283–289.

Neal, E., and C. Cheeseman. 1996. *Badgers.* Poyser, London.

Neal, E. G., and T. J. Roper. 1991. The environmental impact of badgers (*Meles meles*) and their setts. *Symposia of the Zoological Society of London* 63:89–106.

Neiswenter, S. A., D. B. Pence, and R. C. Dowler. 2006. Helminths of sympatric striped, hog-nosed, and spotted skunks in west-central Texas. *Journal of Wildlife Diseases* 42:511–517.

Nellis, C. H., and L. B. Keith. 1976. Population dynamics of coyotes in central Alberta, 1964–68. *Journal of Wildlife Management* 40:380–399.

Nelson, J. L., B. L. Cypher, C. D. Bjurlin, and S. Creel. 2007. Effects of habitat on competition between kit foxes and coyotes. *Journal of Wildlife Management* 71:1467–1475.

Nelson, S. H., A. D. Evans, and R. B. Bradbury. 2005. The efficacy of collar-mounted devices in reducing the rate of predation of wildlife by domestic cats. *Applied Animal Behaviour Science* 94:273–285.

Nelson, S. H., A. D. Evans, and R. B. Bradbury. 2006. The efficacy of an ultrasonic cat deterrent. *Applied Animal Behaviour Science* 96:83–91.

Nelson, T. A., D. G. Gregory, and J. R. Laursen. 2003. Canine heartworms in coyotes in Illinois. *Journal of Wildlife Diseases* 39:593–599.

Nemiroff, L., and J. Patterson. 2007. Design, testing and implementation of a large-scale urban dog waste composting program. *Compost Science and Utilization* 15:237–242.

New, C. J., M. D. Salman, M. King, J. M. Scarlett, P. H. Kass, and J. M. Hutchison. 2000. Characteristics of shelter-relinquished animals and their owners compared with animals and their owners in U.S. pet-owning households. *Journal of Applied Animal Welfare Science* 3:179–201.

New, J. C., W. J. Kelch, J. M. Hutchinson, M. D. Salman, M. King, J. M. Scarlett, and P. H. Kass. 2004. Birth and death rate estimates of cats and dogs in U.S. households and related factors. *Journal of Applied Animal Welfare Science* 7:229–241.

Newman, T. J., P. J. Baker, and S. Harris. 2002. Nutritional condition and survival of red foxes with sarcoptic mange. *Canadian Journal of Zoology* 80:154–161.

Newman, T. J., P. J. Baker, E. Simcock, G. Saunders, P. C. L. White, and S. Harris. 2003. Changes in red fox habitat preference and rest site fidelity following a disease-induced population decline. *Acta Theriologica* 48:79–91.

Newton-Fisher, N., S. Harris, P. White, and G. Jones. 1993. Structure and function of red fox *Vulpes vulpes* vocalisations. *Bioacoustics* 5:1–31.

Ng, S. J., J. W. Dole, R. M. Sauvajot, S. P. D. Riley, and T. J. Valone. 2004. Use of highway undercrossings by wildlife in southern California. *Biological Conservation* 115:499–507.

Nicht, M. 1969. Ein Beitrag zum Vorkommen des Steinmarders, *Martes foina* (Erxleben 1770), in der Großstadt (Magdeburg). *Zeitschrift für Jagdwissenschaft* 15:1–6.

Niebylski, M., H. Savage, R. Nasci, and G. Graig, Jr. 1994. Blood hosts of *Aedes albopictus* in the United States. *Journal of the American Mosquito Control Association—Mosquito News* 10:447–450.

Nielsen, C. K., and A. Woolf. 2001. Bobcat habitat use relative to human dwellings in southern Illinois. Pages 40–44 *in* A. Woolf and C. K. Nielsen, eds. *Proceedings of a Symposium on the Ecology and Management of Bobcats.* Illinois Department of Natural Resources, Champaign.

Nilon, C. H., and R. C. Pais. 1997. Terrestrial vertebrates in urban ecosystems: Developing hypotheses for the Gwynns Falls Watershed in Baltimore, Maryland. *Urban Ecosystems* 1:247–257.

Ninomiya, H., M. Ogata, and T. Makino. 2003. Notoedric mange in free-ranging masked palm civets (*Paguma larvata*) in Japan. *Veterinary Dermatology* 14:339–344.

Noah, D. L., M. G. Smith, J. C. Gotthardt, J. W. Krebs, D. Green, and J. E. Childs. 1996. Mass human exposure to rabies in New Hampshire: Exposures, treatment, and cost. *American Journal of Public Health* 86:1149–1151.

Nogales, M., A. Martín, B. R. Tershy, C. J. Donlan, D. Veitch, N. Puerta, B. Wood, and J. Alonso. 2004. A review of feral cat eradication on islands. *Conservation Biology* 18:310–319.

Nogalski, A., L. Jankiewicz, G. Ćwik, J. Karski, and L. Matuszewski. 2007. Animal related injuries treated at the department of trauma and emergency medicine, Medical University of Lublin. *Annals of Agricultural and Environmental Medicine* 14:57–61.

North, S. 1963. *Rascal: A Memoir of a Better Era.* E. P. Dutton, New York.

North, S. 1966. *Raccoons Are the Brightest People.* E. P. Dutton, New York.

Northeastern Illinois Planning Commission. 1992. *Strategic Plan for Land Resource Management.* Chicago.

Noss, R. F., H. B. Quigley, M. G. Hornocker, T. Merrill, and P. C. Paquet. 1996. Conservation Biology and Carnivore Conservation in the Rocky Mountains. *Conservation Biology* 10:949–963.

Novak, M., M. E. Obbard, J. G. Jones, R. Newman, A. Booth, A. J. Satterthwaite, and G. A. Novikov. 1962. *Carnivorous Mammals of the Fauna of the USSR.* Israel Program for Scientific Translations, Jerusalem.

Nowak, K. I. J., R. Rowntree, E. G. McPherson, S. M. Sisinni, E. Kerkmann, and J. C. Stevens. 1996. Measuring and analyzing urban tree cover. *Landscape and Urban Planning* 36:49–57.

Nowak, R. 1978. Evolution and taxonomy of coyotes and related *Canis*. Pages 3–16 *in* M. Bekoff, ed. Coyotes: *Biology, Behavior, and Management.* Academic Press, New York.

Nowak, R. M. 1991. *Walker's Mammals of the World.* Johns Hopkins University Press, Baltimore.

Nowak, R. M., and J. L. Paradiso. 1983. *Walker's Mammals of the World.* 4th ed. Johns Hopkins University Press, Baltimore.

Nowell, K., and P. Jackson, eds. 1996. *Wild Cats.* Status Survey and Conservation Action Plan. IUCN Felid Specialist Group, IUCN, Gland, Switzerland.

Nunes, C. M., D. de A. Martines, S. Fikaris, and L. H. Queiróz. 1997. Evaluation of dog population in an urban area of southeastern Brazil. *Revista de Saúde Pública* 31:308–309. [In Portuguese.]

Nutter, F. B., J. F. Levine, and M. K. Stoskopf. 2004a. Reproductive capacity of free-roaming domestic cats and kitten survival rate. *Journal of the American Veterinary Medical Association* 225:1399–1402.

Nutter, F. B., M. K. Stoskopf, and J. F. Levine. 2004b. Time and financial costs of programs for live trapping feral cats. *Journal of the American Veterinary Medical Association* 225:1403–1405.

Nyhus, P. J., and R. Tilson. 2004. Characterizing human-tiger conflict in Sumatra, Indonesia: Implications for conservation. *Oryx* 38:68–74.

O'Donnell, M. A., and A. J. DeNicola. 2006. Den site selection of lactating female raccoons following removal and exclusion from suburban residences. *Wildlife Society Bulletin* 34:366–370.

O'Donoghue, M., S. Boutin, C. J. Krebs, G. Zuleta, D. L. Murray, and E. J. Hofer. 1998. Functional responses of coyotes and lynx to the snowshoe hare cycle. *Ecology* 79:1193–1208.

Oli, M. K., I. R. Taylor, and M. E. Rogers. 1994. Snow leopard *Panthera uncia* predation of livestock: An assessment of local perceptions in the Annapurna Conservation Area, Nepal. *Biological Conservation* 68:63–68.

Oliver, A. J., and S. H. Wheeler. 1978. Toxicity of the anticoagulant pindone to the European rabbit, *Oryctolagus cuniculus* and the sheep, *Ovis aries. Australian Wildlife Research* 5:135–142.

Olsen, S. J. 1985. *Origins of the Domestic Dog: The Fossil Record.* University of Arizona Press, Tucson.

Organ, J. F., D. J. Decker, L. H. Carpenter, W. F. Siemer, and S. J. Riley. 2006. *Thinking Like a Manager: Reflections on Wildlife Management.* Wildlife Management Institute, Washington, D.C.

Ortega-Pacheco, A., J. C. Rodriguez-Buenfil, M. E. Bolio-Gonzalez, C. H. Sauri-Arceo, M. Jiménez-Coello, and C. L. Forsberg. 2007. A survey of dog populations in urban and rural areas of Yucatan, Mexico. *Anthrozoos* 20:261–274.

Ough, W. D. 1982. Scent marking by captive raccoons. *Journal of Mammalogy* 63:318–319.

Ovsyanikov, N. G. 1993. *Behavior and Social Organization of the Arctic Fox.* Isd-vo TSNIL Glavachoti RF, Moscow, USSR. [In Russian.]

Padley, W. D. 1990. Home ranges and social interactions of mountain lions in the Santa Ana Mountains, California. Master's thesis, California State Polytechnic University, Pomona.

Page, L. K., S. D. Gehrt, and N. P. Robinson. 2008. Land-use effects on prevalence of raccoon roundworm (*Baylisascaris procyonis*). *Journal of Wildlife Diseases* 44:594–599.

Page, L. K., S. D. Gehrt, K. K. Titcombe, and N. P. Robinson. 2005. Measuring prevalence of raccoon roundworm (*Baylisascaris procyonis*): A comparison of techniques. *Wildlife Society Bulletin* 33:1406–1412.

Page, L. K., R. K. Swihart, and K. R. Kazacos. 1998. Raccoon latrine structure and its potential role in transmission of *Baylisascaris procyonis* to vertebrates. *American Midland Naturalist* 140:180–185.

Page, R. J. C. 1981. Dispersal and population density of the red fox (*Vulpes vulpes*) in an area of London. *Journal of Zoology* 194:485–491.

Page, R. J. C., J. Ross, and D. H. Bennett. 1992. A study of the home ranges, movements and behaviour of the feral cat population at Avonmouth Docks. *Wildlife Research* 19:263–277.

Pal, S. K. 2001. Population ecology of free-ranging urban dogs in West Bengal, India. *Acta Theriologica* 46:69–78.

Pal, S. K. 2003. Urine marking by free-ranging dogs (*Canis familiaris*) in relation to sex, season, place and posture. *Applied Animal Behaviour Science* 80:45–59.

Pal, S. K. 2005. Parental care in free-ranging dogs, *Canis familiaris*. *Applied Animal Behaviour Science* 90:31–47.

Pal, S. K. 2008. Maturation and development of social behaviour during early ontogeny in free-ranging dog puppies in West Bengal, India. *Applied Animal Behaviour Science* 111:95–107.

Pal, S. K., B. Ghosh, and S. Roy. 1998. Dispersal behaviour of free-ranging dogs (*Canis familiaris*) in relation to age, sex, season and dispersal distance. *Applied Animal Behaviour Science* 61:123–132.

Pal, S. K., B. Ghosh, and S. Roy. 1999. Inter- and intra-sexual behaviour of free-ranging dogs (*Canis familiaris*). *Applied Animal Behaviour Science* 62:267–278.

Palomares, F., and T. M. Caro. 1999. Interspecific killing among mammalian carnivores. *American Naturalist* 153:492–508.

Palomares, F., P. Gaona, P. Ferreras, and M. Delibes. 1995. Positive effects on game species of top predators by controlling smaller predator populations: An example with lynx, mongooses, and rabbits. *Conservation Biology* 9:295–305.

Panaman, R. 1981. Behaviour and ecology of free-ranging female farm cats (*Felis catus* L.). *Zeitschrift für Tierpsychologie* 56:59–73.

Parker, G. 1995. *Eastern Coyote: The Story of Its Success*. Nimbus Publishing, Halifax, Nova Scotia.

Parris, K. M., and D. L. Hazell. 2005. Biotic effects of climate change in urban environments: The case of the grey-headed flying-fox (*Pteropus poliocephalus*) in Melbourne, Australia. *Biological Conservation* 124:267–276.

Pate, J., M. J. Manfredo, A. D. Bright, and G. Tischbein. 1996. Coloradans' attitudes toward reintroducing the gray wolf into Colorado. *Wildlife Society Bulletin* 24:421–428.

Patrick, G. R., and K. M. O'Rourke. 1998. Dog and cat bites: Epidemiologic analyses suggest different prevention strategies. *Public Health Reports* 113:252–257.

Patronek, G. J. 1998. Free-roaming and feral cats: Their impact on wildlife and human beings. *Journal of the American Veterinary Medical Association* 212:218–226.

Patronek, G. J., and L. T. Glickman. 1994. Development of a model for estimating the size and dynamics of the pet population. *Anthrozoos* 7:25–42.

Patten, M. A., and D. T. Bolger. 2003. Variation in top-down control of avian reproductive success across a fragmentation gradient. *Oikos* 101:479–488.

Patton, R. F. 1974. Ecological and behavioral relationships of the skunks of Trans Pecos, Texas. Master's thesis, Texas A&M University, College Station.

Paul, M. J., and J. L. Meyer. 2001. Streams in the urban landscape. *Annual Review of Ecology and the Systematics* 32:333–365.

Pauleit, S., R. Ennos, and Y. Golding. 2005. Modeling the environmental impacts of urban land use and land cover change: A study in Merseyside, UK. *Landscape and Urban Planning* 71:295–310.

Pearre, S., Jr., and R. Maass. 1998. Trends in the prey size-based trophic niches of feral and house cats *Felis catus* L. *Mammal Review* 28:125–139.

Pedlar, J. H., and L. Fahrig. 1997. Raccoon habitat use at 2 spatial scales. *Journal of Wildlife Management* 61:102–112.

Pedó, E., A. C. Tomazzoni, S. M. Hartz, and A. U. Christoff. 2006. Diet of crab-eating fox, *Cerdocyon thous* (Linnaeus) (Carnivora, Canidae), in a suburban area of southern Brazil. *Revista Brasileira de Zoologia* 23:637–641.

Pence, D. B., F. D. Matthews, and L. A. Windberg. 1982. Notoedric mange in the bobcat, *Felis rufus*, from south Texas. *Journal of Wildlife Diseases* 18:47–50.

Pence, D. B., M. E. Tewes, D. B. Shindle, and D. M. Dunn. 1995. Notoedric mange in an ocelot (*Felis pardalis*) from southern Texas. *Journal of Wildlife Diseases* 31:558–561.

Penner, L. R., and W. N. Parke. 1954. Notoedric mange in the bobcat, *Lynx rufus*. *Journal of Mammalogy* 35:458.

Pennisi, E. 2006. Crab, raccoon play tag team against turtle. *Science* 311:331.

Perry, B. D., T. M. Kyendo, S. W. Mbugua, J. E. Price, and S. Varma. 1995. Increasing rabies vaccination coverage in urban dog populations of high human population density suburbs: A case study in Nairobi, Kenya. *Preventive Veterinary Medicine* 22:137–142.

Peters, G., and W. C. Wozencraft. 1989. Acoustic communication by fissiped carnivores. Pages 14–56 *in* J. L. Gittleman, ed. *Carnivore Behavior, Ecology, and Evolution*. Cornell University Press, Ithaca, NY.

Peterson, R. O. 1977. Wolf ecology and prey relationships on Isle Royale. *United States National Park Service Monograph Series* 11:1–210.

Pianka, E. R. 1973. The structure of lizard communities. *Annual Review of Ecology and Systematics* 20:53–74.

Pickett, S. T. A., M. L. Cadenasso, J. M. Grove, C. H. Nilon, R. V. Pouyat, W. C. Zipperer, and R. Constanza. 2001. Urban ecological systems: Linking terrestrial ecological, physical, and socioeconomic components of metropolitan areas. *Annual Review of Ecology and Systematics* 32:127–157.

Pickett, S. T. A., M. L. Cadensasso, J. M. Grove, P. M. Groffman, L. E. Band, C. G. Boone, W. R. Burch, Jr., et al.

Wilson. 2008. Beyond urban legends: An emerging framework of urban ecology, as illustrated by the Baltimore Ecosystem Study. *BioScience* 58:139–150.

Pierce B. M., and V. C. Bleich. 2003. Mountain Lion. Pages 744–757 in G. A. Feldhamer, B. C. Thompson, and J. A. Chapman, eds. *Wild Mammals of North America.* Johns Hopkins University Press, Baltimore.

Pierce, B. M., V. C. Bleich, and R. T. Bowyer. 2000a. Selection of mule deer by mountain lions and coyotes: Effects of hunting style, body size, and reproductive status. *Journal of Mammalogy* 81:462–472.

Pierce, B. M., V. C. Bleich, and R. T. Bowyer. 2000b. Social organization of mountain lions: Does a land-tenure system regulate population size? *Ecology* 81:1533–1543.

Pils, C. M., and M. A. Martin. 1978. Population dynamics, predator–prey relationships and management of the red fox in Wisconsin. *Department of Natural Resources, Madison, Wisconsin, Technical Bulletin* 105:1–56.

Pimentel, D., S. McNair, J. Janecka, J. Wightman, C. Simmonds, C. O'Connell, E. Wong, et al. 2001. Economic and environmental threats of alien plant, animal, and microbe invasions. *Agriculture, Ecosystems and Environment* 84:1–20.

Pimm, S. L., L. Dollar, and O. L. Bass. 2006. Genetic rescue of the Florida panther. *Animal Conservation* 9:115–122.

Polis, G. A., W. B. Anderson, and R. D. Holt. 1997. Towards an integration of landscape and food web ecology: The dynamics of spatially subsidized food webs. *Annual Review of Ecology and Systematics* 28:289–316.

Polis, G. A., and R. D. Holt. 1992. Intraguild predation: The dynamics of complex trophic interactions. *Trends in Ecology and Evolution* 7:151–154.

Polis, G. A., C. A. Myers, and R. D. Holt. 1989. The ecology and evolution of intraguild predation: Potential competitors that eat each other. *Annual Review of Ecology of Systematics* 20:297–330.

Pontier, D., E. Fromont, F. Courchamp, M. Artois, and N. G. Yoccoz. 1998. Retroviruses and sexual size dimorphism in domestic cats (*Felis catus* L.). *Proceedings of the Royal Society of London B* 265:167–173.

Pontier, D., and E. Natoli. 1996. Male reproductive success in the domestic cat (*Felis catus* L.): A case history. *Behavioural Processes* 37:85–88.

Pontier, D., and E. Natoli. 1999. Infanticide in rural male cats (*Felis catus* L.) as a reproductive mating tactic. *Aggressive Behavior* 25:445–449.

Pontier, D., L. Say, F. Debais, J. Bried, J. Thioulouse, T. Micol, and E. Natoli. 2002. The diet of feral cats (*Felis catus* L.) at five sites on the Grande Terre, Kerguelen Archipelago. *Polar Biology* 25:833–837.

Power, M. E. 1992. Top-down and bottom-up forces in food webs: Do plants have primacy? *Ecology* 73:733–746.

Prange, S., and S. D. Gehrt. 2004. Changes in mesopredator-community structure in response to urbanization. *Canadian Journal of Zoology* 82:1804–1817.

Prange, S., and S. D. Gehrt. 2007. Skunk response to simulated coyote activity: A test of the mesopredator release hypothesis. *Journal of Mammalogy* 88:1040–1049.

Prange, S., S. D. Gehrt, and E. P. Wiggers. 2003. Demographic factors contributing to high raccoon densities in urban landscapes. *Journal of Wildlife Management* 67:324–333.

Prange, S., S. D. Gehrt, and E. P. Wiggers. 2004. Influences of anthropogenic resources on raccoon (*Procyon lotor*) movements and spatial distribution. *Journal of Mammalogy* 85:483–90.

Presutti, R. J. 2001. Prevention and treatment of dog bites. *American Family Physician* 63:1567–1572.

Prigioni, C., and A. Sommariva. 1997. Ecologia della faina, *Martes foina* (Erxleben, 1777) nell'ambiente urbano di Cavalese (Trento). *Report Centro Ecologia Alpina* 11:1–22. [In Italian.]

Proulx, G. 1988. Control of urban wildlife predation by cats through public education. *Environmental Conservation* 15:358–359.

Pulliam, H. R. 1988. Sources, sinks, and population regulation. *American Naturalist* 132:652–662.

Pybus, M. 1988. Rabies and rabies control in striped skunks (*Mephitis mephitis*) in three prairie regions of western North America. *Journal of Wildlife Diseases* 24:434–449.

Pyšek, P. 1998. Alien and native species in Central European urban floras: A quantitative comparison. *Journal of Biogeography* 25:155–163.

Pyšek, P., Z. Chocholouskova, A. Pysek, V. Jarosik, M. Chytry, and L. Tichy. 2004. Trends in species diversity and composition of urban vegetation over three decades. *Journal of Vegetation Science* 15:781–788.

Quinn, T. 1997a. Coyote (*Canis latrans*) food habits in three urban habitat types of western Washington. *Northwest Science* 71:1–5.

Quinn, T. 1997b. Coyote (*Canis latrans*) habitat selection in urban areas of western Washington via analysis of routine movements. *Northwest Science* 71:289–297.

Rabinowitz, A. R., and M. R. Pelton. 1986. Day-bed use by raccoons. *Journal of Mammalogy* 67:766–769.

Ralls, K., B. L. Cypher, and L. K. Spiegel. 2007. Social monogamy in kit foxes: Formation, association, duration, and dissolution of mated pairs. *Journal of Mammalogy* 88:1439–1446.

Ralls, K., K. Pilgrim, P. J. White, E. E. Paxinos, and R. C. Fleischer. 2001. Kinship, social relationships and den-sharing in kit foxes. *Journal of Mammalogy* 82:858–866.

Ralls, K., and P. J. White. 1995. Predation on San Joaquin kit foxes by larger canids. *Journal of Mammalogy* 76:723–729.

Ramsden, R., and P. Presidente. 1975. *Paragonimus kellicotti* infection in wild carnivores in southwestern Ontario: Prevalence and gross pathologic features. *Journal of Wildlife Diseases* 11:136–141.

Randa, L. A., and J. A. Yunger. 2006. Carnivore occurrence along an urban-rural gradient: A landscape-level analysis. *Journal of Mammalogy* 87:1154–1164.

Randler, C. 2006. Disturbances by dog barking increase vigilance in coots *Fulica atra*. *European Journal of Wildlife Research* 52:265–270.

Rasker, R., and A. Hackman. 1996. Economic development and the conservation of large carnivores. *Conservation Biology* 10:991–1002.

Ratnaswamy, M. J., and R. J. Warren. 1998. Removing raccoons to protect sea turtle nests: Are there implications for ecosystem management? *Wildlife Society Bulletin* 26:846–850.

Ratnaswamy, M. J., R. J. Warren, M. T. Kramer, and M. D. Adam. 1997. Comparisons of lethal and nonlethal techniques to reduce raccoon depredation of sea turtle nests. *Journal of Wildlife Management* 61:368–376.

Ratnayeke, S., G. A. Tuskan, and M. R. Pelton. 2002. Genetic relatedness and female spatial organization in a solitary carnivore, the raccoon, *Procyon lotor*. *Molecular Ecology* 11:1115–1124.

Rauer, G., P. Kacznsky, and F. Knauer. 2003. Experiences with aversive conditioning of habituated brown bears in Austria and other European countries. *Ursus* 14:315–224.

Ray, J. C., K. H. Redford, R. S. Steneck, and J. Berger, eds. 2005. *Large Carnivores and the Conservation of Biodiversity*. Island Press, Washington, D.C.

Rees, W., and M. Wackernagel. 1996. Urban ecological footprints: Why cities cannot be sustainable and why they are a key to sustainability. *Enviromental Impact Assessment Review* 16:223–248.

Reiter, D. K., M. W. Brunson, and R. H. Schmidt. 1999. Public attitudes toward wildlife damage management and policy. *Wildlife Society Bulletin* 27:746–58.

Rembiesa, C., and D. J. Richardson. 2003. Helminth parasites of the house cat, *Felis catus*, in Connecticut, U.S.A. *Comparative Parasitology* 70:115–119.

Renn, O. 1992. *Concepts of Risk: A Classification*. Praeger, Westport, CT.

Reperant, L. A., D. Hegglin, C. Fischer, L. Kohler, J.-M. Weber, and P. Deplazes. 2007. Influence of urbanization on the epidemiology of intestinal helminths of the red fox (*Vulpes vulpes*) in Geneva, Switzerland. *Parasitology Research* 101:605–611.

Revilla, E., M. Delibes, A. Travaini, and F. Palomares. 1999. Physical and population parameters of Eurasian badgers (*Meles meles*) from Mediterranean Spain. *Zeitschrift für Säugetierkunde* 64:269–276.

Rezaeinasab, M., I. Rad, A. R. Bahonar, H. Rashidi, A. Fayaz, S. Simani, A. A. Haghdoost, F. Rad, and M. A. Rad. 2007. The prevalence of rabies and animal bites during 1994 to 2003 in Kerman province, southeast of Iran. *Iranian Journal of Veterinary Research* 8:343–350.

Rheindt, F. E. 2003. The impact of roads on birds: Does song frequency play a role in determining susceptibility to noise pollution? *Journal of Ornithology* 144:295–306.

Richards, D. T., S. Harris, and J. W. Lewis. 1993. Epidemiol-ogy of *Toxocara canis* in red foxes (*Vulpes vulpes*) from urban areas of Bristol. *Parasitology* 107:167–173.

Richards, D. T., S. Harris, and J. W. Lewis. 1995. Epidemiological studies on intestinal helminth parasites of rural and urban red foxes (*Vulpes vulpes*) in the United Kingdom. *Veterinary Parasitology* 59:39–51.

Richardson, P. R. K. 1987. Aardwolf: The most specialized myrmecophagous mammal? *South African Journal of Science* 83:643–646.

Ricketts, R., and M. Imhoff. 2003. Biodiversity, urban areas, and agriculture: Locating priority ecoregions for conservation. *Conservation Ecology* 8:1. www.consecol.org/vol8iss2/art1.

Riffell, S. K., K. J. Gutzwiller, and S. H. Anderson. 1996. Does repeated human intrusion cause cumulative declines in avian richness and abundance? *Ecological Applications* 6:492–505.

Riley, S. J., and D. J. Decker. 2000a. Risk perception as a factor in wildlife stakeholder acceptance capacity for cougars in Montana. *Human Dimensions of Wildlife* 5:50–62.

Riley, S. J., and D. J. Decker. 2000b. Wildlife stakeholder acceptance capacity for cougars in Montana. *Wildlife Society Bulletin* 28:931–939.

Riley, S. J., D. J. Decker, L. H. Carpenter, J. F. Organ, W. F. Siemer, G. F. Mattfeld, and G. Parsons. 2002. The essence of wildlife management. *Wildlife Society Bulletin* 30:585–593.

Riley, S. P. D. 1999. Spatial organization, food habits, and disease ecology of bobcats (*Lynx rufus*) and gray foxes (*Urocyon cinereoargenteus*) in national park areas in urban and rural Marin County, California. Ph.D. diss., University of California, Davis.

Riley, S. P. D. 2001. Spatial and resource overlap of bobcats and gray foxes in urban and rural zones of a national park. Pages 32–39 in A. Woolf and C. K. Nielsen, eds. *Proceedings of a Symposium on the Ecology and Management of Bobcats*. Illinois Department of Natural Resources, Champaign.

Riley, S. P. D. 2006. Spatial ecology of bobcats and gray foxes in urban and rural zones of a national park. *Journal of Wildlife Management* 70:1425–1435.

Riley, S. P. D., C. Bromley, R. H. Poppenga, F. A. Uzal, L. Whited, and R. M. Sauvajot. 2007. Anticoagulant exposure and notoedric mange in bobcats and mountain lions in urban southern California. *Journal of Wildlife Management* 71:1874–1884.

Riley, S. P. D., G. T. Busteed, L. B. Kats, T. L. Vandergon, L. F. S. Lee, R. G. Dagit, J. L. Kerby, R. N. Fisher, and R. M. Sauvajot. 2005. Effects of urbanization on the distribution and abundance of amphibians and invasive species in Southern California streams. *Conservation Biology* 19:1894–1907.

Riley, S. P. D., J. Foley, and B. Chomel. 2004. Exposure to feline and canine pathogens in bobcats (*Lynx rufus*) and gray foxes (*Urocyon cinereoargenteus*) in urban and rural zones of a national park in California. *Journal of Wildlife Diseases* 40:11–22.

Riley, S. P. D., J. Hadidian, and D. A. Manski. 1998. Population density, survival, and rabies in raccoons in an urban national park. *Canadian Journal of Zoology* 76:1153–1164.

Riley, S. P. D., J. P. Pollinger, R. M. Sauvajot, E. C. York, C. Bromley, T. K. Fuller, and R. K. Wayne. 2006. A Southern California freeway is a physical and social barrier to gene flow in carnivores. *Molecular Ecology* 15:1733–1741.

Riley, S. P. D., R. M. Sauvajot, T. K. Fuller, E. C. York, D. A. Kamradt, C. Bromley, and R. K. Wayne. 2003. Effects of urbanization and habitat fragmentation on bobcats and coyotes in Southern California. *Conservation Biology* 17:566–576.

Robel, R. J., N. A. Barnes and L. B. Fox. 1990. Raccoon populations: Does human disturbance increase mortality? *Transactions of the Kansas Academy of Science* 93:22–27.

Robel, R. J., A. D. Dayton, R. R. Henderson, R. L. Meduna, and C. W. Spaeth. 1981. Relationships between husbandry methods and sheep losses to canine predators. *Journal of Wildlife Management* 45:894–911.

Roberts, S. B. 2007. *Ecology of White-Tailed Deer and Bobcats on Kiawah Island, South Carolina: Implications for Suburban Habitat Preservation.* University of Georgia, Athens.

Robertson, C. P. J., P. J. Baker, and S. Harris. 2000. Ranging behaviour of juvenile red foxes and its implication for management. *Acta Theriologica* 45:525–535.

Robertson, C. P. J., and S. Harris. 1995a. The behaviour after release of captive-reared fox cubs. *Animal Welfare* 4:295–306.

Robertson, C. P. J., and S. Harris. 1995b. The condition and survival after release of captive-reared fox cubs. *Animal Welfare* 4:281–294.

Robertson, I. D., and R. C. Thompson. 2002. Enteric parasitic zoonoses of domesticated dogs and cats. *Microbes and Infection* 4:867–873.

Robinson, N. A., and C. A. Marks. 2001. Genetic structure and dispersal of red foxes (*Vulpes vulpes*) in urban Melbourne. *Australian Journal of Zoology* 49:589–601.

Robinson, R. A., and R. N. Pugh. 2002. Dogs, zoonoses, and immunosuppression. *Journal of the Royal Society for the Promotion of Health* 122:95–98.

Rochlitz, I. 2003a. Study of factors that may predispose domestic cats to road traffic accidents. Part 1. *Veterinary Record* 153:549–553.

Rochlitz, I. 2003b. Study of factors that may predispose domestic cats to freeway traffic accidents. Part 2. *Veterinary Record* 153:585–588.

Rochlitz, I. 2004. Clinical study of cats injured and killed in road traffic accidents in Cambridgeshire. *Journal of Small Animal Practice* 45:390–394.

Rodden, M., F. Rodrigues, and S. Bestelmeyer. 2004. Maned wolf. Pages 38–43 *in* C. Sillero-Zubiri, M. Hoffman, and D. W. Macdonald, eds. *Canids: Foxes, Wolves, Jackals, and Dogs.* Status Survey and Conservation Action Plan.

IUCN/SSC Canid Specialist Group. Gland, Switzerland and Cambridge.

Roder, J. D. 2001. *Veterinary Toxicology: The Practical Veterinarian.* Butterworth Heinemann, Boston.

Roelke-Parker, J. E., L. Munson, C. Packer, R. Kock, S. Cleaveland, M. Carpenter, S. J. O'Brien, et al. 1996. A canine distemper virus epidemic in Serengeti lions (*Panthera leo*). *Nature* 379:441–445.

Roemer, G. W., T. J. Coonan, L. Munson, and R. K. Wayne. 2004. Island fox. Pages 97–105 *in* C. Sillero-Zubiri, M. Hoffman, and D. W. Macdonald, eds. *Canids: Foxes, Wolves, Jackals, and Dogs.* Status Survey and Conservation Action Plan. IUCN/SSC Canid Specialist Group. Gland, Switzerland, and Cambridge, UK.

Roetzer, T., M. Wittenzeller, H. Haechel, and J. Nekovar. 2000. Phenology in central Europe: Differences and trends of spring phenophases in urban and rural areas. *International Journal of Biometeorology* 44:60–66.

Rogers, C. M., and M. J. Caro. 1998. Song sparrows, top carnivores, and nest predation: A test of the mesopredator release hypothesis. *Oecologia* 116:227–233.

Rollo, I. M. 1980. Drugs used in the chemotherapy of Helminthiasis. Pages 1013–1037 *in* L. Goodman and A. Gilman, eds. *The Pharmacological Basis of Therapeutics.* 6th ed. Macmillan, New York.

Romanowski, J. 1985. The diet of stray dogs in the suburbs of Warsaw. *Revue d'Ecologie (Terre et la Vie)* 40:203–204.

Roper, T. J. 1992. Badger (*Meles meles*) setts: Architecture, internal environment, and function. *Mammal Review* 22:43–53.

Rosatte, R. C. 1984. Seasonal occurrence and habitat preference of rabid skunks in southern Alberta. *Canadian Veterinary Journal* 25:142–144.

Rosatte, R. C. 1985. The study of rabies in urban wildlife. *Environmental Health Review* 29:5–6.

Rosatte, R. C. 1986. A strategy for urban rabies control: Social change implications. Ph.D. diss., Walden University, University Microfilms International Publishers, Ann Arbor, MI.

Rosatte, R. C. 1987. Striped, spotted, hooded, and hognosed skunk. Pages 599–613 *in* M. Nowak, J. A. Baker, M. E. Obbard, and B. Malloch, eds. *Wild Furbearer Management and Conservation in North America.* Ontario Ministry of Natural Resources, Toronto.

Rosatte, R. C. 1988. Rabies in Canada: History, epidemiology and control. *Canadian Veterinary Journal* 29:362–365.

Rosatte, R. C. 2000. Management of raccoons (*Procyon lotor*) in Ontario, Canada: Do human intervention and disease have significant impact on raccoon populations? *Mammalia* 64:369–390.

Rosatte, R. C., M. Allan, R. Warren, P. Neave, T. Babin, L. Buchanan, D. Donovan, K. Sobey, C. Davies, F. Muldoon, and A. Wandeler. 2005. Movements of two rabid raccoons, *Procyon lotor*, in Eastern Ontario. *Canadian Field-Naturalist* 119:453–454.

Rosatte, R. C., D. Donovan, M. Allan, L. Bruce, T. Buchanan, K. Sobey, C. Davies, A. Wandeler and F. Muldoon. 2007a. Rabies in vaccinated raccoons from Ontario, Canada. *Journal of Wildlife Diseases* 43:300–301.

Rosatte, R. C., D. Donovan, M. Allan, L. Howes, A. Silver, K. Bennett, C. MacInnes, C. Davies, A. Wandeler, and B. Radford. 2001. Emergency response to raccoon rabies introduction into Ontario. *Journal of Wildlife Diseases* 37:265–279.

Rosatte, R. C., and J. R. Gunson. 1984a. Dispersal and home range of striped skunks (*Mephitis mephitis*) in an area of population reduction in southern Alberta. *Canadian Field-Naturalist* 98:315–319.

Rosatte, R. C., and J. R. Gunson. 1984b. Presence of neutralizing antibodies to rabies virus in striped skunks from areas free of skunk rabies in Alberta. *Journal of Wildlife Diseases* 20:171–176.

Rosatte, R. C, D. R. Howard, J. B. Campbell, C. D. MacInnes. 1990. Intramuscular vaccination of skunks and raccoons against rabies. *Journal of Wildlife Diseases* 26:225–230.

Rosatte, R. C., P. M. Kelly-Ward, and C. D. MacInnes. 1987. A strategy for controlling rabies in urban skunks and raccoons. Pages 54–60 *in* L. W. Adams and D. L. Leedy, eds. *Integrating Man and Nature in the Metropolitan Environment*. National Institute for Urban Wildlife, Columbia, MD.

Rosatte, R. C., and S. Larivière. 2003. Skunks (*Genera Mephitis, Spilogale, and Conepatus*). Pages 692–707 *in* G. A. Feldhamer, B. C. Thompson, and J. A. Chapman, eds. *Wild Mammals of North America: Biology, Management, and Conservation*. Johns Hopkins University Press, Baltimore.

Rosatte, R. C., and K. F. Lawson. 2001. Acceptance of baits for delivery of oral rabies vaccine to raccoons. *Journal of Wildlife Diseases* 37:730–739.

Rosatte, R. C., E. MacDonald, K. Sobey, D. Donovan, L. Bruce, M. Allan, A. Silver, et al. 2007b. The elimination of raccoon rabies from Wolfe Island, Ontario: Animal density and movements. *Journal of Wildlife Diseases* 43:242–250.

Rosatte, R. C., and C. D. MacInnes. 1989. Relocation of city raccoons. Pages 87–92 *in Ninth Great Plains Wildlife Damage Control Workshop Proceedings*. USDA Forest Service, Ft. Collins, CO.

Rosatte, R. C., C. D. MacInnes, M. J. Power, D. H. Johnston, P. Bachmann, C. P. Nunan, C. Wannop, M. Pedde, and L. Calder. 1993. Tactics for the control of wildlife rabies in Ontario, Canada. *Reviews of the Science and Technical Office of International Epizootics* 12:95–98.

Rosatte, R. C., C. D. MacInnes, R. T. Williams, and O. Williams. 1997. A proactive prevention strategy for raccoon rabies in Ontario, Canada. *Wildlife Society Bulletin* 25:110–116.

Rosatte, R. C., M. J. Power, D. Donovan, J. C. Davies, M. Allan, P. Bachmann, B. Stevenson, A. Wandeler, and F. Muldoon. 2007c. Elimination of Arctic variant rabies in red foxes, metropolitan Toronto. *Emerging Infectious Diseases* 13:25–27.

Rosatte, R. C., M. J. Power, and C. D. MacInnes. 1991. Ecology of urban skunks, raccoons, and foxes in metropolitan Toronto. Pages 31–38 *in* L. W. Adams and D. L. Leedy, eds. *Wildlife Conservation in Metropolitan Environments*. National Institute for Urban Wildlife, Columbia, MD.

Rosatte, R. C., M. J. Power, and C. D. MacInnes. 1992a. Trap-vaccinate-release and oral vaccination techniques for rabies control in urban skunks, raccoons, and foxes. *Journal of Wildlife Diseases* 28:562–571.

Rosatte, R. C., M. J. Power, and C. D. MacInnes. 1992b. Density, dispersion, movements and habitat of skunks (*Mephitis mephitis*) and raccoons (*Procyon lotor*) in Metropolitan Toronto. Pages 932–944 *in* D. McCullough and R. Barrett, eds. *Wildlife 2001: Populations*. Elsevier Science, London.

Rosatte, R. C., M. J. Pybus, and J. R. Gunson. 1986. Population reduction as a factor in the control of skunk rabies in Alberta. *Journal of Wildlife Diseases* 22:459–467.

Rosatte, R. C., K. Sobey, D. Donovan, M. Allan, L. Bruce, T. Buchanan, and C. Davies. 2007d. Raccoon density and movements in areas where population reduction programs were implemented to control rabies. *Journal of Wildlife Management* 71:2373–2378.

Rosatte, R. C., K. Sobey, D. Donovan, L. Bruce, M. Allan, A. Silver, K. Bennett, et al. 2006. Behavior, movements and demographics of rabid raccoons in Ontario, Canada: Management implications. *Journal of Wildlife Diseases* 42:589–605.

Rosatte, R. C., R. R. Tinline, and D. H. Johnston. 2007e. Rabies control in wild carnivores. Pages 595–634 *in* A. Jackson and W. Wunner, eds. *Rabies*. 2nd ed. Academic Press, San Diego, CA.

Rosatte, R. C., A. Wandeler, F. Muldoon, and D. Campbell. 2007f. Porcupine quills in raccoons as an indicator of rabies, distemper, or both diseases: Disease management implications. *Canadian Veterinary Journal* 48:299–300.

Røskaft, E., B. Händel, T. Bjerke, and B. P. Kaltenborn. 2007. Human attitudes towards large carnivores in Norway. *Wildlife Biology* 13:172–185.

Ross, P. I., and M. G. Jalkotzy. 1992. Characteristics of a hunted population of cougars in southwestern Alberta. *Journal of Wildlife Management* 56:417–426.

Ross, P. I., M. G. Jalkotzy, and P. Daoust. 1995. Fatal traumas sustained by cougars while attacking prey in southern Alberta. *Canadian Field-Naturalist* 109:261–263.

Ross, P. I, M. G. Jalkotzy, and M. Festa-Bianchet. 1997. Cougar predation on bighorn sheep in southwestern Alberta during winter. *Canadian Journal of Zoology* 74:771–775.

Roundtree, G. H. 2004. Comparative study of the home range and habitat usage of red foxes and gray foxes in an urban setting: A preliminary report. Pages 238–244 *in* W. W. Shaw, L. K. Harris, and L. VanDruff, eds.

Proceedings of the 4th International Symposium on Urban Wildlife Conservation, May 1–5 1999. Tucson, AZ.

Roy, L. D., and M. J. Dorrance. 1976. *Methods of Investigating Predation on Domestic Livestock: A Manual for Investigating Officers.* Alberta Agriculture Plant Industry Laboratory, Edmonton.

Rubin, H. D., and A. M. Beck. 1982. Ecological behaviour of free-ranging urban pet dogs. *Applied Animal Ethology* 8:161–168.

Ruell, E. W., S. P. D. Riley, M. R. Douglas, J. P. Pollinger, and K. R. Crooks. 2009. Estimating bobcat population sizes and densities in a fragmented urban landscape using noninvasive capture-recapture sampling. *Journal of Mammalogy* 90:129–135.

Ruth, T. K., J. M. Packard, D. S. Neighbor, and J. R. Skiles. 1991. Mountain lion use of an area of high recreational development in Big Bend National Park, Texas. Page 20 *in* C. E. Braun, ed. *Mountain Lion-Human Interaction Symposium and Workshop.* Colorado Division of Wildlife, Denver.

Ruxton, G. D., S. Thomas, and J. W. Wright. 2002. Bells reduce predation of wildlife by domestic cats (*Felis catus*). *Journal of Zoology* 256:81–83.

Rypula, K., M. Wlodarczyk, P. Chorbinski, and K. Ploneczka. 2004. Viral infections in cats in Wroclaw city. *Medycyna Weterynaryjna* 60:841–844.

Saberwal, V. K., J. P. Gibs, R. Chellam, and A. T. J. Johnsingh, 1994. Lion-human conflict in the Gir Forest, India. *Conservation Biology* 8:501–507.

Sacks, J. J., L. Sinclair, J. Gilchrist, G. C. Golab, and R. Lockwood. 2000. Breeds of dogs involved in fatal human attacks in the United States between 1979 and 1998. *Journal of the American Veterinary Medical Association* 217:836–840.

Saeed, I. S., and C. M. O. Kapel. 2006. Population dynamics and epidemiology of *Toxocara canis* in Danish red foxes. *Journal of Parasitology* 92:1196–1201.

Saeki, M., and D. W. Macdonald. 2004. The effects of traffic on the raccoon dog (*Nyctereutes procyonoides viverrinus*) and other mammals in Japan. *Biological Conservation* 118:559–571.

Saitoh, T., and K. Takahashi. 1998. The role of vole populations in prevalence of the parasite (*Echinococcus multilocularis*) in foxes. *Research on Population Ecology* 40:97–105.

Samuel, W. M., M. J. Pybus, and A. A. Kocan. 2001. *Parasitic Diseases of Wild Mammals.* 2nd ed. Manson Publishing, London.

Sandell, M. 1989. The mating tactics and spacing patterns of solitary carnivores. Pages 164–182 *in* J. L. Gittleman, ed. *Carnivore Behavior, Ecology, and Evolution.* Vol. 1. Cornell University Press, Ithaca, NY.

Sanderson, G. C. 1987. Raccoon. Pages 486–499 *in* M. Novak, J. A. Baker, M. E. Obbard, and B. Malloch, eds. *Wild Furbearer Management and Conservation in North America.* Ministry of Natural Resources, Ontario, Canada.

Sargeant, A. B., and S. H. Allen. 1989. Observed interactions between coyotes and red foxes. *Journal of Mammalogy* 70:631–633.

Sargeant, A. B., S. H. Allen, and J. O. Hastings. 1987. Spatial relations between sympatric coyotes and red foxes in North Dakota. *Journal of Wildlife Management* 51:285–293.

Sargeant, A. B., R. J. Greenwood, J. Piehl, and W. Bicknell. 1982. Recurrence, mortality, and dispersal of prairie striped skunks, *Mephitis mephitis*, and implications to rabies epizootiology. *Canadian Field-Naturalist* 96:312–316.

Saunders, G. 1992. Urban foxes (*Vulpes vulpes*): Implications of their behaviour for rabies control. Master's thesis, University of Bristol.

Saunders, G., B. Coman, J. Kinnear, and M. Braysher. 1995. *Managing Vertebrate Pests: Foxes.* Australian Government Publishing Service, Canberra, Australia Capital Territory.

Saunders, G., C. Lane, S. Harris, and C. Dickman. 2006. *Foxes in Tasmania: A Report on an Incursion of an Invasive Species.* Invasive Animals Cooperative Research Centre, University of Canberra, Australia.

Saunders, G., P. C. L. White, and S. Harris. 1997. Habitat utilisation by urban foxes (*Vulpes vulpes*) and the implications for rabies control. *Mammalia* 61:497–510.

Saunders, G., P. C. L. White, S. Harris, and J. M. V. Rayner. 1993. Urban foxes (*Vulpes vulpes*): Food acquisition, time, and energy budgeting of a generalized predator. *Symposia of the Zoological Society of London* 65:215–234.

Savolainen, P., T. Leitner, A. N. Wilton, E. Matisoo-Smith, and J. Lundeberg. 2004. A detailed picture of the origin of the Australian dingo, obtained from the study of mitochondrial DNA. *Proceedings of the National Academy of Sciences* 101:12387–12390.

Savolainen, P., Y. Zhang, J. Luo, J. Lundeberg, and T. Leitner. 2002. Genetic evidence for an east Asian origin of domestic dogs. *Science* 298:1610–1613.

Say, L., F. Bonhomme, E. Desmarais, and D. Pontier. 2003. Microspatial genetic heterogeneity and gene flow in stray cats (*Felis catus* L.): A comparison of coat colour and microsatellite loci. *Molecular Ecology* 12:1669–1674.

Say, L., and D. Pontier. 2004. Spacing pattern in a social group of stray cats: Effects on male reproductive success. *Animal Behaviour* 68:175–180.

Say, L., D. Pontier, and E. Natoli. 1999. High variation in multiple paternity of domestic cats (*Felis catus* L.) in relation to environmental conditions. *Proceedings of the Royal Society of London B* 266:2071–2074.

Say, L., D. Pontier, and E. Natoli. 2001. Influence of oestrus synchronization on male reproductive success in the domestic cat (*Felis catus* L.). *Proceedings of the Royal Society of London B* 268:1049–1053.

Scarlett, J. M. 1995. Companion animal epidemiology. *Preventive Veterinary Medicine* 25:151–159.

Schaefer, R. J., S. G. Torres, and V. C. Bleich. 2000. Survivorship and cause-specific mortality in sympatric populations of mountain sheep and mule deer. *California Fish and Game* 86:127–135.

Schaller, G. B., Hu Jinchu, Pan Wenshi, and Zhu Jing. 1985. *The Giant Pandas of Wolong*. University of Chicago Press, Chicago.

Schinner, J R. 1969. Ecology and life history of the raccoon (*Procyon lotor*) within the Clifton suburb of Cincinnati, OH. Master's thesis, University of Cincinnati.

Schinner, J. R., and D. L. Cauley. 1974. The ecology of urban raccoons in Cincinnati, Ohio. Pages 125–130 in J. H. Noyes and D. R. Progulske, eds. *Wildlife in an Urbanizing Environment*. Environmental Planning and Resource Development Series No. 28, Holdsworth Resource Center, Amherst, MA.

Schmidt, P. M., R. R. Lopez, and B. A. Collier. 2007. Survival, fecundity, and movements of free-roaming cats. *Journal of Wildlife Management* 71:915–919.

Schneider, D. G., L. D. Mech, and J. R. Tester. 1971. Movement of female raccoons and their young as determined by radio-tracking. *Animal Behavior Monographs* 4:1–43.

Schoener, T. W. 1983. Field experiments on interspecific competition. *American Naturalist* 122:240–285.

Schöffel, I., E. Schein, U. Wittstadt, and J. Hentsche. 1991. Zur Parasitenfauna des Rotfuchses in Berlin. *Berliner und Münchener Tierärztliche Wochenschrift* 104:153–157.

Schowalter, D., and J. Gunson. 1982. Parameters of population and seasonal activity of striped skunks (*Mephitis mephitis*), in Alberta and Saskatchewan. *Canadian Field-Naturalist* 96:409–420.

Schrecengost, J. D., J. C. Kilgo, D. Mallard, H. S. Ray, and K. V. Miller. 2008. Seasonal food habits of the coyote in the South Carolina coastal plain. *Southeastern Naturalist* 7:135–144.

Schubert, C. A., T. D. Nudds, I. K. Barker, R. C. Rosatte, and C. D. MacInnes. 1998. Effect of canine distemper on an urban raccoon population: An experiment. *Ecological Applications* 8:379–387.

Schueler, T. R. 1994. The importance of imperviousness. *Watershed Protection Techniques* 1:100–111.

Schusler, T. M., L. C. Chase, and D. J. Decker. 2000. Community-based management: Sharing responsibility when tolerance for wildlife is exceeded. *Human Dimensions of Wildlife* 5:34–49.

Schwartz, A., and H. E. Henderson. 1991. Amphibians and Reptiles of the West Indies: Descriptions, Distributions, and Natural History. University of Press of Florida, Gainesville.

Schwartz, C. C., J. Swenson, and S. D. Miller. 2003. Large carnivores, moose, and humans: A changing paradigm of predator management in the 21st century. *Alces* 39:41–63.

Schweiger, A., R. W. Ammann, D. Candinas, P.-A. Clavien, J. Eckert, B. Gottstein, N. Halkic, et al. 2007. Human alveolar echinococcosis after fox population increase, Switzerland. *Emerging Infectious Diseases* 13:878–882.

Scott, K. C., J. K. Levy, and P. C. Crawford. 2002. Characteristics of free-roaming cats evaluated in a trap-neuter-return program. *Journal of the American Veterinary Medical Association* 221:1136–1138.

Seaman, D. E., and R. A. Powell. 1996. An evaluation of the accuracy of kernel density estimators for home range analysis. *Ecology* 77:2075–2085.

Sechrest, W., T. M. Brooks, G. A. B. da Fonseca, W. R. Konstant, R. A. Mittermeier, A. Purvis, A. B. Rylands, and J. L. Gittleman. 2002. Hotspots and the conservation of evolutionary history. *Proceedings of the National Academy of Science* 99:2067–2071.

Seidensticker, J. A., M. G. Hornocker, W. V. Wiles, and J. P. Messick. 1973. Mountain lion social organization in the Idaho Primitive Area. *Wildlife Monograph* 35.

Seidensticker, J. A., J. T. Johnsingh, R. Ross, G. Sanders, and M. B. Webb. 1988. Raccoons and rabies in Appalachian mountain hollows. *National Geographic Research* 4:359–370.

Serpell, J. A. 1988. The domestication and history of the cat. Pages 151–158 in D. C. Turner and P. Bateson, eds. *The Domestic Cat: The Biology of Its Behaviour*. Cambridge University Press, Cambridge.

Serpell, J. A. 1995. From paragon to pariah: Some reflections on human attitudes to dogs. Pages 245–256 in J. Serpell, ed. *The Domestic Dog: Its Evolution, Behaviour and Interactions with People*. Cambridge University Press, Cambridge.

Serpell, J. A. 2000. Domestication and history of the cat. Pages 179–192 in D. C. Turner and P. Bateson, eds. *The Domestic Cat: The Biology of Its Behaviour*. 2nd ed. Cambridge University Press, Cambridge.

Shargo, E. S. 1988. Home range, movements, and activity patterns of coyotes (*Canis latrans*) in Los Angeles suburbs. Master's thesis, University of California, Los Angeles.

Sharp, W. M., and L. H. Sharp. 1956. Nocturnal movements and behavior of wild raccoons at a winter feeding station. *Journal of Mammalogy* 37:170–177.

Shaw, H. G. 1980. *Ecology of the Mountain Lion in Arizona*. Final Report, P-R Project W-78-R, Work Plan 2, Job 13. Arizona Game and Fish Department, Phoenix.

Sherif, M. 1936. *The Psychology of Social Norms*. Harper, New York.

Sherif, M., and C. W. Sherif. 1969. *Social Psychology*. Harper & Row, New York.

Shindle, D., D. Land, M. Cunningham, and M. Lotz. 2001. *Florida Panther Genetic Restoration*. Annual Report 2000–01. Florida Fish and Wildlife Conservation Commission, Tallahassee.

Shinobu, G., M. Tooru, and F. Tsuyoshi. 1999. Habitat Fragmentation for Raccoon Dog Caused by Land Use of Transportation in Urban Ecological Networks. Environmental Systems Research.

Shivik, J. A., and D. J. Martin. 2001. Aversive and disruptive stimulus applications for managing predation. *Proceedings Wildlife Damage Management Conference* 9:111–119.

Shochat, E., P. S. Warren, S. H. Faeth, N. E. McIntyre, and D. Hope. 2006. From patterns to emerging processes in mechanistic urban ecology. *Trends in Ecology and Evolution* 21:186–191.

Sieber, O. J. 1984. Vocal communication in raccoons. *Behaviour* 90:80–113.

Sieber, O. J. 1985. Acoustic recognition between mother and cubs in raccoons (*Procyon lotor*). *Behaviour* 96:130–163.

Siemer, W. F., and D. J. Decker. 2006. *An Assessment of Black Bear Impacts in New York*. Human Dimensions Research Unit Series Publication 06-6. Department of Natural Resources, Cornell University Ithaca, NY.

Siemer, W. F., D. J. Decker, P. Otto, and M. L. Gore. 2007. *Working through Black Bear Management Issues: A Practitioners' Guide*. Northeast Wildlife Damage Management Research and Outreach Cooperative, Ithaca, NY.

Sillero-Zubiri, C., and M. K. Laurenson. 2001. Interactions between carnivores and local communities: Conflict or co-existence? Pages 282–312 *in* J. L. Gittleman, S. M. Funk, D. Macdonald, and R. K. Wayne, eds. *Carnivore Conservation*. Cambridge University Press, Cambridge.

Sillero-Zubiri, C., and D. W. Macdonald, eds. 1997. *The Ethiopian Wolf*. Status Survey and Conservation Action Plan. IUCN Canid Specialist Group. Gland, Switzerland.

Sillero-Zubiri, C., and J. Marino. 2004. Ethiopian wolf. Pages 167–174 *in* C. Sillero-Zubiri, M. Hoffman, and D. W. Macdonald, eds. *Canids: Foxes, Wolves, Jackals, and Dogs*. Status Survey and Conservation Action Plan. IUCN/SSC Canid Specialist Group. Gland, Switzerland and Cambridge, U.K.

Sims, V., K. L. Evans, S. E. Newson, J. A. Tratalos, and K. J. Gaston. 2008. Avian assemblage structure and domestic cat densities in urban environments. *Diversity and Distributions* 14:387–399.

Sjöberg, L. 1998. Worry and risk perception. *Risk Analysis* 18:85–93.

Skerget, M., C. Wenisch, F. Daxboeck, R. Krause, R. Haberl, and D. Stuenzner. 2003. Cat or dog ownership and seroprevalence of ehrlichiosis, Q fever, and cat-scratch disease. *Emerging Infectious Diseases* 9:1337–1340.

Skinner, C., P. Skinner, and S. Harris. 1991a. The past history and recent decline of badgers *Meles meles* in Essex: An analysis of some of the contributory factors. *Mammal Review* 21:67–80.

Skinner, C., P. Skinner, and S. Harris. 1991b. An analysis of some of the factors affecting the current distribution of badger *Meles meles* setts in Essex. *Mammal Review* 21:51–65.

Slabbekoorn, H., and M. Peet. 2003. Birds sing at a higher pitch in urban noise. *Nature* 424:267.

Slate, D. 1985. Movement, activity, and home range patterns among members of a high density suburban raccoon population. Master's thesis, Rutgers: The State University of New Jersey.

Slate, D., C. E. Rupprecht, J. A. Rooney, D. Donovan, D. H. Lein, and R. B. Chipman. 2005. Status of oral rabies vaccination in wild carnivores in the United States. *Virus Research* 111:68–76.

Slater, M. R. 2001. The role of veterinary epidemiology in the study of free-roaming dogs and cats. *Preventive Veterinary Medicine* 48:273–286.

Slater, M. R. 2002. *Community Approaches to Feral Cats: Problems, Alternatives, and Recommendations*. The Humane Society Press, Washington D.C.

Slater, M. R., A. Di Nardo, O. Pediconi, P. Dalla Villa, L. Candeloro, B. Allessandrini, and S. Del Papa. 2008. Free-roaming dogs and cats in central Italy: Public perceptions of the problem. *Preventive Veterinary Medicine* 84:27–47.

Slovic, P. 1987. Perception of risk. *Science* 236:280–285.

Slovic, P. 1993. Perceived risk, trust, and democracy. *Risk Analysis* 13:675–682.

Smith, A. E., S. R. Craven, and P. D. Curtis. 2001. *Managing Canada Geese in Urban Environments*. Jack Berryman Institute Publication 16. Cornell Cooperative Extension, Ithaca, NY.

Smith, D. 1982. *Brocky, the Bathwick Badger*. Privately published.

Smith, D. W., L. D. Mech, M. Meagher, W. E. Clark, R. Jaffe, M. K. Phillips, and J. A. Mack. 2000. Wolf-bison interactions in Yellowstone National Park. *Journal of Mammalogy* 81:1128–1135.

Smith, D. W., R. O. Peterson, and D. B. Houston. 2003. Yellowstone after wolves. *BioScience* 53:330–340.

Smith, G. C., B. Gangadharan, Z. Taylor, M. K. Laurenson, H. Bradshaw, G. Hide, J. M. Hughes, et al. 2003. Prevalence of zoonotic important parasites in the red fox (*Vulpes vulpes*) in Great Britain. *Veterinary Parasitology* 118:133–142.

Smith, G. C., and S. Harris. 1989. The control of rabies in urban fox populations. Pages 209–224 *in* R. J. Putman, ed. *Mammals as Pests*. Chapman and Hall, London.

Smith, G. C., and S. Harris. 1991. Rabies in urban foxes (*Vulpes vulpes*) in Britain: The use of a spatial stochastic simulation model to examine the pattern of spread and evaluate the efficacy of different control régimes. *Philosophical Transactions of the Royal Society of London B* 334:459–479.

Smith, G. C., and D. Wilkinson. 2003. Modeling control of rabies outbreaks in red fox populations to evaluate culling, vaccination, and vaccination combined with fertility control. *Journal of Wildlife Diseases* 39:278–286.

Smith, H. T., and R. M. Engeman. 2002. An extraordinary raccoon, *Procyon lotor*, density at an urban park. *Canadian Field-Naturalist* 116:636–639.

Smith, T. E., R. R. Duke, M. J. Kutilek, and H. T. Harvey. 1986. *Mountain Lions* (Felis concolor) *in the Vicinity of Carlsbad Caverns National Park, New Mexico, and Guadalupe Mountains National Park, Texas*. U.S. Department of Interior, National Park Service, Santa Fe, NM.

Smith-Patten, B. D., and M. A. Patten. 2008. Diversity,

seasonality, and context of mammalian roadkills in the southern Great Plains. *Environmental Management* 41:844–852.

Somers, M. J., and J. A. J. Nel. 2004. Movement patterns and home range of Cape clawless otters (*Aonyx capensis*), affected by high food density patches. *Journal of Zoology* 262:91–98.

Sorace, A., and M. Gustin. 2009. Distribution of generalist and specialist predators along urban gradients. *Landscape and Urban Planning* 90:111–118.

Soulé, M. E., A. C. Alberts, and D. T. Bolger. 1992. The effects of habitat fragmentation on chaparral plants and vertebrates. *Oikos* 63:39–47.

Soulé, M. E., D. T. Bolger, A. C. Alberts, R. Sauvajot, J. Wright, M. Sorice, and S. Hill. 1988. Reconstructed dynamics of rapid extinctions of chaparral-requiring birds in urban habitat islands. *Conservation Biology* 2:75–92.

Soulsbury, C. D., P. J. Baker, G. Iossa, and S. Harris. 2008a. Fitness costs of dispersal in red foxes (*Vulpes vulpes*). *Behavioral Ecology and Sociobiology* 62:1289–1298.

Soulsbury, C. D., G. Iossa, P. J. Baker, N. C. Cole, S. M. Funk, and S. Harris. 2007. The impact of sarcoptic mange *Sarcoptes scabiei* on the British fox *Vulpes vulpes* population. *Mammal Review* 37:278–296.

Soulsbury, C. D., G. Iossa, P. J. Baker, and S. Harris. 2008b. Environmental variation at the onset of independent foraging affects full-grown body mass in the red fox. *Proceedings of the Royal Society of London B* 275:2411–2418.

Souter, M. A., and M. D. Miller. 2007. Do animal-assisted activities effectively treat depression? A meta-analysis. *Anthrozoos* 20:167–180.

South Coast Wildlands. 2008. South Coast Missing Linkages: A Wildland Network for the South Coast Ecoregion. Available online at http://www.scwildlands.org.

Sovada, M. A., C. C. Roy, J. B. Bright, and J. R. Gillis. 1998. Causes and rates of mortality of swift foxes in western Kansas. *Journal of Wildlife Management* 62:1300–1306.

Sovada, M. A., A. B. Sargeant, and J. W. Grier. 1995. Differential effects of coyotes and red foxes on duck nest success. *Journal of Wildlife Management* 59:1–9.

Spiegel, L. K. 1996. Studies of San Joaquin kit fox in undeveloped and oil-developed areas: An overview. Pages 1–14 *in* L. K. Spiegel, ed. *Studies of the San Joaquin Kit Fox in Undeveloped and Oil-Developed Areas*. California Energy Commission, Sacramento.

Spiegel, L. K., T. C. Dao, and M. Bradbury. 1996. Spatial ecology and habitat use of San Joaquin kit foxes in undeveloped and oil-developed lands of Kern County, California. Pages 93–114 *in* L. K. Spiegel, ed. *Studies of the San Joaquin Kit Fox in Undeveloped and Oil-Developed Areas*. California Energy Commission, Sacramento.

Stahl, P., J. M. Vandel, V. Herrenschmidt, and P. Migot, 2001. Predation on livestock by an expanding reintroduced lynx population: Long-term trend and spatial variability. *Journal of Applied Ecology* 38:674–687.

Stanton, W. 1997. Letters. *New Scientist* 2099:56.

Steen-Ash, S. 2004. Intraspecific spatial dynamics of urban stray cats. Pages 222–227 *in* W. W. Shaw, L. K. Harris, and L. Vandruff, eds. *Proceedings of the 4th International Symposium on Urban Wildlife Conservation*. School of Natural Resources, College of Agriculture and Life Sciences, University of Arizona, Tucson.

Stevens, T. H., T. A. More, and R. J. Glass. 1994. Public attitudes about coyotes in New England. *Society and Natural Resources* 7:57–66.

Stieger, C., D. Hegglin, G. Schwarzenbach, A. Mathis, and P. Deplazes. 2002. Spatial and temporal aspects of urban transmission of *Echinococcus multilocularis*. *Parasitology* 124:631–640.

Stockwell, C. A., A. P. Hendry, and M. T. Kinnison. 2003. Contemporary evolution meets conservation biology. *Trends in Ecology and Evolution* 18:94–101.

Stone, W., J. C. Okoniewski, and J. R. Stedelin. 1999. Poisoning of wildlife with anticoagulant rodenticides in New York. *Journal of Wildlife Diseases* 35:187–193.

Stoner, D. C., M. L. Wolfe, and D. M. Choate. 2006. Cougar exploitation levels in Utah: Implications for demographic structure, population recovery, and metapopulation dynamics. *Journal of Wildlife Management* 70:1588–1600.

Storm, G. L. 1972. Daytime retreats and movements of skunks on farmlands in Illinois. *Journal of Wildlife Management* 36:31–45.

Storm, G. L., R. D. Andrews, R. L. Phillips, R. A. Bishop, D. B. Siniff, and J. R. Tester. 1976. Morphology, reproduction, dispersal, and mortality of midwestern red fox populations. *Wildlife Monographs* 49:1–82.

Stoskopf, M. K., and F. B. Nutter. 2004. Analyzing approaches to feral cat management—one size does not fit all. *Journal of the American Veterinary Medical Association* 225:1361–1364.

Stout, R. J., R. C. Stedman, D. J. Decker, and B. A. Knuth. 1993. Perceptions of risk from deer-related vehicle accidents: Implications for public preferences for deer herd size. *Wildlife Society Bulletin* 21:237–249.

Stratford, J. A., and W. D. Robinson. 2005. Distribution of Neotropical migratory bird species across an urbanizing landscape. *Urban Ecosystems* 8:59–77.

Stuewer, F. W. 1943. Raccoons: Their habits and management in Michigan. *Ecological Monographs* 13:203–257.

Sturla, K. 1993. Role of breeding regulation laws in solving the dog and cat overpopulation problem. *Journal of the American Veterinary Medical Association* 202:928–932.

Suarez, A. V., D. T. Bolger, and T. J. Case. 1998. The effects of habitat fragmentation and invasion on the native ant community in coastal Southern California. *Ecology* 79:2041–2056.

Sukopp, H. 2002. On the early history of urban ecology in Europe. *Preslia Praha* 74:373–393.

Sundqvist, A.-K., S. Björnerfeldt, J. A. Leonard, F. Hailer, Å. Hedhammar, H. Ellegren, and C. Vilà. 2006. Unequal contribution of sexes in the origin of dog breeds. *Genetics* 172:1121–1128.

Sunquist, M. E. 1974. Winter activity of striped skunks (*Mephitis mephitis*) in east-central Minnesota. *American Midland Naturalist* 92:434–446.

Sunquist, M. E., G. G. Montgomery, and G. L. Storm. 1969. Movements of a blind raccoon. *Journal of Mammalogy* 50:145–147.

Sunquist, M. E., and F. C. Sunquist. 1989. Ecological constraints on predation by large felids. Pages 283–301 *in* J. L. Gittleman, ed. *Carnivore Behavior, Ecology, and Evolution.* Cornell University Press, Ithaca, NY.

Suzán, G., and G. Ceballos. 2005. The role of feral mammals on wildlife infectious disease prevalence in two nature reserves within Mexico City limits. *Journal of Zoo and Wildlife Medicine* 36:479–484.

Svensson, A. M., and J. Rydell. 1998. Mercury vapour lamps interfere with the bat defence of tympanate moths (*Operophtera* spp; Geometridae). *Animal Behaviour* 55:223–226.

Sweanor, L. L., K. A. Logan, J. W. Bauer, B. Millsap, and W. M. Boyce. 2008. Puma and human spatial and temporal use of a popular California state park. *Journal of Wildlife Management* 72:1076–1084.

Sweanor, L. L., K. A. Logan, and M. G. Hornocker. 2000. Cougar dispersal patterns, metapopulation dynamics, and conservation. *Conservation Biology* 14:798–808.

Sweanor, L. L., K. A. Logan, and M. G. Hornocker. 2005. Puma responses to close encounters with researchers. *Wildlife Society Bulletin* 33:905–911.

Sweitzer, R. A., S. H. Jenkins, and J. Berger. 1997. Near-extinction of porcupines by mountain lions and consequences of ecosystem change in the Great Basin desert. *Conservation Biology* 11:1407–1417.

Taborsky, M. 1988. Kiwis and dog predation: Observations in Waitangi State Forest. *Notornis* 35:197–202.

Takatsuki, S., H. Hirasawa, and E. Kanda. 2007. A comparison of the point-frame method with frequency methods in fecal analysis of an omnivorous mammal, the raccoon dog. *Mammal Study* 32:1–5.

Tanaka, H. 2005. Seasonal and daily activity patterns of Japanese badgers (*Meles meles anakuma*) in Western Honshu, Japan. *Mammal Study* 30:11–17.

Tanaka, H., A. Yamanaka, and K. Endo. 2002. Spatial distribution and sett use of Japanese badger, *Meles meles anakuma. Mammal Study* 27:15–22.

Taylor, M. E. 1989. Locomotor adaptations by carnivores. Pages 382–409 *in* J. L. Gittleman, ed. *Carnivore Behavior, Ecology, and Evolution.* Cornell University Press, Ithaca, NY.

Teagle, W. G. 1967. The fox in the London suburbs. *London Naturalist* 46:44–68.

Teagle, W. G. 1969. The badger in the London area. *London Naturalist* 48:48–75.

Teel, T. L., R. S. Krannich, and R. H. Schmidt. 2002. Utah stakeholders' attitudes toward selected cougar and black

bear management practices. *Wildlife Society Bulletin* 30:2–15.

Tembrock, G. 1957. Zur ethologie des Rotfuchses (*Vulpes vulpes* (L.), unter besonderer Berücksichtigung der Fortpflanzung. *Zoologische Garten* 23:289–532.

Tenter, A. M., A. R. Heckeroth, and L. M. Weiss. 2000. *Toxoplasma gondii:* From animals to humans. *International Journal for Parasitology* 30:1217–1258.

Terborgh, J., K. Feeley, M. Silman, P. Nunez, and B. Balukjian. 2006. Vegetation dynamics of predator-free land-bridge islands. *Journal of Ecology* 94:253–263.

Terborgh J., L. Lopez L, P. Nuñez, M. Rao, G. Shahabuddin, G. Orihuela, M. Riveros, et al. 2001. Ecological meltdown in predator-free forest fragments. *Science* 294:1923–1926.

Ternent, M. A., and D. L. Garshelis. 1999. Taste-aversion conditioning to reduce nuisance activity by black bears in a Minnesota military reservation. *Wildlife Society Bulletin* 27:720–728.

Tester, U. 1986. Vergleichende Nahrungsuntersuchung beim Steinmarder *Martes foina* (Erxleben, 1777) in großstädtischem und ländlichem Habitat. *Säugetierkundliche Mitteilungen* 33:37–52.

Tevis, L., Jr. 1947. Summer activities of California raccoons. *Journal of Mammalogy* 28:323–332.

Theberge, J. B., and C. H. R. Wedeles. 1989. Prey selection and habitat partitioning in sympatric coyote and red fox populations, southwest Yukon. *Canadian Journal of Zoology* 67:1285–1290.

Thompson, C. M., and E. M. Gese. 2007. Food webs and intraguild predation: Community interactions of a native mesocarnivore. *Ecology* 88:334–346.

Thompson, D. J., and J. A. Jenks. 2005. Long distance dispersal by a subadult male cougar from the Black Hills, South Dakota. *Journal of Wildlife Management* 69:818–820.

Tigas, L. A., D. H. Van Vuren, and R. M. Sauvajot. 2002. Behavioral responses of bobcats and coyotes to habitat fragmentation and corridors in an urban environment. *Biological Conservation* 108:299–306.

Tigas, L. A., D. H. Van Vuren, and R. M. Sauvajot. 2003. Carnivore persistence in fragmented habitats in urban Southern California. *Pacific Conservation Biology* 9:144–151.

Timm, R. M., R. O. Baker, J. R. Bennett, and C. C. Coolahan. 2004. Coyote attacks: An increasing suburban problem. *Transactions of the North American Wildlife and Natural Resources Conference* 69:67–88.

Timm, R. M., and G. E. Connolly. 2001. Sheep-killing coyotes a continuing dilemma for ranchers. *California Agriculture* 55:26–31.

Torres, S. G., T. M. Mansfield, J. E. Foley, T. Lupo, and A. Brinkhaus. 1996. Mountain lion and human activity in California: Testing speculations. *Wildlife Society Bulletin* 24:451–460.

Torrey, E. F., and R. H. Yolken. 2003. *Toxoplasma gondii* and schizophrenia. *Emerging Infectious Diseases* 9:1375–1380.

Totton, S., R. Tinline, R. Rosatte, and L. Bigler. 2002. Contact rates of raccoons (*Procyon lotor*) at a communal feeding site in rural eastern Ontario. *Journal of Wildlife Diseases* 38:313–319.

Trapp, G., and D. L. Hallberg. 1975. Ecology of the gray fox (*Urocyon cinereoargenteus*): A review. Pages 164–178 in M. W. Fox, ed. *The Wild Canids: Their Systematics, Behavioral Ecology, and Evolution*. Van Nostrand Reinhold, New York.

Treves, A., and K. U. Karanth. 2003. Human-carnivore conflict and perspectives on carnivore management worldwide. *Conservation Biology* 17:1491–99.

Trewhella, W. J., S. Harris, and F. E. McAllister. 1988. Dispersal distance, home-range size, and population density in the red fox (*Vulpes vulpes*): A quantitative analysis. *Journal of Applied Ecology* 25:423–434.

Trewhella, W. J., S. Harris, G. C. Smith, and A. K. Nadian. 1991. A field trial evaluating bait uptake by an urban fox (*Vulpes vulpes*) population. *Journal of Applied Ecology* 28:454–466.

Tsukada, H., K. Hamazaki, S. Ganzorig, T. Iwaki, K. Konno, J. T. Lagapa, K. Matsuo, et al. 2002. Potential remedy against *Echinococcus multilocularis* in wild red foxes using baits with anthelmintic distributed around fox breeding dens in Hokkaido, Japan. *Parasitology* 125:119–129.

Tsukada, H., Y. Morishima, N. Nonaka, Y. Oku, and M. Kamiya. 2000. Preliminary study of the role of red foxes in *Echinococcus multilocularis* transmission in the urban area of Sapporo, Japan. *Parasitology* 120:423–428.

Turkowski, F. J., and L. D. Mech. 1968. Radio-tracking the movements of a young male raccoon. *Journal of the Minnesota Academy of Science* 35:33–38.

Turner, D. C., and C. Mertens. 1986. Home range size, overlap and exploitation in domestic farm cats (*Felis catus*). *Behaviour* 99:22–45.

Turner J. W., Jr., M. L. Wolfe, and J. F. Kirkpatrick. 1992. Seasonal mountain lion predation on a feral horse population. *Canadian Journal of Zoology* 70:929–934.

Twitchell, A. R., and H. H. Dill. 1949. One hundred raccoons from one hundred and two acres. *Journal of Mammalogy* 30:130–133.

Ulrich, R. S., R. F. Simons, E. F. Losito, M. A. Miles, and M. Zelson. 1991. Stress recovery during exposure to natural and urban environments. *Journal of Environmental Psychology* 11:201–230.

United Nations. 2006. *World Urbanization Prospects: The 2005 Revision*. New York: Population Division, Department of Economic and Social Affairs, United Nations.

United Nations Population Fund (UNFPA). 2007. *State of World Population 2007: Unleashing the Potential of Urban Growth*. UNFPA, New York.

U.S. Department of Agriculture. 2006. *Cooperative Rabies Management Program National Report 2006*. United States Department of Agriculture, Animal and Plant Health Inspection Service, Wildlife Services. 138 pp.

U.S. Fish and Wildlife Service. 1998. *Recovery Plan for Upland Species of the San Joaquin Valley, California*. U.S. Fish and Wildlife Service, Region 1, Portland, OR.

Uzal, F. A., R. S. Houston, S. P. D. Riley, R. Poppenga, J. Odani, and W. Boyce. 2007. Notoedric mange in two free-ranging mountain lions (*Puma concolor*). *Journal of Wildlife Diseases* 43:274–278.

Valentine, R. L., C. A. Bache, W. H. Gutenmann, and D. J. Lisk. 1988. Tissue concentrations of heavy metal and polychlorinated biphenyls in raccoons in central New York. *Bulletin of Environmental Contamination and Toxicology* 40:711–716.

Van Apeldoorn, R. C., J. Vink, and T. Matyáštík. 2006. Dynamics of a local badger (*Meles meles*) population in the Netherlands over the years 1983–2001. *Mammalian Biology* 71:25–38.

Van Deelen, T. R., and T. E. Gosselink. 2006. Coyote survival in a row-crop agricultural landscape. *Canadian Journal of Zoology* 84:1630–1636.

van der Zee, F. F., J. Wiertz, C. J. F. Ter Braak, and R. C. van Apeldoorn. 1992. Landscape change as a possible cause of the badger *Meles meles* L. decline in The Netherlands. *Biological Conservation* 61:17–22.

Van Dyke, F. G., R. H. Brocke, H. G. Shaw, B. B. Ackerman, T. P. Hemker, and F. G. Lindzey. 1986. Reactions of mountain lions to logging and human activity. *Journal of Wildlife Management* 50:95–102.

Van Valkenburgh, B. 1989. Carnivore dental adaptations and diet: A study of trophic diversity within guilds. Pages 410–436 in J. L. Gittleman, ed. *Carnivore Behavior, Ecology, and Evolution*. Cornell University Press, Ithaca, NY.

van Wijngaarden, A., and J. van de Peppel. 1964. The badger, *Meles meles* (L.), in the Netherlands. *Lutra* 6:1–60.

Vaske, J. J., and D. Whittaker. 2004. Normative approaches to natural resources. Pages 283–294 in M. Manfredo, J. J. Vaske, B. L. Bruyere, D. R. Field, and P. J. Brown, eds. *Society and Natural Resources: A Summary of Knowledge*. Modern Litho: Jefferson, MO.

Veeramani, A., E. A. Jayson, and P. S. Easa. 1996. Man-wildlife conflict: Cattle lifting and human casualties in Kerala. *Indian Forester* 10:897–902.

Vergara, V. 2001. Two cases of infanticide in a red fox, *Vulpes vulpes*, family in southern Ontario. *Canadian Field-Naturalist* 115:170–173.

Verts, B. J. 1967. *The Biology of the Striped Skunk*. University of Illinois Press, Urbana.

Verts, B. J., L. N. Carraway, and A. Kinlaw. 2001. *Spilogale gracilis*. *Mammalian Species* 674:1–10.

Vesey-Fitzgerald, B. 1965. *Town Fox, Country Fox*. Andre Deutsch, London.

Vickers, T. W. 2007. Attitudes toward, and acceptance of, mountain lions by urban and rural southern Californians, and comparison with residents of other western states. Master's thesis, School of Veterinary Medicine, University of California, Davis.

Vigne, J.-D., J. Guilaine, K. Debue, L. Haye, and P. Gérard. 2004. Early taming of the cat in Cyprus. *Science* 304:259.

Vilà, C., J. E. Maldonado, and R. K. Wayne. 1999. Phylogenetic relationships, evolution, and genetic diversity of the domestic dog. *American Genetic Association* 90:71–77.

Vilà, C., P. Savolainen, J. E. Maldonado, I. R. Amorim, J. E. Rice, R. L. Honeycutt, K. A. Crandall, J. Lundeberg, and R. K. Wayne. 1997. Multiple and ancient origins of the domestic dog. *Science* 276:1687–1689.

Vilà, C., J. Seddon, and H. Ellegren. 2005. Genes of domestic mammals augmented by backcrossing with wild ancestors. *Trends in Genetics* 21:214–218.

Vilà, C., and R. K. Wayne. 1999. Hybridization between wolves and dogs. *Conservation Biology* 13:195–198.

Voight, D. R., and B. D. Earle. 1983. Avoidance of coyotes by red fox families. *Journal of Wildlife Management* 47:852–857.

von der Lippe, M., and I. Kowarik. 2008. Do cities export biodiversity? Traffic as dispersal vector across urban-rural gradients. *Diversity and Distributions* 14:18–25.

von Schantz, T. 1981. Female cooperation, male competition, and dispersal in the red fox *Vulpes vulpes*. *Oikos* 37:63–68.

von Schantz, T. 1984a. Spacing strategies, kin selection, and population regulation in altricial vertebrates. *Oikos* 42:48–58.

von Schantz, T. 1984b. "Non-breeders" in the red fox *Vulpes vulpes*: A case of resource surplus. *Oikos* 42:59–65.

Vos, A., E. Pommerening, L. Neubert, S. Kachel, and A. Neubert. 2002. Safety studies of the oral rabies vaccine SAD B19 in striped skunk (*Mephitis mephitis*). *Journal of Wildlife Diseases* 38:428–431.

Wade-Smith, J., and B. J. Verts. 1982. *Mephitis mephitis*. *Mammalian Species* 173:1–7.

Wade-Smith, R. A., and M. E. Richmond. 1978a. Reproduction in captive striped skunks (*Mephitis mephitis*). *American Midland Naturalist* 100:452–455.

Wade-Smith, R. A., and M. E. Richmond. 1978b. Induced ovulation, development of the corpus luteum, and tubal transport in the striped skunk (*Mephitis mephitis*). *American Journal of Anatomy* 153:123–142.

Walton, L. R., H. D. Cluff, P. C. Paquet, and M. A. Ramsay. 2001. Movement patterns of barren-ground wolves in the central Canadian arctic. *Journal of Mammalogy* 82:867–876.

Wandeler, P., S. M. Funk, C. R. Largiadèr, S. Gloor, and U. Breitenmoser. 2003. The city-fox phenomenon: Genetic consequences of a recent colonization of urban habitat. *Molecular Ecology* 12:647–656.

Wang, H., and T. K. Fuller. 2003. Food habits of four sympatric carnivores in southeastern China. *Mammalia* 67:513–519.

Wang, S. W., and D. W. Macdonald. 2006. Livestock predation by carnivores in Jigme Singye Wangchuk National Park, Bhutan. *Biological Conservation* 129:558–565.

Wania, A., I. Kühn, and S. Klotz. 2006. Plant richness patterns in agricultural and urban landscapes in Central Germany: Spatial gradients of species richness. *Landscape and Urban Planning* 75:97–110.

Warner, R. E. 1985. Demography and movements of free-ranging domestic cats in rural Illinois. *Journal of Wildlife Management* 49:340–346.

Warren, R. J., M. J. Conroy, W. E. James, L. A. Baker, and D. R. Diefenbach. 1990. Reintroduction of bobcats on Cumberland Island, Georgia: A biopolitical lesson. *Transactions of the 55th North American Wildlife and Natural Resources Conference* 55:580–589.

Warrick, G. D., H. O. Clark, Jr., P. A. Kelly, D. F. Williams, and B. L. Cypher. 2007. Use of agricultural lands by San Joaquin kit foxes. *Western North American Naturalist* 67:270–277.

Wassmer, D. A., D. D. Guenther, and J. N. Layne. 1988. Ecology of the bobcat in south-central Florida. *Bulletin of the Florida State Museum, Biological Sciences* 33:159–228.

Way, J. G. 2002. Radiocollared coyote crosses Cape Cod Canal. *Northeast Wildlife* 57:63–65.

Way, J. G. 2003. Description and possible reasons for an abnormally large group size of adult eastern coyotes observed during summer. *Northeastern Naturalist* 10:335–342.

Way, J. G. 2007. A comparison of body mass of *Canis latrans* (coyotes) between eastern and western North America. *Northeastern Naturalist* 14:111–124.

Way, J. G., P. J. Auger, I. M. Ortega, and E. G. Strauss. 2001. Eastern coyote denning behaviour in an anthropogenic environment. *Northeast Wildlife* 56:18–30.

Way, J. G., S. M. Cifuni, D. L. Eatough, and E. G. Strauss. 2006. Rat poison kills a pack of Eastern Coyotes, *Canis latrans*, in an urban area. *Canadian Field-Naturalist* 120:478–480.

Way, J. G., I. M. Ortega, and P. J. Auger. 2002. Eastern coyote home range, territoriality, and sociality in urbanized Cape Cod. *Northeast Wildlife* 57:1–18.

Way, J. G., I. M. Ortega, and E. G. Strauss. 2004. Movement and activity patterns of eastern coyotes in a coastal, suburban environment. *Northeastern Naturalist* 11:237–254.

Weaver, H. W., J. T. Anderson, J. W. Edwards, and T. Dotson. 2003. Physical and behavioral characteristics of nuisance and non-nuisance black bears in southern West Virginia. *Proceedings Southeast Association of Fish and Wildlife Agencies* 57:308–316.

Webbon, C. C., P. J. Baker, and S. Harris. 2004. Faecal density counts for monitoring changes in red fox numbers in rural Britain. *Journal of Applied Ecology* 41:768–779.

Weber, J.-M., and L. Dailly. 1998. Food habits and ranging behaviour of a group of farm cats (*Felis catus*) in a Swiss mountainous area. *Journal of Zoology* 245:234–237.

Webster, J. P., C. F. A. Brunton, and D. W. Macdonald. 1994. Effect of *Toxoplasma gondii* upon neophobic behaviour in wild brown rats, *Rattus norvegicus*. *Parasitology* 109:37–43.

Webster, W. 1967. *Filaroides mephitis* N. sp. (Metastrongyloidea: Filaroididae) from lungs of eastern Canadian skunks. *Canadian Journal of Zoology* 45:145–147.

Wegan, M. T., 2007. Aversive conditioning, population estimation, and habitat preference of black bears (*Ursus americanus*) on Fort Drum military installation in northern New York. Master's thesis, Cornell University, Ithaca, NY.

Weggler, M., and B. Leu. 2001. A source population of black redstarts (*Phoenicurus ochruros*) in villages with a high density of feral cats (*Felis catus*). *Journal für Ornithologie* 142:273–283.

Westbrook, W. H., and R. D. Allen. 1979. Animal field research. Pages 148–167 *in* R. D. Allen and W. H. Westbrook, eds. *The Handbook of Animal Welfare*. Garland STPM Press, New York.

Western Governors' Association (WGA). 2008. *Wildlife Corridors Initiative: Transportation Working Group Report*. Western Governors' Association, Denver, Colorado. www.westgov.org/wga/publicat/wildlife08.pdf.

White, D., Jr., K. C. Kendall, and H. D. Picton. 1998. Grizzly bear feeding activity at alpine army cutworm moth aggregation sites in northwest Montana. *Canadian Journal of Zoology* 76:221–227.

White, J. 1999. *An Assessment of Habitat Manipulation as a Fox Control Strategy*. Final Report to the National Feral Animal Control Program. Bureau of Rural Sciences, Australia.

White, J. G., R. Gubiani, N. Smallman, K. Snell, and A. Morton. 2006. Home range, habitat selection and diet of foxes (*Vulpes vulpes*) in a semi-urban riparian environment. *Wildlife Research* 33:175–180.

White, L. A., and S. D. Gehrt. 2009. Coyote attacks on humans in North America and Canada. *Human Dimensions of Wildlife* 14:419–432.

White, P. A. 1992. Social organization and activity patterns of the arctic fox (*Alopex lagopus pribilofensis*) on St. Paul Island, Alaska. Master's thesis, University of California, Berkeley. 139 pp.

White, P. C. L., P. J. Baker, J. C. R. Smart, S. Harris, and G. Saunders. 2003. *Control of Foxes in Urban Areas: Modelling the Benefits and Costs*. Symposium on Urban Wildlife, Third International Wildlife Management Congress. Christchurch, New Zealand.

White, P. C. L., and S. Harris. 1994. Encounters between red foxes (*Vulpes vulpes*): Implications for territory maintenance, social cohesion, and dispersal. *Journal of Animal Ecology* 63:315–327.

White, P. C. L., S. Harris, and G. C. Smith. 1995. Fox contact behaviour and rabies spread: A model for the estimation of contact probabilities between urban foxes at different population densities and its implications for rabies control in Britain. *Journal of Applied Ecology* 32:693–706.

White, P. C. L., G. Saunders, and S. Harris. 1996a. Spatiotemporal patterns of home range use by foxes (*Vulpes vulpes*) in urban environments. *Journal of Animal Ecology* 65:121–125.

White, P. C. L., and S. J. Whiting. 2000. Public attitudes towards badger culling to control bovine tuberculosis in cattle. *Veterinary Record* 147:179–84.

White, P. J., W. H. Berry, J. J. Eliason, and M. T. Hanson. 2000. Catastrophic decrease in an isolated population of kit foxes. *Southwestern Naturalist* 45:204–211.

White, P. J., and R. A. Garrott. 1997. Factors regulating kit fox populations. *Canadian Journal of Zoology* 75:1982–1988.

White, P. J., and K. Ralls. 1993. Reproduction and spacing patterns of kit foxes relative to changing prey availability. *Journal of Wildlife Management* 57:861–867.

White, P. J., C. A. Vanderbilt White, and K. Ralls. 1996b. Functional and numerical responses of kit foxes to a short-term decline in mammalian prey. *Journal of Mammalogy* 77:370–376.

Whitney, L. F. 1933. The raccoon—some mental attributes. *Journal of Mammalogy* 14:108–144.

Whittaker, D., M. J. Manfredo, P. J. Fix, R. Sinnott, S. Miller, and J. J. Vaske, 2001. Understanding beliefs and attitudes about an urban wildlife hunt near Anchorage, Alaska. *Wildlife Society Bulletin* 29:1114–1124.

Whittaker, D. G., and F. G. Lindzey. 1999. Effect of coyote predation on early fawn survival in sympatric deer species. *Wildlife Society Bulletin* 27:256–262.

Wieczorek Hudenko, H., D. J. Decker, and W. F. Siemer. 2008. *Stakeholder Insights into the Human-Coyote Interface in Westchester County, New York*. Human Dimensions Research Unit Series Publ. 08-1. Department of Natural Resources, Cornell University, Ithaca, NY.

Wiertz, J., and J. Vink. 1986. The present status of the badger *Meles meles* (L., 1758) in the Netherlands. *Lutra* 29:21–53.

Wilford, C. L. 2004. [Untitled]. *Journal of the American Veterinary Medical Association* 224:1749–1752.

Wilkinson, D., and G. C. Smith. 2001. A preliminary survey for changes in urban fox (*Vulpes vulpes*) densities in England and Wales, and implications for rabies control. *Mammal Review* 31:107–110.

Williams, C. K., G. Ericsson, and T. Heberlein. 2002. A quantitative summary of attitudes toward wolves and their reintroduction (1972–2000). *Wildlife Society Bulletin* 30:575–584.

Williams, E. S., and I. K. Barker. 2001. *Infectious Diseases of Wild Mammals*. 3rd ed. Manson Publishing, London.

Williams, J. C., G. V. Byrd, N. B. Konyukhov, R. Barrett, J. Cooper, and T. Gaston. 2003. Whiskered auklets *Aethia pygmaea*, foxes, humans and how to right a wrong. *Marine Ornithology* 31:175–180.

Williams, J. S., J. J. McCarthy, and H. D. Picton. 1995. Cougar habitat use and food habits on the Montana Rocky Mountain Front. *Intermountain Journal of Sciences* 1:16–26.

Willingham, A. L., N. W. Ockens, C. M. O. Kapel, and J. Monrad. 1996. A helminthological survey of wild red foxes (*Vulpes vulpes*) from the metropolitan area of Copenhagen. *Journal of Helminthology* 70:259–263.

Wilmers, C. C., R. L. Crabtree, D. W. Smith, K. M. Murphy, W. M. Getz. 2003. Trophic facilitation by introduced predators: Gray wolf subsidies to scavengers in Yellowstone National Park. *Journal of Animal Ecology* 72:909–916.

Wilson, D. E., and D. M. Reeder. 2005. *Mammal Species of the World: A Taxonomic and Geographic Reference.* 3rd ed. Johns Hopkins University Press, Baltimore.

Wilson, G., S. Harris, and G. McLaren. 1997. *Changes in the British Badger Population, 1988 to 1997.* People's Trust for Endangered Species, London.

Wilson, M. A. 1997. The wolf in Yellowstone: Science, symbol, or politics? Deconstructing the conflict between environmentalism and wise use. *Society and Natural Resources* 10:453–468.

Wilson, P. J., W. J. Jakubas, and S. Mullen. 2004. *Genetic Status and Morphological Characteristics of Maine Coyotes as Related to Neighboring Coyote and Wolf Populations.* Final Report to the Maine Outdoor Heritage Fund Board, Grant 011-3-7. Maine Department of Inland Fisheries and Wildlife, Bangor. 58 pp.

Wincentz, T.-L. 2004. Population dynamics of urban and rural red foxes (*Vulpes vulpes*) in Denmark. Master's thesis, University of Copenhagen, Denmark.

Windberg, L. A., H. L. Anderson, and R. M. Engeman. 1985. Survival of coyotes in southern Texas. *Journal of Wildlife Management* 49:301–307.

Winter, L., and G. E. Wallace. 2006. *Impacts of Feral and Free-Ranging Cats on Bird Species of Conservation Concern: A Five-State Review of New York, New Jersey, Florida, California, and Hawaii.* American Bird Conservancy, Washington D.C.

Witmer, G. M., and D. S. DeCalesta. 1986. Resource use by unexploited sympatric bobcats and coyotes in Oregon. *Canadian Journal of Zoology* 64:2333–2338.

Witter, M. S., and I. C. Cuthill. 1993. The ecological costs of avian fat storage. *Philosophical Transactions of the Royal Society of London B* 340:73–92.

Wittmann, K., J. J. Vaske, M. J. Manfredo, and H. C. Zinn. 1998. Standards for lethal control of problem wildlife. *Human Dimensions of Wildlife* 3:29–48.

Wolch, J. R., Gullo, A., and Lassiter, U. 1997. Changing attitudes toward California's cougars. *Society and Animals* 5:9–116.

Wolf, K. N., F. Elvinger, and J. L. Pilcicki. 2003. Infra-red triggered photography and tracking plates to monitor oral rabies vaccine bait contact by raccoons in culverts. *Wildlife Society Bulletin* 31:387–391.

Wolfe, A., S. Hogan, D. Maguire, C. Fitzpatrick, L. Vaughan, D. Wall, T. J. Hayden, and G. Mulcahy. 2001. Red foxes (*Vulpes vulpes*) in Ireland as hosts for parasites of potential zoonotic and veterinary significance. *Veterinary Record* 149:759–763.

Wood, W. F. 1999. The history of skunk defensive secretion research. *Chemical Educator* 4:44–50.

Woodroffe, R., J. W. McNutt, and M. G. L. Mills. 2004. African wild dog. Pages 174–183 *in* C. Sillero-Zubiri, M. Hoffman, and D. W. Macdonald, eds. *Canids: Foxes, Wolves, Jackals and Dogs.* Status Survey and Conservation Action Plan. IUCN/SSC Canid Specialist Group. Gland, Switzerland and Cambridge, UK.

Woods, M., R. A. McDonald, and S. Harris. 2003. Predation of wildlife by domestic cats *Felis catus* in Great Britain. *Mammal Review* 33:174–188.

Woollard, T., and S. Harris. 1990. A behavioural comparison of dispersing and non-dispersing foxes (*Vulpes vulpes*) and an evaluation of some dispersal hypotheses. *Journal of Animal Ecology* 59:709–722.

Woolridge, D. R. 1980. Chemical aversion conditioning of polar and black bears. *International Conference of Bear Research and Management* 4:167–173.

Wozencraft, W. C. 1989. The phylogeny of the recent Carnivora. Pages 495–535 *in* J. L. Gittleman, ed. *Carnivore Behavior, Ecology, and Evolution.* Cornell University Press, Ithaca, NY.

Wozencraft, W. C. 2005. Order Carnivora. Pages 532–628 *in* D. E. Wilson and D. M. Reeder, eds. *Mammal Species of the World: A Taxonomic and Geographic Reference.* Vol. 1. 3rd ed. Johns Hopkins University Press, Baltimore.

Wright, A., A. H. Fielding, and C. P. Wheater. 2000. Predicting the distribution of Eurasian badger (*Meles meles*) setts over an urbanized landscape: A GIS approach. *Photogrammetric Engineering and Remote Sensing* 66:423–428.

Wulff, R. 2003. [Untitled]. *Journal of the American Veterinary Medical Association* 223:607.

Wulff, R. 2005. Comments on welfare of feral cats and wildlife. *Journal of the American Veterinary Medical Association* 226:30.

Yamaguchi, N., H. L. Dugdale, and D. W. Macdonald. 2006. Female receptivity, embryonic diapause, and superfetation in the European badger (*Meles meles*): Implications for the reproductive tactics of males and females. *Quarterly Review of Biology* 81:33–48.

Yamane, A. 1998. Male reproductive tactics and reproductive success of the group-living feral cat (*Felis catus*). *Behavioural Processes* 43:239–249.

Yamane, A., J. Emoto, and N. Ota. 1997. Factors affecting feeding order and social tolerance to kittens in the group-living feral cat (*Felis catus*). *Applied Animal Behaviour Science* 52:119–127.

Yom-Tov, Y., S. Ashkenazi, and O. Viner. 1995. Cattle predation by the golden jackal *Canis aureus* in the Golan Heights, Israel. *Biological Conservation* 73:19–22.

Young, R. P., J. Davison, I. D. Trewby, G. J. Wilson, R. J. Delahay, and C. P. Doncaster. 2006. Abundance of hedgehogs (*Erinaceus europaeus*) in relation to the density and distribution of badgers (*Meles meles*). *Journal of Zoology* 269:349–356.

Zabel, C. J., and S. J. Taggart. 1989. Shift in red fox, *Vulpes vulpes*, mating system associated with El Niño in the Bering Sea. *Animal Behaviour* 38:830–838.

Zeveloff, S. I. 2002. *Raccoons: A Natural History.* Smithsonian Institution Press, Washington, D.C.

Zhang, X., M. A. Friedl, C. B. Schaaf, A. H. Strahler, and A. Schneider. 2004. The footprint of urban climates on vegetation phenology. *Geophysical Research Letters* 31. doi:10.1029/2004GL020137.

Zimmermann, B., P. Wabakken, and M. Dötterer. 2001. Human-carnivore interactions in Norway: How does the re-appearance of large carnivores affect people's attitudes and levels of fear? *Forest, Snow, and Landscape Research* 76:137–153.

Zinn, H. C., M. J. Manfredo, and D. J. Decker. 2008. Human conditioning to wildlife: Steps toward theory and research. *Human Dimensions of Wildlife* 13:388–399.

Zinn, H. C., M. J. Manfredo, J. J. Vaske, and K. Wittmann. 1998. Using normative beliefs to determine acceptability of wildlife management actions. *Society and Natural Resource* 11:649–662.

Zinn, H. C., and C. L. Pierce. 2002. Values, gender, and concern about potentially dangerous wildlife. *Environment and Behavior* 34:239–256.

Zint, M., A. Kraemer, H. Northway, and M. Lim. 2002. Evaluation of the Chesapeake Bay Foundation's conservation education program. *Conservation Biology* 16:641–649.

Zoellick, B. W., C. E. Harris, B. T. Kelly, T. P. O'Farrell, T. T. Kato, and M. E. Koopman. 2002. Movements and home ranges of San Joaquin kit foxes relative to oil-field development. *Western North American Naturalist* 62:151–159.

Zoellick, B. W., T. P. O'Farrell, P. M. McCue, C. E. Harris, and T. T. Kato. 1987. *Reproduction of the San Joaquin Kit Fox on Naval Petroleum Reserve No. 1, Elk Hills, California, 1980–1985.* U.S. Department of Energy Topical Report No. EGG 10282-2144.

INDEX